Plants, People and Practices

UPOV and the UPOV Convention are increasingly relevant and important. They have technical, social and normative legitimacy and have standardised numerous concepts and practices related to plant varieties and plant breeding. In this book, Jay Sanderson provides the first sustained and detailed account of the Convention. Building upon the idea that it has an open-ended and contingent relationship with scientific, legal, technical, political, social and institutional actors, the author explores the Convention's history, concepts and practices. Part I examines the emergence of the UPOV Convention during the 1950s and its expanding legitimacy in relation to plant variety protection and plant breeding more broadly. Part II explores the Convention's key concepts and practices, including plant breeder, plant variety, plant names (denomination), characteristics, protected material, essentially derived varieties (EDV) and farm-saved seed. This book is an invaluable resource for academics, policy-makers, agricultural managers and researchers in this field.

JAY SANDERSON is an Associate Professor at USC Law School (USC Australia), a member of the Australian Centre for Intellectual Property in Agriculture (ACIPA) and an adjunct with the Law Futures Centre, Griffith University Law School, Queensland. He has published widely on issues of intellectual property, plants and agri-food, and has been cited by Australia's Productivity Commission and Advisory Council on Intellectual Property. He is the co-editor of *The Intellectual Property and Food Project: From Rewarding Innovation and Creation to Feeding the World* (with Charles Lawson, 2013).

Cambridge Intellectual Property and Information Law

As its economic potential has rapidly expanded, intellectual property has become a subject of front-rank legal importance. Cambridge Intellectual Property and Information Law is a series of monograph studies of major current issues in intellectual property. Each volume contains a mix of international, European, comparative and national law, making this a highly significant series for practitioners, judges and academic researchers in many countries.

A list of books in the series can be found at the end of this volume.

Plants, People and Practices

The Nature and History of the UPOV Convention

Jay Sanderson
USC Law School
USC Australia

CAMBRIDGE
UNIVERSITY PRESS

University Printing House, Cambridge CB2 8BS, United Kingdom

One Liberty Plaza, 20th Floor, New York, NY 10006, USA

477 Williamstown Road, Port Melbourne, VIC 3207, Australia

4843/24, 2nd Floor, Ansari Road, Daryaganj, Delhi – 110002, India

79 Anson Road, #06–04/06, Singapore 079906

Cambridge University Press is part of the University of Cambridge.

It furthers the University's mission by disseminating knowledge in the pursuit of education, learning, and research at the highest international levels of excellence.

www.cambridge.org
Information on this title: www.cambridge.org/9781107126497

10.1017/9781316411216

First published 2017

A catalogue record for this publication is available from the British Library.

Library of Congress Cataloging-in-Publication data
Names: Sanderson, Jay (Law teacher), author.
Title: Plants, people and practices : the nature and history of the UPOV convention / Dr. Jay Sanderson.
Other titles: Nature and history of the UPOV convention | Cambridge intellectual property and information law.
Description: New York, NY : Cambridge University Press, 2017. | Series: Cambridge intellectual property and information law
Identifiers: LCCN 2016043401 | ISBN 9781107126497 (Hardback)
Subjects: LCSH: International Union for the Protection of New Varieties of Plants–History. | Plant varieties–Patents. | Plant breeders–Legal status, laws, etc.
Classification: LCC K3876 .S25 2017 | DDC 346.04/86–dc23
LC record available at https://lccn.loc.gov/2016043401

ISBN 978-1-107-12649-7 Hardback

Contents

Acknowledgements

As with any book, I have benefitted enormously from the support and help of many people. Indeed, the list of family, friends and colleagues who have supported or helped me along the way is long. However, a special debt is owed to the following people. Brad Sherman, previously based at Griffith Law School and now at the TC Beirne School of Law (UQ), for his guidance, ideas and support. He is passionate about research and a wonderful mentor and friend. Leanne Wiseman, who gave me my first research work – as a 'not so young' undergraduate law student – continues to be a mentor and collaborator. Over the years, she has also become a good friend.

I have also benefitted from discussions and comradeship with my colleagues from the Australian Centre for Intellectual Property in Agriculture (ACIPA), particularly Charles Lawson but also Carol Ballard, Matthew Rimmer, Stephen Hubicki, Berris Charnley and Brendan Tobin. Thanks also to Peter Button, Yolanda Huerta and Bernard Le Buanec from UPOV. Peter Button, in particular, has been generous with his time and patience, particularly with my initial attempts to simplify some of the technical aspects of UPOV and the UPOV Convention. Translations of the Diplomatic Conference of 1957–1961, 1972 were funded by ACIPA and done by Monique Coombs. Numerous other colleagues at the Griffith Law School and the USC Law School have contributed to this book: providing a fertile research environment (William MacNeil, Kieran Tranter, Charles Lawson and Leanne Wiseman), talking about transgenics and science fiction (Tim Peters), engaging with ideas and humour, and being man enough to wear brightly coloured pants (Edwin Bikundo), or being passionate about all things research and theoretical (Brendon Murphy). The editorial and production staff at Cambridge University Press were nothing but supportive throughout the review and publication process, and I am thankful for their efforts. It goes without saying that all faults lie with me and that nothing contained in the book reflects the views of anyone but the author.

Finally the book would not have been possible without a loving and supportive family. Jenny and Cam (or as I like to call them, Mum and Dad) have supported everything that I have done or attempted to do. Isaac, Ruby and Baxter have grown a lot in the time it took me to complete this book. They give me perspective, joy and purpose. Most importantly, I owe so much to Jacqueline: where I am pragmatic and rational, she is creative and passionate; where I am words and thoughts, she is art and love... 'all in your back pocket'.[1]

[1] grey space
tender humour
restricted rage
dust
art
and
love
all in your back pocket.
Kylie Johnson, *Count Me the Stars* (Pier 9, 2008)

1 Introduction

1.1 Introduction

This is a book about the UPOV Convention.[1] More specifically it is about the UPOV Convention's history, key concepts and the practices that inform, sustain and sometimes challenge both the UPOV Convention and UPOV.[2] There are numerous motives for this book. As the only international treaty concerned with intellectual property protection of new plant varieties, the UPOV Convention plays a key social, political and economic role in the development of new plant varieties. Encouraging the development of new plant varieties is important in order to overcome problems associated with food security, climate change and environmental degradation. The development of new plant varieties can help to deal with problems of food availability and affordability, as well as economic and social development. This is particularly important in light of growing populations, the scarcity of arable land and reduced genetic diversity. Generally speaking, by providing an exclusive right, granted to the plant breeder of a new plant variety, to control the exploitation of their plant variety the UPOV Convention provides a mechanism for national or regional legislators to offer incentives for those engaged in plant breeding. In doing so it sets out the conditions for grant, the plant material to be protected, uses to which protection applies and various exceptions to the rights granted.

Whilst the issue of food security is complicated, there is no denying the importance of producing modified, new or improved plants. In *Feeding the World: An Economic History of Agriculture, 1800–2000*, Giovanni Federico examines the factors that have affected the performance of

[1] International Convention for the Protection of New Plant Varieties (1991) 815 UNTS 89, being the original agreement of 2 December 1961 and revised on 10 November 1972, 23 October 1978 and 19 March 1991 (UPOV 1991).

[2] The International Union for the Protection of New Varieties of Plants ('UPOV') was established in 1961 and is based in Geneva, Switzerland. UPOV, and its legitimacy, is the topic of Chapter 3.

agriculture and the ability to produce food between 1800 and 2000.[3] Federico identifies a broad range of factors including the development of new or modified plants as central to agricultural development, particularly where the development of new plants has largely been directed towards public goods such as food availability and affordability. As the central international treaty dealing with plant varieties, the UPOV Convention, and the UPOV system more generally, plays an important role in international policy aimed at promoting the development of new plants, and in the continued development of agriculture throughout the world.

Another reason why it is necessary to comprehensively examine the UPOV Convention is because member countries to the WTO have an obligation to provide some form of protection over plant varieties in their national laws. Article 27(3)(b) of the Agreement on Trade-Related Aspects of Intellectual Property Rights (TRIPS)[4] sets out an obligation to provide some form of protection for plant varieties. Specially, Article 27(3)(b) states that '[m]embers shall provide for the protection of plant varieties either by patents or by an effective *sui generis* system or by any combination thereof'.[5] Article 27(3)(b) has the dual effect of bringing into question UPOV's status as an 'effective *sui generis*' legal scheme, as well as intensifying the relationship between the UPOV Convention (as a possible 'effective' *sui generis* system), patent law and alternative *sui generis* plant variety protection schemes.[6] Part of the difficulty for WTO members looking to meet their obligations is that Article 27(3)(b) adopted a compromise position in which the UPOV Convention is not explicitly referenced.[7] So, while there are currently 74 members (72 Member States and 2 intergovernmental organisations, covering

[3] G. Federico, *Feeding the World: An Economic History of Agriculture, 1800–2000* (Princeton University Press, 2005).

[4] *Marrakech Agreement Establishing the World Trade Organization*, April 15, 1994, Annex 1C: *Agreement on Trade-Related Aspects of Intellectual Property Rights Including Trade*, 33 I.L.M. 81, 84 (TRIPS).

[5] TRIPs, Art. 27(3)(b). Currently, least-developed countries are not required to provide any form of plant variety protection until 2021.

[6] The relationship between plant variety rights and patents is one of the main themes of M. Llewelyn and M. Adcock, *European Plant Intellectual Property* (Hart Publishing, 2006). See also WIPO-UPOV, Symposium on the Co-Existence of Patents and Plant Breeders' Rights in the Promotion of Biotechnological Developments (Geneva, October 25, 2002). For a discussion of alternative schemes see, e.g., C. Correa (with S. Shashikant and F. Meienberg), *Plant Variety Protection in Developing Countries: A Tool for Designing a Sui Generis Plant Variety Protection System, An Alternative to UPOV 1991* (APBREBES, 2015).

[7] D. Gervais, *The TRIPs Agreement: Drafting History and Analysis* (Sweet and Maxwell, 1998).

93 States) to the UPOV Convention, there are a number of countries – including Thailand, Nepal, Sri Lanka and Bangladesh – that have, for the time being at least, chosen not to join UPOV.[8] Instead, these countries have decided to provide plant variety rights protection in ways that are different to that prescribed by the UPOV Convention. India, for example, chose to meet its obligations to protect plant varieties outside the UPOV Convention with the Protection of Plant Varieties and Farmers' Rights Act 2001. That said, even those countries opting for a *sui generis* form of plant variety protection base a large portion of their national laws on the UPOV Convention; often UPOV 1978, which is felt to be better suited (balanced) to the interests of farmers because it treats farm-saved seed more liberally, and does not extend protection to harvested material, products derived from harvested material and essentially derived varieties (EDVs).[9]

While there is a degree of flexibility in TRIPS over how to 'effectively' protect plant varieties, increasingly trade and economic agreements are removing this flexibility: requiring signatories to accede to, or ratify, the UPOV Convention (specifically UPOV 1991).[10] Signatories to trade agreements that are not UPOV Members have been asked to join UPOV or if countries are already Members to UPOV 1978, they may 'agree' to sign up to the stricter provisions of UPOV 1991. For example, while the North American Free Trade Agreement (NAFTA) gave Mexico the choice between UPOV 1978 and UPOV 1991, by signing up to the Trans Pacific Partnership (TPP) in 2015, Mexico has 'agreed' to accede to UPOV 1991 (within three years of the TPP coming into force) and will need to implement stronger national plant variety protection laws that are UPOV 1991 compliant, as opposed to UPOV 1978.[11] In 2015, the African Regional Intellectual Property Organisation (ARIPO) initiated the procedure to become a member of UPOV, and has drafted plant variety protection laws consistent with UPOV 1991.[12] While this is an

[8] In 2002, India attempted to join UPOV, although the proposed national laws were deemed to be incompatible with the UPOV Convention by the UPOV Council, particularly in relation to farmers' rights: see D. Matthews, *Intellectual Property, Human Rights and Development: The Role of NGOs and Social Movements* (Edward Elgar, 2011), pp. 196–198.

[9] D. Jefferson, 'Development, farmers' rights, and the Ley Monsanto: The struggle over the ratification of UPOV 1991 in Chile' (2014–2015) 55 *IDEA* 31.

[10] These topics are discussed in more detail in Chapter 10 (farm-saved seed), Chapter 8 (protected material) and Chapter 9 (EDVs).

[11] Although, the future of the TPP is now uncertain: on 24 January United States President, Donald Trump, signed an Executive Order withdrawing the U.S. from the TPP.

[12] Members of ARIPO adopted the Arusha Protocol for the Protection of New Varieties of Plants ARIPO on 6 July 2015: see ARIPO, Protocol for the Protection of New Varieties of Plants Adopted, www.aripo.org/news-events-publications/news/item/69-aripo-protocol-for-the-protection-of-new-varieties-of-plants-adopted.

early step in the process of ARIPO members implementing national plant variety protection – and strictly speaking there is no obligation to sign, ratify or accede to the ARIPO's draft law (the 'Arusha Protocol') – it is now likely that all members will need to implement national plant variety protection that is compliant with UPOV 1991. While Kenya is already a UPOV member, it is likely that more ARIPO member states will join UPOV. For example, on 22 October 2015, Tanzania acceded to UPOV, and on 22 November 2015 became the seventy-fourth member of UPOV.[13] ARIPO argues that national laws consistent with UPOV 1991 will 'provide Member States with a regional plant variety protection system that recognizes the need to provide growers and farmers with improved varieties of plants in order to ensure sustainable Agricultural production'.[14] Many African countries are also likely to join UPOV and implement UPOV-based plant variety protection because of various trade and economic agreements with the United States, Europe and being part of the Southern African Development Community (SADC).[15]

For the reasons outlined above it is important to examine both UPOV and the UPOV Convention thoroughly. While others have studied the UPOV Convention,[16] this book is the first sustained investigation into UPOV and the UPOV Convention. As Helfer suggests, 'The importance of food security to human survival and the widespread interest in intellectual property in genetic materials suggest that PVP [plant variety rights] should be the subject of widespread interest by scholars and policymakers. In fact, nothing could be further from the truth.'[17] Before I set out the approach and structure of this book, I want to outline some of the main concerns and controversies over the UPOV Convention.

[13] See, e.g., H. Haugen, 'Inappropriate processes and unbalanced outcomes: Plant variety protection in Africa goes beyond UPOV 1991 requirements' (2015) 18(5) *The Journal of World Intellectual Property* 196; B. De Jonge, 'Plant variety protection in Sub-Saharan Africa: Balancing commercial and smallholder farmers' interests' (2014) 7(3) *Journal of Politics and Law* 100.

[14] ARIPO, Protocol for the Protection of New Varieties of Plants.

[15] For a discussion of 'Africa's road to UPOV', see C. Oguamanam, 'Breeding apples for oranges: Africa's misplaced priority over plant breeders' rights' (2015) 18(5) *The Journal of World Intellectual Property* 165.

[16] See, e.g., M. Llewelyn and M. Adcock, *European Plant Intellectual Property*; M. Janis, H. Jervis and R. Peet, *Intellectual Property Law of Plants* (Oxford University Press, 2014); G. Dutfield, 'Turning plant varieties into intellectual property: The UPOV Convention' in G. Tansey and T. Rajotte (eds.), *The Future Control of Food: A Guide to International Negotiations and Rules on Intellectual Property, Biodiversity, and Food Security* (Earthscan/IDRC, 2008), pp. 27–47.

[17] L. Helfer, 'The demise and rebirth of plant variety protection: A comment on technological change and the design of plant variety protection regimes' (2007) 82 *Chicago Kent Law Review* 1619.

1.2 Concerns and Controversies

Adopted in 1961, the UPOV Convention has been revised three times: in 1972 (UPOV 1972), 1978 (UPOV 1978) and 1991 (UPOV 1991).[18] Currently, UPOV has 74 members, a further 16 States and one intergovernmental organisation have initiated the procedure for acceding to the UPOV Convention, and a further 24 States and one intergovernmental organisation have sought assistance with the development of laws based on the UPOV Convention.[19] By joining UPOV, Member States agree to enact national plant variety protection laws that are UPOV-compliant.[20] The purpose of the UPOV Convention is 'to provide and promote an effective system of plant variety protection, with the aim of encouraging the development of new varieties of plants, for the benefit of society'.[21]

The UPOV Convention has provided the legal framework used to protect new plant varieties for almost 50 years, yet it occupies a peculiar place in intellectual property law. Viewed as unimportant by some and too technical, outmoded or obsolete by others, the UPOV Convention seldom receives the kind of attention bestowed on intellectual property law heavyweights such as copyright and patents. In this way the UPOV Convention occupies a peculiar place in intellectual property law and is often given cursory treatment, or ignored altogether, in textbooks and legal journals. As Sherman puts it, '[Plant variety rights] have been treated as outsiders that are begrudgingly tolerated, but not liked.'[22] Part of the reason for the legal fraternity's apparent aversion to the UPOV Convention is that it is *sui generis* and specifically targeted towards the needs of plant breeding. More specifically the technical nature of plant variety rights means that many plant variety rights schemes are managed by agricultural department rather that intellectual property offices, and the assessment of the technical requirements of plant variety rights is carried out by plant breeders and scientists rather than lawyers or patent attorneys. As a consequence, and despite the fact that making a plant variety application is a (relatively) straightforward process, particularly for those people the scheme is intended for (i.e. plant breeders) and does

[18] The UPOV Convention is to be reviewed periodically: UPOV 1961, Art. 27; UPOV 1978, Art. 27; UPOV 1991, Art. 38.

[19] UPOV, *Members of the International Union for the Protection of New Varieties of Plants* (15 April 2016)), www.upov.int/export/sites/upov/members/en/pdf/pub423.pdf.

[20] UPOV, *Guidance for Members of UPOV on How to Ratify, or Accede to, the 1991 Act of the UPOV Convention*, UPOV/INF/14/1 (UPOV, Geneva, 2009).

[21] UPOV, *UPOV Mission Statement*, www.upov.int/en/about/.

[22] B. Sherman, 'Taxonomic property' (2008) 67(3) *Cambridge Law Journal* 560.

not need the skills of a lawyer or patent attorney, the plant variety rights scheme is viewed as an exclusive, selective 'club'.[23]

One of the persistent concerns over the UPOV Convention is that it is, and perhaps has always been, unnecessary and obsolete. In this way, the UPOV Convention tends to be viewed unfavourably when compared with patent law and advancing science and technology. As early as the 1980s, Cornish argued that the changing nature of science and technology meant that plant breeders would attain more adequate and appropriate protection through patent law and ultimately questioned whether 'the [UPOV] regime had a viable future' because it obstructed the 'logical framework of protection', namely the use of patents for the protection of plant material.[24] Often the claim that the UPOV Convention is obsolete and outmoded is expressed either explicitly or implicitly in the argument that the UPOV Convention is rooted to a particular technological paradigm. While this view is tied to the fact that the UPOV Convention is industry specific and *sui generis*, this 'out-dated world-view' argument also means that the UPOV Convention is seen to be ahistorical and immutable. Consequently, the UPOV Convention has been labelled outmoded, a Neanderthal and obsolete.[25]

The UPOV Convention, like other forms of intellectual property, suffers from the problem of proof. The purpose of the UPOV Convention is to 'provide and promote' effective plant variety protection and encourage the 'development of new varieties of plants for the benefit of society'. However, not everyone believes that it does this. And over the years, a number of countries – including Thailand, Nepal, Sri Lanka and Bangladesh – have not joined UPOV or implemented UPOV-compliant national, plant variety protection. The fact that countries are willing to implement plant variety rights outside of the UPOV scheme is significant, as member countries to the WTO have an obligation to provide a minimum level of intellectual property protection in their national laws. As we have already seen, Article 27(3)(b) of TRIPS sets out an obligation to provide some form of protection for plant varieties: specially, '[m]embers shall provide for the protection of plant varieties either by patents or by an effective *sui generis* system or by any combination

[23] G. Dutfield, *Food, Biological Diversity and Intellectual Property: The Role of the International Union for the Protection of New Varieties of Plants (UPOV)* (Quakers United Nations Office, 2011).

[24] W. Cornish, *Intellectual Property: Patents, Copyright, Trade Marks and Allied Rights* (Sweet and Maxwell, 1989), p. 148 at note 37.

[25] See, e.g., C. Fowler, *Unnatural Selection: Technology, Politics and Plant Evolution* (Gordon and Breach, 1994), p. 152; M. Janis and S. Smith, 'Technological change and the design of plant variety protection regimes' (2007) 82 *Chicago Kent Law Review* 1557.

thereof'.[26] The question of whether the UPOV Convention is 'effective' is a vexed one. The language of 'effective' is used by both UPOV and TRIPS, although not necessarily used explicitly in the same context. The argument is that by providing intellectual property protection, in the form of a targeted and specific scheme such as the UPOV Convention, plant breeders will be encouraged to develop new plant varieties.

Since 2000, UPOV has taken some initiatives to show the benefits of the UPOV Convention and the convenience of adopting a ready-made system of plant variety protection.[27] In 2005, for example, UPOV published its *Report on The Impact of Plant Variety Protection* ('Impact Report').[28] In its *Impact Report*, UPOV argues that the introduction of the UPOV system of plant variety protection and UPOV membership have a range of benefits. Specifically, UPOV argues that the introduction of the UPOV system of plant variety protection and UPOV membership increases breeding activities, availability of improved varieties, the number of new varieties and the number of foreign varieties, and diversifies the types of plant breeders (private breeders, researchers) and encourages competitiveness.[29] On the basis of its study, UPOV confidently and unequivocally concluded that it had 'demonstrated that in order to enjoy the full benefits which plant variety protection is able to generate, both implementation of the UPOV Convention and membership of UPOV are important'.[30]

Not all of the evidence on UPOV-based plant variety protection is positive. For example: in a report commissioned by the World Bank, Louwaars et al. examined the effects of intellectual property rights, including plant variety protection, in developing countries by focusing on five countries: China, Colombia, Kenya, India and Uganda.[31] The researchers questioned the need for UPOV-based plant variety protection and suggested that private seed industries owed relatively little to national plant variety rights protection. They also speculated that plant variety rights may in fact reduce the effectiveness of traditional rights and practices for farmers to save, exchange or sell saved seed.

[26] TRIPs, Art. 27(3)(b).

[27] See, e.g., J. Sanderson, 'Why UPOV is relevant, transparent and looking to the future: a conversation with Peter Button' (2013) 8(8) *Journal of Intellectual Property Law & Practice* 615, 622; UPOV, Council: Forty-Seventh Ordinary Session, C/47/4 (12 August 2013).

[28] UPOV, *Report on the Impact of Plant Variety Protection* (UPOV, 2005).

[29] Ibid.

[30] Ibid.

[31] N. Louwaars et al., *Impacts of Strengthened Intellectual Property Rights Regimes on the Plant Breeding Industry in Developing Countries: A Synthesis of Five Case Studies* (World Bank, 2005). See also R. Tripp, N. Louwaars and D. Eaton, 'Plant variety protection in developing countries: a report from the field' (2007) 32 *Food Policy* 354.

A lack of evidence is a problem for both those looking to support and oppose UPOV-based schemes. Because plant variety rights schemes generally interact with other factors (such as other intellectual property and legal schemes, agricultural markets, increased globalisation and a reduction of public expenditure for agricultural research and seed production), it is difficult to confidently conclude on the possible contributions and concerns that plant variety rights protection might offer plant breeding. Therefore, many of the concerns about the UPOV Convention are largely theoretical and speculative as there is little or no ethnographic or empirical evidence to support them. It must be pointed out, however, that the problem of proof is not unique to the UPOV Convention. Evidence in support of intellectual property is notoriously difficult to measure, distil and substantiate. As Merges states,

> Estimating costs and benefits, modeling them over time, projecting what would happen under counterfactuals (such as how many novels or pop songs really would be written in the absence of copyright protection, and who would benefit from such a situation) – these are all overwhelmingly complicated tasks. And this problem poses a major problem for utilitarian theory. The sheer practical difficulty of measuring or approximately all the variables involved means that the utilitarian program will always be at best aspirational.[32]

Yet other concerns are directed at UPOV (or 'the Union') rather than the UPOV Convention. More specifically some commentators, policy makers and advocates have accused UPOV of lacking transparency, being unaccountable and, in its relationships with developing countries, misusing its influence and power.[33] UPOV has also been criticised for being self-serving and biased, with some civil society organisations and governments urging developing and least developed countries to avoid UPOV.[34] During the late 1990s, for example, GRAIN argued that UPOV was biased towards industrial agriculture and, even though it was not in their interests, had pressured many developing countries into not joining UPOV.[35] More recently, UPOV has been accused of ignoring its own Convention by approving ARIPO's Draft Protocol for the

[32] R. Merges, *Justifying Intellectual Property* (Harvard University Press, 2011), pp. 2–3.

[33] In 2011, for example, Graham Dutfield likened UPOV to an exclusive, selective 'club' and emphasised concerns over transparency, democratic accountability and lack of public debate: Dutfield, *Food, Biological Diversity and Intellectual Property*, pp. 12–14.

[34] See Correa, *Plant Variety Protection in Developing Countries*, p. 4; GRAIN/GAIA, 'Ten reasons not to join UPOV', *Global Trade and Biodiversity in Conflict* (1988) 2, www.grain.org/article/entries/1-ten-reasons-not-to-join-upov; J. Ekpere, *TRIPs, Biodiversity and Traditional knowledge: OAU Model Law on Community Rights and Access to Genetic Resources* (UNCTAD/ICTSD, 2000).

[35] See GRAIN, 'UPOV on the warpath', *Seedling* (June 1999), www.grain.org/es/article/entries/257-upov-on-the-war-path; GRAIN/GAIA, 'Ten reasons not to join UPOV'.

Protection of New Varieties of Plants on 11 April 2014.[36] At times, these criticisms of and challenges to UPOV have manifested in public protest: for example, farmers, indigenous communities and civil society groups have protested against UPOV and the UPOV Convention in Thailand (2013), Costa Rica (2014), Ghana (2014) and Chile (2014).[37]

Finally, there are a range of concerns over the UPOV Convention that can be broadly construed as relating to farmers' rights, conservation and biodiversity. The proliferation of forums and institutions dealing with plants and plant genetic resources has meant that there are differing interests related to plants, plant genetic resources and people. For example, the Convention on Biological Diversity (CBD) and the 2010 Nagoya Protocol on Access to Genetic Resources and Equitable Sharing of Benefits Arising from their Utilization (Nagoya Protocol) impose an obligation to equitably share the benefits of genetic resources.[38] And the International Treaty on Plant Genetic Resources for Food and Agriculture (Plant Treaty) aims to guarantee food security through the conservation, exchange and sustainable use of the world's plant genetic resources for food and agriculture and has in so doing recognised farmers' rights.[39] A large part of civil society groups' opposition to UPOV and the UPOV Convention is based on the view that the UPOV Convention prohibits the rights of farmers to save, exchange or sell farm-saved seed. For example, in her book *Agrobiodiversity and the Law: Regulating Genetic Resources, Food Security and Cultural Diversity*,[40]

[36] UPOV, 'UPOV council holds its thirty-first extraordinary session', Press Release 96 (11 April 2014), www.upov.int/export/sites/upov/news/en/pressroom/pdf/pr96.pdf. Specifically, it has been suggested that in ratifying ARIPO's Draft Protocol, UPOV ignored Articles 34 and 30(2) of UPOV 1991: See, e.g., S. Gura, 'UPOV breaking its own rules to tie-in African countries' (11 April 2014) APBREBES, www.apbrebes.org/press-release/upov-breaking-its-own-rules-tie-african-countries; Alliance for Food Sovereignty in Africa, 'ARIPO sells out African farmers, seals secret deal on plant variety protection' (8 July 2015), http://afsafrica.org/aripo-sells-out-african-farmers-seals-secret-deal-on-plant-variety-protection/. The Draft Protocol was replaced with the Arusha Protocol for the Protection of New Varieties of Plants on 6 July 2015: see also Haugen, 'Inappropriate processes and unbalanced outcomes'.

[37] See Jefferson, 'Development, farmers' rights, and the Ley Monsanto'; GRAIN, 'Thai farmers and civic groups protest UPOV lobby' (19 November 2013), www.grain.org/bulletin_board/entries/4833-thai-farmers-and-civic-groups-protest-upov-lobby; APBREBES, 'Massive protests in Ghana over UPOV style Plant Breeders' Rights Bill' (11 February 2014), www.apbrebes.org/news/massive-protests-ghana-over-upov-style-plant-breeders-bill.

[38] Convention on Biological Diversity (1992) 1760 UNTS 79 ('CBD').

[39] International Treaty on Plant Genetic Resources for Food and Agriculture (2001) 2400 UNTS 303 ('Plant Treaty').

[40] J. Santilli, *Agrobiodiversity and the Law: Regulating Genetic Resources, Food Security and Cultural Diversity* (Routledge, 2012). For a discussion of some of these issues as they relate to customary law, see B. Tobin, *Indigenous Peoples, Customary Law and Human Rights–Why Living Law Matters* (Routledge, 2014).

Juliana Santilli considers the loss of biodiversity by analysing the impacts that international and national legal instruments have on farming systems and on the local small-scale farmers who conserve and manage them. In so doing, Santilli suggests that the exchange of seeds through local networks (e.g. farmer exchanges) is an essential component of agrobiodiversity, and that the implementation of UPOV 1991 by developing countries 'should be considered only after the implementation of the [Plant Treaty] ... which recognises farmers' rights, so that inconsistencies between these two instruments are avoided'.[41] By linking the UPOV Convention to farmers' rights, conservation and biodiversity commentators have raised a number of questions: Do plant variety rights encourage or hinder biodiversity and agrobiodiversity? Do plant variety rights recognise the work of farmers? Do plant variety rights recognise the relationship between 'conventional' breeders and farmers? Can plant variety rights meet the needs of developing countries?

1.3 Overmining and Undermining

Before I outline the structure of this book, it is worth chronicling the tone of many of the concerns and controversies over the UPOV Convention. In doing so, I will also say something about my own approach. One of the shortcomings of many of the accusations levelled at UPOV and the UPOV Convention is that the complexity and context of plant variety protection is conflated and subordinated to other agendas. Those opposed to the UPOV Convention tend to mobilise and use the label of intellectual property or UPOV's association with the WTO to rationalise and evoke the idea that the UPOV Convention is part of a grand plan to monopolise plants and farming and facilitate large-scale commercial farming and food production practices at the expense of poor or small-scale farmers. While such portrayals are, on some level at least, justified and worth making, they tend to reduce UPOV and the UPOV Convention to a caricature of itself. They tend to be made from a theoretical or philosophical position. This problem was acknowledged at the Diplomatic Conference of 1991, where the then President of the Council of UPOV lamented:

> We sometimes hear it suggested that the interests of plant breeders and of agriculturists are fundamentally opposed; that the protection of plant varieties only benefits the plant breeders and is always contrary to the interests of the persons who must pay to purchase the seed or propagating material of their

[41] Santilli, *Agrobiodiversity and the Law*, p. 337.

chosen protected variety! I sometimes have the impression that even the most open-minded people feel bound to defend one or the other entrenched position wherever they are talking about patents or plant breeders' rights. This kind of discussion often leaves me with the feeling that those involved are more interested in defending their institutional or sectoral interests than in having an in-depth, balanced discussion about the contents and rationale of the one or the other form of protection . . . However, I only expect this kind of posturing, or position-taking, from those who have forgotten that the only way of achieving a good result in a discussion is to consider carefully the true pros and cons of an argument.[42]

Another way to think about many of these concerns and controversies over the UPOV Convention is through what philosopher Graham Harman calls overmining and undermining. Overmining is a process of 'reduc[ing] upward rather than downward' so that an object (in this case the UPOV Convention and its concepts and practices) becomes part of a bigger event, affecting other objects.[43] An example of overmining in the context of the UPOV Convention comes from the concerns over farmers' rights, conservation and biodiversity and simple binaries such as developing countries versus developed countries or large-scale farming versus small-holder farms. By contrast, undermining is a process of reducing or dissolving the UPOV Convention down into its component parts (such as genes or DNA).[44] An example of undermining is the concern that the UPOV Convention is outmoded and obsolete because of its focus on physically observable (i.e. phenotypic) characteristics. Yet while it is one thing to reformulate plant varieties in terms of molecular structures and breeding techniques, it does not necessarily follow that the identification and description of plant varieties for the purposes of the UPOV Convention should be reduced to the same level. The process of overmining and undermining makes the UPOV Convention an easy target; however, it misses some of the nuance and context within which the UPOV Convention operates. So in order to take the UPOV Convention seriously, my starting point is to adopt a contextual or constitutive approach and identify and analyse the various heterogeneous elements or actors that have been assembled to establish, amend and maintain the UPOV Convention.[45] The significance of assemblages to the study of society, politics, science and law is nothing new and is evident in a number of theorists

[42] UPOV, *Records of the Diplomatic Conference for the Revision of the International Convention for the Protection of New Varieties of Plants*, Publication No. 337(E) (UPOV, 1992), pp. 161–162 [2].

[43] G. Harman, *The Quadruple Object* (Zero Books, 2011), p. 10.

[44] Ibid., pp. 7–19.

[45] M. DeLanda, *A New Philosophy of Society: Assemblage Theory and Social Complexity* (Continuum, 2006).

and academics who have used various terms – societies (Whitehead),[46] associations or networks (Latour),[47] togetherness (Stengers)[48] and mangle of practice (Pickering)[49] – to describe the dynamic, open-ended and contingent interactions that weave together heterogeneous elements or actors. As we will see throughout the remainder of this book, actors in the 'UPOV-assemblage' include at the very least plants, plant breeders, farmers, scientists, techniques, practices, politicians, laws and regulations, and the uses to which plant varieties are put.

There are a number of examples of this kind of contextual approach in scholarship on intellectual property law, where broad, interdisciplinary analysis has been used to examine the relationships between the law and biotechnology. Acknowledging the contextual, situated or relational nature of plant patents, Fowler argues that the Plant Patent Act 1930 is 'historically situated' and its origins and workings cannot be 'found in the simple unfolding of legal logic, an inevitable response to a major scientific advance. Nor can it be understood as the institutionalization of custom or as a simple expression of public opinion.'[50] And, in a paper titled 'Taxonomic property', Sherman writes about the way in which plant inventions are described, defined and differentiated in botany and horticulture. He argues that standard discussion and debate of plant intellectual property is often based on 'an image of the plant invention as a physical entity that is isolated and removed from the environment in which it is protected'.[51] As an example, Sherman points to academics that ascribe difference between science's shift towards genotype and law's apparent fixation on phenotype.[52] According to Sherman, one problem with this kind of reductionist thinking is that it is 'restricted both as a result of the way that they look at the plant invention and also as a consequence of the way that they think about the nature of the relationship between law and science'.[53] Inspired by the work of Whitehead, Sherman evokes the notion of 'informed materials' to argue that plant inventions embody taxonomic practices, and thus incorporate the

[46] A. Whitehead, *Process and Reality* (The Free Press, 1978).
[47] B. Latour, *Reassembling the Social: An Introduction to Actor-Network-Theory* (Oxford University Press, 2005).
[48] I. Stengers, *Power and Invention: Situating Science* (University of Minnesota Press, translated P. Bains, 1997), p. 89.
[49] A. Pickering, *The Mangle of Practice: Time, Agency and Science* (University of Chicago, 1995).
[50] C. Fowler, 'The Plant Patent Act of 1930: A sociological history of its creation' (2000) 82 *Journal of the Patent and Trademark Office Society* 621, 622.
[51] Sherman, 'Taxonomic property', 565.
[52] See, e.g., Janis and Smith, 'Technological change'.
[53] Sherman, 'Taxonomic property', 561.

informational and material environments in which they are generated. Sherman argues, in other words, that 'plant intellectual property is a hybrid of law, botany and horticulture: it is, in a sense, a form of taxonomic property'.[54]

Intellectual property law is increasingly being analysed by non-legal scholars. A collection of essays titled *Making and Unmaking Intellectual Property Law* includes contributions from anthropologists, literary theorists, historians, science studies practitioners and cultural critics, and reflects a cross-disciplinary approach to law, science and technology.[55] Broadly speaking, the collection acknowledges that the law is much more than legal doctrine and that it is, therefore, necessary to engage with the law without privileging the legal profession's point of view.[56] Indeed, the editors propose that '[i]t is only through this kind of attention to cultural production in the shadow of IP that one can find and analyse the interesting slippages between practice and its legal conceptualization'.[57] One of the chapters in the collection, written by Daniel Kelves, titled 'New Blood, New Fruits: Protection for Breeders and Originators, 1789–1930', examines the central role that business practice played in the development of the United States Plant Patent Act of 1930.[58] Although plant breeders did not generally seek patent protection in the eighteenth century – because of the belief that living organisms did not satisfy the requirements of patentability – they did protect their plant improvements using a combination of 'alternative arrangement[s]' such as trademarks, trade names and reputation.[59] And, in the end, this played a crucial role in developing protection over asexually reproduced plants from 1930 and then over sexually reproduced plants from 1970.

Acknowledging that the UPOV Convention is not isolated but is interconnected, relational and contextual, and that it accommodates and translates a range of actors is not only important to situate the UPOV Convention within broader legal, political and scientific frameworks but also for the way in which specific principles and concepts – such as plant breeder, plant variety, the scope of protection, farm-saved seed and essential derivation – are considered. Indeed, as we will see in the

[54] Ibid., 583.
[55] M. Biagioli, P. Jaszi and M. Woodmansee (eds.), *Making and Unmaking Intellectual Property: Creative Production in Legal and Cultural Perspectives* (The University of Chicago Press, 2011).
[56] Ibid., p. 11. [57] Ibid., p. 11.
[58] D. Kelves, 'New blood, new fruits: Protection for breeders and originators, 1789–1930' in M. Biagioli, P. Jazi and M. Woodmansee (eds.), *Making and Unmaking of Intellectual Property*, pp. 253–267.
[59] Ibid., p. 254.

remaining chapters, the key concepts and principles of the UPOV Convention continue to be a reworking of historical, botanical, legal, scientific and commercial particulars. By examining the history and nature of the UPOV Convention, this book aims to draw attention to its rich, complex and open-ended nature. And hopefully in doing so will help to avoid artificial or exaggerated controversies and criticisms: to know more, not less; to multiply, not subtract, information.[60] Simply stated, then, the aim of this book is to add another perspective to the scholarship on UPOV, the UPOV Convention and plant variety protection more generally.

Finally, I want to be clear about what the book is not. The book does not engage directly with doctrinal or normative issues. The book does not attempt to provide an answer to the question posed by Article 27(3)(b) of TRIPS on whether the UPOV system is an 'effective *sui generis* system' of protection or plant varieties.[61] Nor does it consider how the aims and objectives of the UPOV Convention, or plant variety rights protection more broadly, can be improved. Attempts to provide an answer to these kinds of questions have been published elsewhere and include, for example, analysis of the enforcement of plant variety rights,[62] discussions about whether the scheme provides an effective incentive to invest in plant breeding and an examination of the impact that protection has on plant breeding programs,[63] overviews of plant variety rights in different countries[64] and identification of alternative *sui generis* plant variety protection.[65]

1.4 Structure of This Book

This book has two Parts. Part I examines the emergence of the UPOV Convention during the 1950s (Chapter 2) and the expanding legitimacy of UPOV in and around plant variety protection (Chapter 3).

[60] B. Latour, 'Why has critique run out of steam? From matters of fact to matters of concern' (2004) 30(2) *Critical Inquiry* 225, 231–2.

[61] TRIPs, Art. 27(3)(b).

[62] See, e.g., G. Würtenberger, 'Questions on the law of evidence in plant variety infringement proceedings' (2006) 1(7) *Journal of Intellectual Property Law and Practice* 458; GRAIN, 'The end of farm-saved seed? Industry's wish list for the next revision of UPOV' (2007), www.grain.org/briefings/?id=202; Advisory Council on Intellectual Property, *A Review of Enforcement of Plant Breeder's Rights: Final Report* (2010).

[63] UPOV, *Report on the Impact of Plant Variety Protection.* For other attempts at assessing plant variety rights schemes, see T. Dhar and J. Foltz, 'The impact of intellectual property rights in the plant and seed industry' in J. Kesan (ed.) *Agricultural Biotechnology and Intellectual Property: Seeds of Change* (CABI Publishing Co, 2007); Tripp, Louwaars and Eaton, 'Plant variety protection in developing countries', 32.

[64] See, e.g., Janis, Jervis and Peet, *Intellectual Property Law of Plants* (including Argentina, Australia, Brazil, Canada, China, Europe, India, Japan, Korea and Mexico).

[65] See, e.g., Correa, *Plant Variety Protection in Developing Countries.*

In examining the emergence of the UPOV Convention during the 1950s, Chapter 2 provides an account of the growing economic importance of plant breeding and plant varieties, and the dissatisfaction with existing protection over plants during this time. It demonstrates that although the creation of a completely separate legal regime to regulate plant varieties marked a radical change in the international intellectual property landscape, the UPOV Convention was not the creation of a completely new legal regime, with new concepts and legal rights. The UPOV Convention was, and still is, a mixture of scientific, technical, commercial, institutional, legal and political particulars. Chapter 3 tries to establish UPOV's legitimacy in and around plant variety protection. It provides an examination of UPOV's expanding membership, and its clearly defined objectives and hierarchical structure, which set out clearly delineated functions and spheres of competence and technical abilities. Perhaps most importantly, Chapter 3 shows how UPOV and the UPOV Convention have played, and continue to play, a crucial role in the standardisation and normalisation of concepts and practices in and around plant variety protection. It becomes clear, therefore, that UPOV has legal, technical, social and normative legitimacy in and around plant variety protection.

Having considered the origins to the UPOV Convention and the legitimacy of UPOV, Part II of this book shifts the focus to the UPOV Convention's key concepts and practices. The concept of a plant breeder is the subject of Chapter 4. It shows that the UPOV Convention endowed plant breeders – long thought of as discoverers, practitioners and technicians – with a legal right over new plant varieties. Chapter 4 also points out that during the 1980s, as a consequence of engagement with concerns over the misappropriation of biological resources and other treaties such as the CBD and Plant Treaty, the place of the discoverer in intellectual property law was challenged. As a consequence, a definition of a plant breeder as a person who 'bred, or discovered and developed, a new variety' was introduced to UPOV 1991 to clarify the place of discoverers in the UPOV Convention. Chapter 4 finishes by considering the question of whether farmers can be plant breeders for the purpose of the UPOV Convention.

Taking the plant variety as its focus, Chapter 5 considers how the definition of plant variety has been rendered more legal and political. First, the definition of plant variety in UPOV 1991 'retains sufficient flexibility to accommodate the various forms that the existing types of variety will take, at the same time allowing for and efficiently satisfying expectations and clearing up the assortment of situations that caused the inclusion of an express definition to be desirable'.[66] Second, there has been an expansion and proliferation of the kinds of plant varieties included in the UPOV Convention as well as *sui generis* plant variety protection schemes. Broadly speaking, the introduction of different kinds of varieties has been with one of two intents: to strengthen plant variety rights protection (e.g. EDVs) or to recognise the role played by farmers and local communities in the development of new plant varieties (e.g. farmers' varieties and local domestic varieties which can be found in *sui generis* laws of some developing countries including India and Thailand).[67] The end result of this is that the plant variety concept has been rendered more political and legal, bringing with it the challenges of implementation, application and interpretation.

Chapter 7 examines the way in which plants are described and distinguished for the purposes of plant variety rights protection.[68] The chapter begins by establishing the central and historical role of physically observable characteristics in plant variety protection schemes and then explores some of the reasons why molecular or genotypic information (as a means of describing and distinguishing plants) are not suited to plant variety rights schemes. Most notably, molecular or genotypic information is highly reductionist and, while it is one thing to reformulate plant varieties in terms of molecular structures, it does not necessarily follow that the identification and description of plant varieties has been (or needs to be) reduced to the same level.[69] Indeed, the utilitarian nature of plant variety rights and plant breeding more generally means that end-users – such as farmers, gardeners and consumers – play an important role in the way in which plant varieties are described and distinguished. Chapter 8 considers some of the consequences of extending protection to harvested material and products derived directly from harvested material. Notably, the introduction of the term 'reasonable opportunity' provides a clear example of the increased use of legal language within the UPOV Convention, and has entrenched the use of contractual and licensing arrangements in plant variety protection. Another consequence of extending the scope of protection under the UPOV Convention is that it has given plant

[66] Rosselló, 'The UPOV Convention – The concept of variety and the technical criteria of distinctness, uniformity and stability', 59.

[67] Other countries to introduce concepts of farmer or local plant varieties include Malaysia, the Philippines and Indonesia.

[68] While the deposit of plant material is required for the UPOV Convention, Chapter 7 focuses on the description of plant varieties. UPOV 1991, Article 12. In relation to growing trials also see UPOV, *Experience and Cooperation in DUS Testing*, Document TGP/5 (UPOV, 2008).

[69] For a discussion of technological determinism, see M. Smith and L. Marx (eds.), *Does Technology Drive History? The Dilemma of Technological Determinism* (MIT Press, 1994).

variety rights a new lease of life. While plant variety rights have been criticised as outdated and obsolete, in some respects at least, plant variety rights are better able to deal with the protection of plant-related developments, even in cases in which plant varieties contain specific genetic information.

The topic of Chapter 9 is EDVs. It shows that examining and identifying EDVs is not simply a matter of determining the quantity of genetic similarities between plant varieties. Indeed, it is paradoxical to reduce the assessment of essential derivation to a scientific problem in which essential derivation is measured by way of coefficients, quotients or percentages. The relevant jurisprudence, while limited, suggests that in legal disputes over essential derivation the crucial question is whether there are 'substantial' or 'important' differences between plant varieties. Further, because of the mixture of science (quantification) and law (qualification) EDVs are an 'agreed fact', in which adopted guidelines and arbitration are crucial to the examination and identification of EDVs. It appears then that while science can quantify the ways in which plant varieties are the same, it cannot tell us whether that sameness should have any meaning. Chapter 10 examines how the practice of saving and exchanging seed is embodied in the UPOV Convention. By establishing sides and relying on dichotomies (e.g. farmers vs. industry; developed countries vs. developing countries) the issue of farm-saved seed is often reduced to political, civil or other agendas. Chapter 10 argues that little is known about the extent to which the UPOV Convention influences farmer exchange networks. Yet it is possible that the UPOV Convention and farmer exchange networks are not mutually exclusive, and that there is nothing to suggest that the implementation of UPOV-based plant variety protection will have an adverse effect on farmer exchange networks.

The conclusion, Chapter 11, reiterates that the UPOV Convention is based on open-ended and contingent engagements with scientific, legal, technical, political, social and institutional actors. It also makes a number of more specific observations about the nature of UPOV and the UPOV Convention. First, UPOV and the UPOV Convention have technical, social and normative legitimacy in and around plant variety protection. Second, UPOV and the UPOV Convention have standardised and normalised numerous concepts and practices in and around plant variety protection. In so doing, UPOV and the UPOV Convention inform and underpin the work of many actors including plant breeders, taxonomists, researchers, lawyers, administrator and advocates. Finally, over time, UPOV and the UPOV Convention have been rendered more legal and political. All of this suggests that as more countries join UPOV, and begin the processes of implementing regional or national plant variety protection laws, these countries will inform, shape and sustain future revisions to the UPOV Convention. Similar to the emergence of the UPOV Convention during the 1950s and 1960s, future revisions to the UPOV Convention are not inevitable, nor will they be sudden. Therefore, it is up to UPOV Members to set the agenda and parameters around the UPOV Convention and its key concepts and practices.

Finally, some preliminary notes and explanations. Throughout this book I use plant variety rights, rather than plant breeders' rights;[70] and I use UPOV, rather than 'the Union', to refer to the International Union for the Protection of New Varieties of Plants. Translations of the Diplomatic Conference of 1957–1961, 1972 were funded by the Australian Centre for Intellectual Property in Agriculture (ACIPA) and done by Monique Coombs.[71] All websites are valid as of 15 August 2016. Permission to reproduce the texts of the Acts of the UPOV Convention was given by UPOV, the copyright owner, on 31 March 2016.

[70] Although in some countries, such as Australia, it is referred to as plant breeder's rights.

[71] UPOV, *Actes des Conférence Internationales pour la Protection des Obtentions Végétales 1957–1961, 1972*, Publication No. 316 (UPOV, 1972).

Part I

2 The Emergence of the UPOV Convention

A New Context of Plant Breeding and Dissatisfaction with Existing Forms of Protection

> A man can patent a mousetrap or copyright a nasty song, but if he gives to the world a new fruit that will add millions to the value of earth's annual harvests he will be fortunate if he is rewarded by so much as having his name connected with the result.[1]

2.1 Introduction

The introduction of the UPOV Convention in 1961 was neither sudden nor inevitable. Rather, as this chapter will demonstrate, the emergence of the UPOV Convention was the result of an ongoing dialectic of resistance (various impediments to providing non-legal and legal protection to new varieties of plants) and accommodation (responses to the impediments).[2] More precisely, the introduction of the UPOV Convention was the result of the coalescence of numerous events including the transformation of plant breeding into both a science and a business, dissatisfaction with existing legal and non-legal protection of plant varieties, and, ultimately, the collective will of plant breeder and intellectual property organisations. As a result of the coalescing of these influences and actors, in the period from the early 1900s through to the 1960s, a range of different organisations and committees were called upon to consider, either directly or indirectly, how plant varieties could be protected at both international and national levels. Importantly, too, despite the fact that the emergence of a distinct legal regime to regulate plant varieties marked a radical change in the international intellectual property landscape, the UPOV Convention was not the creation of a completely new

[1] Luther Burbank.

[2] For a discussion, and application, of the dialectic to scientific practice, see A. Pickering and K. Guzik, *The Mangle in Practice: Science, Society and Becoming* (Duke University Press, 2008); A. Pickering, *The Mangle of Practice: Time, Agency and Science* (The University of Chicago Press, 1995).

legal regime, with new concepts and legal rights.[3] Rather, the UPOV Convention was informed by, drew from and distinguished itself from existing non-legal and legal protection of plant varieties.

2.2 A New Context of Plant Breeding

A key development in the push for the legal protection of plant varieties was a new context in plant breeding in which plant breeding was seen as a distinct and valuable endeavour. During the late nineteenth and early twentieth centuries plant breeding was being transformed into both a science and a business, and was distinguished from the growing of crops, plants, trees and flowers – something I will refer broadly to as farming. Up until this point, farmers often cultivated plants and produced propagating material on their own farms which meant that plant breeding was generally viewed as indistinct from farming. This changed, however, during the nineteenth century when plant breeders experimented on plants because of intellectual curiosity, rather than purely for determining how to improve the performance of plants.[4] One of the earliest pioneers of plant breeding was the German naturalist Joseph Gottlieb Kolreuter, who, during the mid-eighteenth century, conducted numerous experiments on plants. Kolreuter, who was more interested in discovering how plants worked rather than making better crops, pioneered the study of artificial fertilisation and plant hybridisation. More well known than Kolreuter, perhaps, it is Gregor Mendel's work on pea plants during the mid-nineteenth century that best illustrates the transformation of plant breeding into a separate science. Although largely ignored until the early 1900s, Mendel's experiments revealed the basic laws of heredity, in which characteristics are passed from one generation to another by individual factors (which we now call genes).[5] Rather than being concerned with the agronomic performance of common pea plants, Mendel

[3] According to William Cornish, the UPOV Convention emerged from a process of 'emulation' in which a wide range of existing forms of protection were used to come up with a targeted international treaty that accommodated the needs of the plant-breeding industry: W. Cornish, 'The international relations of intellectual property' (1993) 52(1) *Cambridge Law Journal* 46. See also G. Bugos and D. Kevles, 'Plants as intellectual property: American practice, law and policy in world context' (1992) 7 *Osiris* 75.

[4] C. Tudge, *Neanderthals, Bandits and Farmers* (Yale University Press, 1998), arguing that people have been controlling their environment as far back as 30,000 years. For a history of plant breeding, see N. Kingsbury, *Hybrid: The History and Science of Plant Breeding* (The University of Chicago Press, 2009); D. Murphy, *Plant Breeding and Biotechnology: Societal Context and the Future of Agriculture* (Cambridge University Press, 2007).

[5] P. Bowler, *Evolution: The History of an Idea* (University of California Press, 2003); R. Moore, 'The rediscovery of Mendel's work' (2001) 27(2) *Bioscene* 13.

chose the common garden pea for his experiments because they were easy to cultivate, grew fast and, perhaps most importantly, their traits – such as flower colour, stem length, seed shape and colour and pod shape and colour – were easily identified and observed.

Whether the advances in science during the late 1880s and early 1900s had an immediate and practical effect on improving plant varieties is disputed. In a book titled *Plants, Patients and the Historian: (Re)membering in the Age of Genetic Engineering*, Paolo Palladino examines plant breeding research in Britain in the period 1910–1940.[6] In doing so Palladino points out that plant breeders did not agree on the utility of science, such as Mendel's genetics, for farming practice and the improvement of plant varieties. Instead, the new science of plant breeding was being mobilised for different purposes. On the one hand, for example, Sir Rowland Biffen, the first director of the Plant Breeding Institute in Cambridge, believed that genetic principles were essential to developing improved plant varieties. On the other hand, John Percival of the Department of Agriculture at University College of Reading insisted that characteristics of interest to farmers, such as yield and strength, were influenced by a complex array of physiologic and environmental factors that could not be reduced to Mendelian principles.[7] In the context of the UPOV Convention, however, the exact nature and scope of the influence of science on plant breeding in the nineteenth and early twentieth centuries are, perhaps, a moot point. The end result was much the same: the distinction between plant breeding and farming was both reinforced and amplified.[8]

The upshot of the changes in plant-breeding methods and practice – whether immediate or deferred, actual or perceived – was the realisation that plant breeding was a valuable endeavour. Importantly for the transformation of plant breeding, prior to the twentieth century there was little variety, and few choices, in the foods eaten and flowers grown. As a consequence of the interest and demand for new plants, the early twentieth century was a time in which it was believed that substantial benefits would be derived from the creation of new and different plants: plants that had bigger fruits, vegetables that tasted fresher or flowers that bloomed for longer. In this way the importance of plant breeding to agriculture and horticulture was beginning to be fully acknowledged,

[6] P. Palladino, *Plants, Patients and the Historian: (Re)membering in the Age of Genetic Engineering* (Manchester University Press, 2002).

[7] Ibid.

[8] See, e.g., D. Goodman, B. Sorj and J. Wilkinson, *From Farming to Biotechnology: A Theory of Agro-Industrial Development* (Basil Blackwell, 1987); S. Krimsky, 'The profit of scientific discovery and its normative implications' (1999–2000) 75 *Chicago-Kent Law Review* 15.

and the attention of both plant breeders and consumers was being focused on the value of plant breeding.[9] And, as biological improvements obtained from plant breeding led to yield increase, improved taste and enhanced beauty, the demand for new plants intensified.

At the same time as the role of the plant breeder was being reconfigured, the piracy of plant innovations was increasingly seen as a problem for plant breeders. The problem of piracy was, in large part, due to plants themselves. More precisely, the self-reproductive nature of plants meant that it was (relatively) easy for competitors to produce and multiply a new plant variety once it had been placed on the market.[10] Indeed, the fact that living things such as plants could reproduce themselves meant that a farmer could make a one-off purchase of a plant and, by saving seed or other propagating material such as bulbs or tubers, be in a position to grow a potentially limitless number of plants on their own farms in the future. Plants were not the only thing to be pirated, however, and competitors sometimes either pirated plant names or engaged in the deceptive use of plant names on old or inferior plant varieties. Referring to the difficulties that he faced trying to safeguard his own hard work and innovations against thieves, Luther Burbank, one of the most well-known and vocal plant breeders in the United States, believed that he had

> been robbed and swindled out of my best work by name thieves, plant thieves and in various ways too well known to the originator ... A plant which has cost thousands of dollars in coin and years of intensest labor and care which is of priceless value to humanity may now be stolen with perfect impunity by any sneaking rascal ... Many times have I named a new fruit or flower and before a stock could be produced some horticultural pirate had either appropriated the name, using it on some old, well-known or inferior variety or stealing the plant and introducing it as their own, or offering a big stock as soon as the originator commences to advertise the new variety.[11]

The problem of plant-related piracy was exacerbated by a number of other changes in plant breeding occurring at the time. First, there was increasing involvement from private plant breeders and plant-breeding

[9] J. Kloppenburg, *First the Seed: The Political Economy of Plant Biotechnology* (Cambridge University Press, 1988); C. Fowler, *Unnatural Selection: Technology, Politics and Plant Evolution* (Gordon and Breach Science Publishers, 1994), p. 49.

[10] G. Würtenberger, 'Questions on the law of evidence in plant variety infringement proceedings' (2006) 1 (7) *Journal of Intellectual Property Law and Practice* 458, 466 (quoting Freda Wuesthoff).

[11] Luther Burbank to Jacob Moore, 4 May 1898, published in *Green's Fruit Grower*, June 1898 (cited in D. Kevles, 'New blood, new fruits: Protection for breeders and originators, 1789–1930' in M. Biagioli, P. Jazi and M. Woodmansee (eds.), *The Making and Unmaking of Intellectual Property: Creative Production in Legal and Cultural Perspective*, pp. 253–267, 261).

companies. Indeed, from the late nineteenth century, private individuals and plant-breeding companies began to play an increasingly important role in the breeding of non-cereal crops, as well as ornamental and decorative plants and flowers (such as roses and carnations) for private growers.[12] Importantly, private individuals and companies involved in plant breeding argued that they would not invest in plant-breeding programmes unless they could be guaranteed an economic return. Second, the ease with which plant developments and names could be copied was exacerbated by the increased trade of plants and propagating material across international and national borders, making it even more difficult for both public and private plant breeders to protect their innovations.

Taken together, the changes in plant-breeding practices and the piracy of plants and plant names meant that plant breeders were more determined than ever to protect their plant innovations. Despite the discontent of plant breeders, however, mechanisms to protect plant breeder innovations were not entirely absent, and during the late nineteenth and early twentieth centuries plant breeders had numerous legal and non-legal ways of protecting their innovations. The problem, however, as we will see in Section 2.3, was that these mechanisms were largely inadequate, or were restricted to a complex and disparate set of national laws and regulations that could not be repeated on an international level.

2.3 The Inadequacies of Existing Non-Legal and Legal Protection of Plants, the 1800s to Early 1900s

One of the oldest ways in which plant breeders have protected their innovations is through secrecy. Plant breeders, for example, went to great lengths to hide parent lines, breeding methods and techniques in the hope that this would increase the economic reward for their work. It was generally felt that if as much information as possible about plant breeding was kept secret, plant breeders had a greater chance of receiving a fair price for their plants.[13] Secrecy was so important to the practice of plant

[12] The increase in 'private' plant breeders has been considered elsewhere. In 1890, there were nearly 600 companies involved in commercial seed production. See Fowler, *Unnatural Selection: Technology, Politics and Plant Evolution*, p. 38; also see G. Dutfield, *Intellectual Property Rights and the Life Science Industries: A Twentieth Century History* (Ashgate, 2003), p. 30.

[13] B. Charnley, 'Why didn't an equivalent to the US Plant Patent Act of 1930 emerge in Britain? Historicising the Boundaries of Un-Patentable Innovation' in C. Lawson and J. Sanderson (eds.), *The Intellectual Property and Food Project: From Feeding the World to Rewarding Innovation and Creation* (Ashgate, 2013), pp. 103–122.

breeding that plant breeders tended to work alone so as to avoid other plant breeders, growers and the public more generally. At the Royal Horticultural Society Conference held in London and Cambridge in 1906, for example, the absence of plant breeders was noted. So, although there were important discussions on how to protect plant varieties, many plant breeders chose to be absent. Explaining the absence of plant breeders from the session on intellectual property, as well as highlighting the need for secrecy in plant breeding more generally, George Paul pondered why so many well-known growers were absent:

> The fact is, these gentlemen do not like to tell us, or to show, what they have done in their experiments, because when once their knowledge becomes public, they have not the slightest chance of receiving any pecuniary reward for their labours. If they were properly protected from being deprived of the due reward of their labours, they would nodoubt be much more willing to come forward and help us, and place their invaluable experience at our disposal.[14]

Another way in which plant breeders protected their innovations during the 1800s and early 1900s was through the careful development and use of reputation. In many ways, plant breeders and seed companies were at the forefront of advertising and the use of marketing to persuade customers to purchase one breeder's plants over another's was common place.[15] To enhance their reputations and make their plants stand out, plant breeders often used colourfully detailed catalogues and, where relevant, advertised any awards that they had won from agricultural and horticultural shows.[16] Again, Luther Burbank provides an example of the importance of reputation to plant breeders during the nineteenth century.[17] Burbank considered his reputation to be so important, and one of his most marketable assets, that he used various methods to market his reputation. In 1893, for example, Burbank published a fifty-two page catalogue called *New Creations in Fruits and Flowers* that highlighted his many achievements, as well as the various 'empirical methods' and scientific processes that were used at his nursery and experimental farm.[18] Eventually, because of the success of his plants including

[14] W. Wilks (ed.), *Report of the Third International Conference on Genetics 1906* (London, 1907), p. 474.

[15] See, e.g., D. Kevles, 'A primer of A, B, seeds: Advertising, branding, and intellectual property in an emerging industry' (2013) 47 *UCDL Review* 657; Bugos and Kevles, 'Plants as intellectual property: American practice, law and policy in world context'.

[16] What Berris Charnley has described as a 'moral economy' within which plant breeders worked and collaborated: B. Charnley, 'Seeds without patents' (2013) 64(1) *Revue économique* 69.

[17] Another is Britain's John Garton (1863–1922).

[18] J. Smith, *The Garden of Invention: Luther Burbank and the Business of Plant Breeding* (Penguin Books, 2010).

the Russet 'Burbank' potato and the Santa Rosa plum, and the reputation that came with that success, Burbank became known as the 'Wizard of Santa Rosa'.

Yet another way that plant breeders protected their innovations during the early twentieth century was through biological or scientific methods. One such method was hybridisation – a process by which two genetically distinct plants were crossed and resulted, after a number of generations, in progeny that possess the most desirable, or exhibit amplification of the desired, characteristics of both plants. Significantly, hybrid plant varieties also tended to exhibit hybrid vigour and have yields higher than open-pollinated plant varieties. The significance of hybridisation to plant breeding is perhaps seen best in corn. In the early 1900s, George Shull, a US plant scientist, examined the inheritance of corn traits by growing pure-bred lines of corn through self-pollination. While the pure-bred lines were less vigorous and productive, Shull noticed that when he crossed the pure-bred lines, the hybrid yields were better than any of the parents or corn plants pollinated in the open fields. In so doing, Shull recognised the potential for using hybridisation to improve crop yields, and by the 1930s and 1940s, the majority of farms in the United States were growing hybrid corn.[19]

Hybrid plants such as corn were not successful merely because of the increased vigour and yield. Seed of second (and later) generations of hybrid varieties lost some of their yield potential and uniformity. This meant that seed saved from a hybrid plant variety, when replanted, often displayed a considerable decrease in the quality and quantity of plants and crops. As a consequence, there was little, if any, benefit in saving and replanting hybrid seed, which discouraged farmers from using their own (saved) crops to propagate the next generation. Farmers had to instead purchase new propagating material each year. Because of the reduced output and quality of second generations, hybridisation served as a way of ensuring biological control over the propagating material of plant varieties, and became an effective way for plant breeders to protect their innovation from unauthorised use and exploitation, as farmers could only reliably access the sought-after trait for one, or possibly a few, generation(s).[20] So, while the improved characteristics of the hybrid varieties helped to justify the technique, it has been suggested that it

[19] See Z. Griliches, 'Hybrid corn: An exploration in the economics of technological change' (1957) 25(4) *Journal of the Econometric Society* 501; D. Duvick, 'Biotechnology in the 1930s: the development of hybrid maize' (2001) 2(1) *Nature Reviews Genetics* 69.

[20] The use of technological protection measures may also prolong the monopoly beyond the life of any plant breeder's right or patent.

was the proprietary character of hybridisation that was most appealing to plant breeders and the seed industry. Indeed, hybridised plants gave plant breeders a level of control and protection over their innovation that was not available using traditional breeding methods. According to Berlan, hybridisation was favoured by plant breeders because it provided an in-built biological protection method that was not available, or even politically possible, at the time.[21] Going even further, Cary Fowler suggested that the protective nature of hybridisation meant that plant breeders would 'divert their efforts to the creation of hybrid varieties, with an inherent biological solution to the problem [of farm-saved seed]'.[22]

Despite the fact that plant breeders made use of a range of non-legal methods – such as secrecy, reputation and hybridisation – to protect their innovations, there were crucial impediments to the level and effectiveness of protection afforded by these methods. First, the self-producing nature of plants undermined the effectiveness of using secrecy and reputation to protect plant breeders' work. Even if plant breeders were able to charge a premium for their plant varieties based on their own reputations, they still faced the very real prospect that their plant varieties would be appropriated by other plant breeders (who would sell the plants) or by farmers (who could save the propagating material). Second, relying on secrecy to protect plant innovations meant that plant breeders were often isolated and untrusting, and did not share their knowledge and skills. As we have already seen, according to Mr Paul, one of the unfortunate consequences of the secretive nature of plant breeding was that plant breeders were often absent from meetings and 'invaluable' knowledge was not available to other plant breeders.[23] Third, biological methods of protection such as hybridisation were limited because not all crops could be bred using these techniques.[24] Most notably, crops such as wheat and barley could not be reliably bred using hybridisation techniques and were therefore still thought to be highly susceptible to piracy, particularly by farmers who could easily save seed from one generation to the next.

In addition to non-legal protection afforded to plant breeders during the nineteenth and early twentieth centuries, there were a number of

[21] J.P. Berlan, 'The Political Economy of Agricultural Genetics', in R. Singh (ed.), *Thinking about Evolution: Historical, Philosophical, and Political Perspectives* (Cambridge University Press, 2000), pp. 510–528, 511.

[22] Fowler, *Unnatural Selection: Technology, Politics and Plant Evolution*, p. 61. See also Kloppenburg, *First the Seed: The Political Economy of Plant Biotechnology*.

[23] Wilks (ed.), *Report of the Third International Conference on Genetics 1906*, p. 474

[24] For a discussion of hybridisation in various crops, see S.S. Banga and S.K. Banga (eds.) *Hybrid Cultivar Development* (Springer, 1998).

legal and regulatory schemes that provided a level of protection to plant breeders.[25] These included plant patents, seed certification and testing systems, trademarks and licences. These legal schemes were generally implemented domestically; so, despite the fact that there were certain similarities between the schemes, there were important differences in both the substance and implementation of the schemes. Although there were limitations of the complex and disparate set of laws and regulations, they are the antecedents to the UPOV Convention, so it is worth considering some of these in more detail.[26]

During the early 1900s, French plant breeders used a complex mix of seed certification, trade mark and plant names to protect their work. In 1922, the French government declared that it would introduce a Register for Newly Bred Plants.[27] Under the proposed register, wheat plant breeders would be granted twelve years protection, and would be entitled to the exclusive use of a plant denomination and official mark. There were, however, a number of problems with the Register for Newly Bred Plants. Most notably, the declaration was not implemented. And although there were twenty-two wheat varieties submitted for registration (over an eight-year period), not one certificate was issued.[28] In addition to the inoperative plant register, French plant breeders could seek protection using a mixture of royalty payments, seed control laws, trade mark law and patents (for a limited number of ornamental varieties such as roses and carnations). The complex framework in which French plant breeders could obtain protection was summarised, in 1960, by the British Committee on Transactions in Seeds in the following way: (i) a breeder may trade mark the name of his variety, which enables him to license others who wish to use the trade mark when selling the variety; (ii) seed of a variety may not be sold until the variety has been officially tested and

[25] The first legislative proposal for the protection of agricultural innovations was a Papal States Edict of 1833. While the edict purported to grant ownership of new inventions and discoveries in technological arts and agriculture, it was never implemented: A. Heitz, 'The history of plant variety protection' in *The First Twenty-Five Years of the International Convention for the Protection of New Varieties of Plants* (UPOV, 1987), pp. 60–61, referring to Bernard Laclavière: see also B. Lavlavière, 'La Protection des Droits des Obtenteurs sur les Nouvelles Espèces ou Variétés des Plantes et la Convention de Paris du 2 Décembre 1961 pour la Protection des Obtentions Végétales' (1962) 168 *Bulletin Technique D'Information des Ingénieurs des Services Agricoles*.

[26] For an overview of these laws, see Committee on Transactions in Seeds, *Plant Breeders' Rights: Report of the Committee on Transactions in Seeds*, Cmnd. 1092 (London, July 1960), pp. 18–24; Heitz, 'The history of plant variety protection', pp. 63–77.

[27] See Heitz, 'The history of plant variety protection'.

[28] See B. Laclavière, 'The French law on the protection of new plant varieties' (1971) 10 *Industrial Property* 44. Also see Heitz, 'The history of plant variety protection' (citing Bernard Laclavière), p. 69.

registered under its name in the Official Catalogue of Varieties; and (iii) seed must be sold under the registered varietal name and in bags labelled with that name, and it is an offence to sell seed under any other name.[29]

A different approach was used in the United States. While the United States Congress considered introducing a 'Bill to amend the laws of patents in the interests of the originators of horticultural products' in 1892,[30] it was not until 1930 that the Plant Patent Act of 1930 extended patent protection to asexually reproduced plants.[31] Under the Plant Patent Act, plant patent holders had the right to exclude others from asexually reproducing the plant and from using, offering for sale or selling the plant or any of its parts. By excluding sexually propagated plants, and plants propagated by tubers such as potatoes, the Plant Patent Act 1930 was largely confined to decorative plants such as roses. The rationale for restricting the Plant Patent Act in this way was to exclude the main sources of food from the patent system. According to Fowler, the exclusion of sexually produced plants, as well as food crops such as tubers, served a vital political purpose, and allowed the nursery industry to convince the American Seed Trade Association that it was necessary to exclude sexually produced plants.[32] Perhaps, most importantly, excluding food crops was seen as a necessary step in the evolution of a broader, more inclusive right for plant breeders. And, at the time, it was argued that it was only a matter of time before protection was expanded to include sexually propagated plants.[33]

Another country to provide legal protection over plant varieties during this period was West Germany. In addition to seed certification, trade mark law could be used by West German plant breeders to protect generic designations.[34] However, due to the limits of seed certification and trade mark law in protecting plant breeders, a draft Seed and Seedling law was introduced into Parliament in the early 1900s. Although this draft law was never passed, the need for stronger protection of plant breeders' work was taken up by Franz and Freda Wuestoff in the 1930s, and after years of debate the Protection of Varieties and the Seeds of Cultivated Plant ('Seed Law') of 1953 was introduced into

[29] Committee on Transactions in Seeds, *Plant Breeders' Rights: Report of the Committee on Transactions in Seeds*, p. 19.

[30] Arguments before the House of Commons on Patents on H.R. 18851, To Amend the Laws of the United States Relating to Patents in the Interest of the Originators of Horticultural Products, 59th Congress, 3–18 (1906).

[31] Now sections 161–164. See P. Moser and P. Rhode, 'Did plant patents create the American rose?' in J. Lerner and S. Stern (eds.), *The Rate and Direction of Inventive Activity Revisited* (The University of Chicago Press, 2011).

[32] Fowler, *Unnatural Selection: Technology, Politics and Plant Evolution.*

[33] Ibid.

[34] Heitz, 'The history of plant variety protection', pp. 72–76.

German law. The purpose of the Seed Law was to promote the creation of useful new varieties of cultivated plants, and to do so the Seed Law granted owners the exclusive right to produce and sell seed of protected plant varieties. It also gave owners of a protected variety a registered designation that could be used by anyone that marketed a protected plant variety.

The complex and disparate use of seed certification, plant patents and trade mark in national laws was largely seen either as inadequate or as a legal experiment that would not be repeated on a broader scale. The United States Plant Patent Act model, for example, was repeated only in a very small number of other countries including Cuba (1937) and South Africa (1952). In addition, many of these laws and regulations applied differently depending on the plant or crop being registered or certified: most noticeably the United States Plant Patent Act was limited to vegetatively produced plant varieties, and many of the certification schemes varied depending on whether the plant breeders were developing food crops such as wheat, or ornamental plants such as roses or carnations. Enforcement of these laws was also a problem, and plants protected using trade mark and certification schemes could be renamed and offered for sale under different names. At the 1927 Congress of the International Institute of Agriculture, the protection of plant denominations was criticised as deficient, and seed licences were necessary to ensure that 'any grower who engaged in reproduction of those breeds for the purpose of sale to pay a royalty to the producer'.[35] Even the payment of royalties, however, was problematic, as the use of contracts and licences did not bind third parties who gained access to plants.

When thinking about how to better protect the work of plant breeders, a number of possibilities were considered. In Britain, for example, at the Third Conference of the Royal Horticultural Society, held in 1906, the possibility of using intellectual property to protect the work of plant breeders was mooted. More precisely, at the fifth session on the third day of the conference, participants considered whether plant breeders could be protected using copyright law. The session was given the title 'Copyright for the Raisers of Novelties', and Mr George Paul, a commercial rose breeder, likened plant breeders to publishers and canvassed the possibility of providing copyright protection to plant breeders due to the 'risk and labour, added to the observations and experience which have taken the best years of one's life to amass'.[36] In so doing, Mr Paul argued that plant breeders should be able to retain the sale of plant stock

[35] Ibid., p. 81.
[36] Wilks, *Report of the Third International Conference on Genetics 1906*, p. 474.

and bring an infringement action 'in the same way that one publisher can against another who pirates his productions'.[37] There was little detail about how this could be achieved, however, and Mr Paul did little more than suggest the passing of 'some resolution'.[38] The vague proposal to protect plant breeders through copyright or other means was opposed on a number of grounds. First, protecting plant breeders under copyright law, or any other scheme, was seen to be problematic because of the nature and use of plants. Professor Hansen, for example, argued that 'seedling is regarded as the gift of God, and it would be hard to patent that'.[39] Second, the proposal was opposed on fairness grounds: that is, the idea that all plant breeders benefit from sharing propagating material. One participant put it the following way: 'Mr. Paul propagates other people's new roses, and other people propagate his.'[40] Third, the proposal was opposed on pragmatic grounds in that it would be difficult, if not impossible, to implement and enforce a law over living organisms. In the end, the chairman of the session on 'Copyright for the Raisers of Novelties' concluded that there was not sufficient support for the idea of protecting plant breeders through intellectual property because while

> 'the point raised by Mr Paul is a most interesting one ... there are evidently two sides to it, as to most things, and, unless there were a very decided majority in favour of it, I do not think it would be wise for us to move in the matter'.[41]

Due to the problem of scope, selectivity and enforcement, there was a push for a more global and harmonised international approach to the protection of plant breeders' rights. In the mid-1900s, the question was not so much about whether protection of plant varieties was needed, but whether plant varieties could be protected under existing patent law, or whether a new and distinct form of protection was needed.

2.4 Towards an International Law on Plant Varieties

During the mid-1900s, calls to provide protection over the work of plant breeders grew louder as a number of organisations and groups, which sought to represent the interests of plant breeding and intellectual property more broadly, emerged and found their voice.[42] Perhaps, the most

[37] Ibid., p. 475. [38] Ibid., p. 474. [39] Ibid., p. 474. [40] Ibid., p. 475.
[41] Ibid., p. 475.
[42] Detailed accounts of the events and organisations involved in the emergence of the UPOV Convention are provided in: Heitz, 'The history of plant variety protection'; G. Dutfield, 'Turning plant varieties into intellectual property: the UPOV Convention' in G. Tansey and T. Rajotte (eds.), *The Future Control of Food: A Guide to International Negotiations and Rules on Intellectual Property, Biodiversity and Food Security*, pp. 27–47;

important of these groups was the International Association of Plant Breeders for the Protection of Plant Varieties (ASSINSEL).[43] ASSINSEL was established in 1938 with the primary objective of introducing an international convention for the protection of new varieties of plants. The First Congress of ASSINSEL, held in 1939, was concluded with a number of resolutions including accepting international trademarks and appellations as a means of protecting plant breeders, and the use of licences (which were to be drawn up by ASSINSEL) for the purposes of multiplication and sale of plant varieties.[44] Another organisation advocating for the expansion of intellectual property law was the International Association for the Protection of Intellectual Property (AIPPI). Formed in 1897, AIPPI's focus was on promoting 'understanding of the need for international protection of industrial and other intellectual property' and 'to encourage further development of the protection of intellectual property'.[45]

In order to achieve their goals of protecting the work of plant breeders and promoting intellectual property more generally, ASSINSEL and AIPPI relied on and mobilised the kind of instrumental and utilitarian justifications that had dominated intellectual property since the 1600s. One justification for granting plant breeders property rights in their plant varieties echoed Lockean notions of property acquisition: that is, plant breeders, like any other worker, have an equitable right in the products of their labour.[46] Another justification for providing plant breeders property rights was based on utilitarian grounds, and the idea that encouraging the creation of new plants is for the greatest good. Using utilitarian arguments, ASSINSEL and AIPPI argued that plant breeders should receive property rights over their plant varieties because it would provide an incentive for the production of new, modified and improved plant varieties. The result of property rights over plants, therefore, is benefit for users of plant varieties and the community more generally, with new and improved varieties of plants leading to better food crops and flowers. It was also argued that providing intellectual property protection to plant varieties was needed to help to combat serious and long-term challenges

M. Blakeney, 'Stimulating agricultural innovation' in K.E. Maskus and J.H. Reichman (eds.), *International Public Goods and Transfer of Technology under a Globalized Intellectual Property Regime* (Cambridge University Press, 2005).

43 ISF is a non-political, non-profit organisation resulting from the merger of two highly respected international organisations: FIS and ASSINSEL.

44 Heitz, 'The history of plant variety protection', p. 82.

45 AIPPI, *Objectives of AIPPI*, www.aippi.hu/en/about-aippi/33-objectives-of-aippi.html.

46 See, e.g., Committee on Transactions in Seeds, *Plant Breeders, Rights: Report of the Committee on Transactions in Seeds*, pp. 14–15.

to the plantbreeding industry such as the development of plant varieties that are resistant to pests and diseases, and the broader issue of food production.[47]

While the momentum of ASSINSEL and AIPPI was delayed by the Second World War, these organisations continued their push for the protection of plant varieties during the 1950s, and were deeply involved in determining how to protect plant varieties. During the mid-to-late nineteenth century the cause of ASSINSEL, and other plant-breeding organisations such as the International Community of Breeders of Asexually Reproduced Ornamental Varieties (CIOPORA) and the International Federation of the Seed Trade (FIS),[48] was helped by the development and expansion of intellectual property law. It was around this time that the Paris Convention for the Protection of Industrial Property 1883 (Paris Convention) and the Berne Convention for the Protection of Literary and Artistic Works 1886 (Berne Convention) were introduced and began to be implemented. While it was generally agreed that plant breeders deserved and needed protection, there was less consensus about the way in which this could be achieved. One option was to include plant varieties under existing patent laws. Another option was to introduce a new form of intellectual property distinct from patent law. Each of these options will be discussed in turn.

2.4.1 *Protecting Plant Varieties Using Existing Patent Law*

Given the well-established nature of patent law, it is not surprising that one of the preferred options for the protection of plant varieties was to have plant varieties included under the auspice of patent law. Further, given the predominant role that the International Bureau of the Union for the Protection of Industrial Property played in intellectual property law in the early part of the twentieth century, it is not surprising that they were called on to oversee early efforts to protect plant innovation.[49] To some extent, initial support for the idea that plant varieties could be protected using existing patent law relied on Article 1(3) of the Paris Convention which states that industrial property applies to 'agricultural and extractive industries and to all manufactured or natural products, for example wines, grains, tobacco leaf, fruit, cattle, minerals, beer, flowers

[47] Ibid.

[48] In 2002, ASSINSEL merged with FIS to form the International Seed Federation (ISF).

[49] The International Bureau of the Union for the Protection of Industrial Property was established by the Paris Convention for the Protection of Industrial Property 1883 to carry out administrative tasks and for conducting studies designed to facilitate the protection of industrial property.

and flour'.[50] Initially, AIPPI advocated for the inclusion of plant varieties under existing patent law, and at the AIPPI Congress of 1952, held in Vienna, the Congress adopted the following text:

The Congress expresses the view that, in order to achieve effective protection for new plant varieties, the legislation of the countries of the Union must:

1. Provide, in so far as it is not yet granted, for patent or equivalent protection for plants that possess important new properties, with a view to their exploitation, provided that their propagation is assured;
2. Place on an equal footing an invention's suitability for use in agriculture, forestry, market gardening and other comparable fields, and an invention's suitability for use in industry as provided in the patent laws of many countries.[51]

While Article 1(3) of the Paris Convention had been interpreted so that plant products including grain, flowers and flour could be considered industrial property, the problem was with the term 'products'. It appears that at the time the consensus was that plant varieties could not be protected under the Paris Convention because they were not considered to be 'products'.[52] As a consequence, it became apparent that it was going to be difficult if not impossible for patent law to provide sufficient protection for plant varieties. In 1954, even AIPPI, which had generally advocated for the inclusion of plant varieties in patent law, contemplated a form of protection distinct from patent. And at the AIPPI Congress of 1954, held in Brussels, it was resolved that 'in each of the countries of the Union, inventions relating to the plant kingdom be assimilated, with respect to legal protection, to industrial inventions and that new plant varieties be also protected'.[53]

In addition to the reservations about including plant varieties within the framework of the Paris Convention, and despite what many contemporary commentators may suggest, initial objections to the patenting of plant varieties were not based on ethical arguments such as the availability and cost of food crops or the control of nature. Instead, the objections to the patenting of plants were based on the nature of plants, the practice

[50] UPOV, *The First Twenty Five Years of the International Convention for the Protection of New Varieties of Plants*, pp. 22–25, 54 (Address by Dr Arpad Bogsch, Secretary-General of UPOV).

[51] Heitz, 'The history of plant variety protection', p. 78.

[52] See e.g., Committee on Transactions in Seeds, *Plant Breeders, Rights: Report of the Committee on Transactions in Seeds*; F. Wuesthoff, 'Patenting of plants' [1956–58] *Industrial Property Quarterly* 12; B. Laclavière, 'The Convention of Paris 2 Dec 1961 for the Protection of New Varieties of Plants and the International Union for the Protection of New Varieties of Plants' [1965] *Industrial Property* 224.

[53] Heitz, 'The history of plant variety protection', p. 80.

of plant breeding and the more pragmatic argument that plants and plant breeding did not qualify for patent protection.[54] One of the main stumbling blocks to using patent law to protect plant developments was the fact that new plants did not fit, or at least were widely seen not to fit, within the requirements of patent law.

Proponents of modifying patent law to accommodate plant varieties had to confront the conceptual problem that new plants were not seen as inventions (but as discoveries) and plant breeders were not seen as inventors (but as discoverers).[55] The distinction between invention/discovery and inventor/breeder was based on the view that inventions were creations whilst discoveries were simply findings of something not created by 'man'.[56] The characterisation of plant breeding as technical rather than intellectual presented a conceptual obstacle to including plant varieties under patent law because it was generally felt that it was difficult to satisfy the requirements needed to obtain patent protection.[57] More specifically the characterisation of plant breeding as a technical endeavour, in which perseverance and a 'breeder's eye' are central to success,[58] meant that it was generally assumed that plant-breeding developments did not meet the threshold requirement of patent law – one that excluded discoveries from the realm of patent law.[59] Similarly, the inventive-step or non-obviousness requirement (that, put generally, is that given the state of the art, new plant varieties are not obvious to other plant breeders) proved too much for plant breeders. And, the work of a

[54] In the early 1900s there were questions over the place of scientific discoveries in intellectual property law, and in the 1920s the League of Nations Committee on Intellectual Cooperation unsuccessfully attempted to gain international agreement for a scheme extending intellectual property to scientific discoveries. See D. Miller, 'Intellectual property and narratives of discovery/invention: The League of Nations' Draft Convention on "Scientific Property" and Its fate' (2008) 46 *History of Science* 299.

[55] C. MacLeod, 'Concepts of invention and the patent controversy in Victorian Britain' in R. Fox (ed.), *Technological Change: Methods and Themes in the History of Technology* (Harwood Academic Publishers, 1998), pp. 137–154.

[56] B. Sherman and L. Bently, *The Making of Modern Intellectual Property: The British Experience, 1760–1911* (Cambridge University Press, 1999), pp. 44–47.

[57] See UPOV, *Industrial Patents and Plant Breeders' Rights – Their Proper Fields and Possibilities for Their Demarcation* (UPOV, 1984), pp. 73–95.

[58] Some authors acknowledge that scientists' practice involves 'magic': see, e.g., A. Cambrosio and P. Keating, '"Going Monoclonal": Art, science, and magic in the day-to-day use of hybridoma technology' (1988) 35 *Social Problems* 244.

[59] In addition, it was argued that patent law could not adequately protect plant breeding because of the difficulty of reporting unlawful propagation. See A. Caudron, 'Plant breeding: A common understanding for public laboratories, breeding firms and users of varieties' in *The First Twenty-Five Years of the International Convention for the Protection of New Varieties of Plants*, p. 43.

plant breeder was generally considered to be 'obvious' and 'non-inventive'.[60]

The cornerstone of plant breeding was the practice of selection and propagation, where plant breeders endeavoured to pick out plants with the best combinations of desirable characteristics and then reproduce them. Grafting, a technique in which plant breeders remove a small piece of branch with dormant buds and attach it to another tree to create a new variety of plant, for example, was seen as a crude technical skill that did not warrant patent protection. The characterisation of plant breeding as a technical skill that involved both perseverance and chance is summed up by Jane Smith in *The Garden of Invention: Luther Burbank and the Business of Plant Breeding*, where she recounts how Luther Burbank noticed an unusual seed ball forming on one of his potato plants:

> Burbank marked the potato with a strip of fabric torn from the hem of his shirt . . . and waited for the seed ball to ripen. When he came one day to check on his plant and discovered that his precious seed ball had fallen from the stem, he spent three days searching on hands and knees to find the tiny thing in the potato patch. He somehow recovered it, carefully saved the twenty-three seeds inside, and waited until spring to plant them.[61]

In some jurisdictions this problem manifested itself as the so-called product of nature doctrine. Most notably, the product of nature doctrine was evident, and clearly elucidated, in US case law during the nineteenth and early twentieth centuries. The product of nature doctrine is generally traced back to the nineteenth-century case of *Ex parte Latimer*[62] and the statement in American law that to patent 'trees of the forest and the plants of the earth' would be 'unreasonable and impossible'.[63] Another significant case in relation to the product of nature doctrine is *Funk Brothers Seed Co* v. *Kalo Inoculant Co*[64] where the United States Supreme Court held a patent to be invalid 'for want of invention'. This was because the claimed invention had been discovered and not invented. In *Funk Brothers*, the United States Supreme Court reasoned that the claimed invention failed to 'disclose an invention or discovery within the

[60] Plant breeding was generally characterised by judgement rather than inventiveness. See R. Magnuson, 'A short discussion on various aspects of plant patents' (1948) 30 *Journal of the Patent Office Society* 493, 496; A. Pottage and B. Sherman, 'Organisms and manufactures: On the history of plant inventions' (2007) 31 *Melbourne University Law Review* 539.

[61] Smith, *The Garden of Invention*, p. 41.

[62] *Ex parte Latimer* [1889] Dec Comm Pat 123.

[63] Ibid., 126. For a discussion of the European perspective, see B. Fleck and C. Baldcok, 'Intellectual property protection for plant-related inventions in Europe' (2003) 4(10) *Nature Reviews Genetics* 834.

[64] 333 U.S.127 (1948) 127.

meaning of the patent statutes'.[65] In so doing, the Supreme Court laid out the product of nature doctrine stating:

> patents cannot issue for the discovery of the phenomena of nature … The qualities of these bacteria, like the heat of the sun, electricity, or the qualities of metals, are part of the storehouse of knowledge of all men … He who discovers a hitherto unknown phenomenon of nature has no claim to a monopoly of it which the law recognises. If there is to be invention from such a discovery, it must come from the application of the law of nature to a new and useful end.[66]

Under the product of nature doctrine, the argument was that plants were essentially a product of nature, the discovery of which could not confer rights on any particular individual. Although the existence of the product of nature doctrine is generally taken for granted, upon closer inspection it is unlikely that there was a *per se* exclusion on plant varieties being patented. Instead, the legal issues were more to do with description, enablement and the level of human intervention necessary to satisfy the requirements of patent law.[67] In *American Fruit Growers Inc* v. *Brogdex Co*[68] the claimed invention related to 'certain new and useful improvements in the art of preparing fresh fruit for market'.[69] The plaintiff, Brogdex, alleged that the American Fruit Growers Association had infringed a number of its products and processes claimed in United States Patent 1,529,461.[70] In the context of the product of nature doctrine, claim 26 of the patent application is most relevant as it claimed a '[f]resh citrus fruit of which the rind or skin carries borax in amount that is very small but sufficient to render the fruit resistant to blue mold decay'. The defendant argued that claim 26 was invalid on the basis that the innovation was not an article of manufacture within the meaning of US patent law. In addressing the validity of the patent, the United States Supreme Court said that the added borax 'only protects the natural article against deterioration … [t]here is no change in the name, appearance or general character of the fruit'.[71] As such, the Supreme Court concluded that the fruit remains a 'fresh orange fit only for the same beneficial uses'.[72] This kind of rationale was also expressed in *Ex parte B*,[73] a case in which a patent was refused for the treatment of seeds or plants with an organic compound that affected the plants' growth

[65] Ibid., 132. [66] Ibid.

[67] See M. Janis and J. Kesan, 'US Plant Variety Protection: Sound and Fury…?' (2002) 39 *Houston Law Review* 727, 734; R. Cook, 'The first plant patent decision' (1937) 19 *Journal of the Patent Office Society* 187, 190.

[68] *American Fruit Growers Inc* v. *Brogdex Co* 283 U.S. 1 (1931). [69] Ibid., 6.

[70] United States Patent Number 1,529,461 ('Art of Preparing Fresh Fruit for Market').

[71] *American Fruit Growers Inc* v. *Brogdex* Co 283 U.S. 1 (1931) 12. [72] Ibid.

[73] *Ex parte B* (Board of Appeal) October 1 1943 (Case No. 146).

characteristics, resulting in higher germination or root development. One ground for rejection was that the claims did not constitute inventions because they did not produce a 'new article' which possesses 'a new or distinctive form, quality, or property'.[74]

In relation to the product of nature doctrine, it is significant that the claimed orange in *Brogdex* and the seed in *Ex parte B* did not differ from other oranges or seeds. Furthermore, relying on earlier decisions, the United States Supreme Court in *American Fruit Growers* v. *Brogdex* stated that '[t]here must be a transformation, a new and different article must emerge having a distinctive name, character or use'.[75] Therefore, under US patent law, while '[t]he laws of nature, physical phenomenon, and abstract ideas [were] held not patentable',[76] it appears that a plant variety was patentable as long as it had undergone some substantial transformation.[77] These decisions of US patent law show that it was difficult to separate out plant developments that were considered discoveries (and were not patentable) from other plant developments that were invented.[78] Pottage and Sherman suggest that part of the problem is that one single plant 'is an index to many different principles or origination, many potential owners, and many different modes of scientific or therapeutic knowledge'.[79]

Another obstacle facing those who wanted to accommodate plant varieties under patent law was plants themselves. More specifically, due

[74] Ibid. [75] *American Fruit Growers Inc* v. *Brogdex Co* 283 U.S. 1 (1931) 13.

[76] *Diamond* v. *Chakrabarty* 447 U.S. (1980) 303, 309.

[77] In 1980, the United States Supreme Court considered the product of nature doctrine in the area of biological organisms in the case of *Diamond* v. *Chakrabarty* 447 U.S. 303 (1980). In a 5-4 decision, the Supreme Court overturned the Patent and Trademarks Office's denial of a patent for a microorganism, paving the way for the use of the patent system for all sorts of life forms and their genes. The Court distinguished earlier cases, differentiating between the 'non-naturally occurring manufacture or composition of matter' (at issue in *Chakrabarty*) and the naturally occurring bacteria considered in *Funk Brothers Seed Co* v. *Kalo Inoculant Co* 333 U.S. 127 (1948). The Supreme Court held that a living organism (in this case a bacterium that was modified so that it could break down petroleum) could be patented as long as it was the product of human activity and not just a discovery of an existing organism. In 1985, the United States Patent and Trademarks Office granted the first utility patent for a plant variety: *Ex parte Hibberd* 227 USPQ 443 (Board of Appeals and Interferences, 1985).

[78] See also *J.E.M. Ag Supply v. Pioneer Hi-Bred International Inc* 534 U.S. 124 (2001) 134 (quoting *Chakrabarty*, 447 U.S. at 311–12). For a discussion, see J. Conley and R. Makowski, 'Back to the future: Rethinking the product of nature doctrine as a barrier to biotechnology patents (Part I)' (2003) 85 *Journal of the Patent and Trade mark Office* 301*Trade mark*; J. Conley and R. Makowski, 'Back to the future: Rethinking the product of nature doctrine as a barrier to biotechnology patents (Part II)' (2003) 85 *Journal of the Patent and Trade mark Office* 371.

[79] Pottage and Sherman, 'Organisms and manufactures: On the history of plant inventions', 541.

to the nature of plants – particularly the way in which plants were reproduced – it was difficult, if not impossible, to reduce plant varieties to a written form such that it would satisfy the requirement of enabling disclosure. This is the doctrinal rule that embodies the idea of patents as a social contract, between the patentee and the public, granting the patentee a temporary monopoly in return for the full disclosure of how to produce the invention. The problem here was that while machines and chemicals could be disclosed with drawings and formulas, the same could not be said about plants. It is simply not possible to reproduce a plant variety from a written form.[80] In order to be able to reproduce a plant variety it was only necessary to have access to the actual plant variety so that the plant material could be used in breeding programmes as a source of heredity. The description problem was summed up by the British Committee on Transactions in Seeds (Engholm Committee), which, in 1960, examined the question of protecting the rights of plant breeders, and stated that:

> In the case of living plants, [patents] could not work in the same way as for methods of manufacture. Fundamentally, a plant is a self-reproducing mechanism. When the breeder has completed his work on a new variety, it is usually possible for others to multiply and use it without necessarily knowing how it was bred. Information about its parentage, or the methods used by the breeder, is in most cases superfluous. It is the variety itself that is important and which the breeder must be able to protect, not the method by which the variety was originated. Even if the method could be fully described and published, the information would seldom be of any help to other people who might wish to attempt to recreate the variety from the breeder's starting material.[81]

Yet another obstacle to protecting plant varieties using patent law was that it was difficult, if not impossible, to identify and demarcate the property that was being protected. Broadly speaking, intellectual property law, including patent law, requires that the subject matter of protection is reproducible. Because plants are living, dynamic organisms they do not always breed true, and thus, have a tendency to 'revert to their original types, notwithstanding the care of the seed-grower'.[82] Plant

[80] P. Rosenberg, Patent Law Fundamentals (Clark Boardman Co., 1975); J. Rossmann, 'The preparation and prosecution of plant patent applications' (1935) 17 Journal of the Patent Office Society 632; R. Allyn, *The First Plant Patents: A Discussion on the New Law and Patent Office Practice* (Educational Foundation, 1934), p. 18; R. Feldman, *Rethinking Patent Law* (Harvard University Press, 2012).

[81] Committee on Transactions in Seeds, *Plant Breeders, Rights: Report of the Committee on Transactions in Seeds*, p. 39.

[82] D. Kevles, 'Patents, protections, and privileges' (2007) 98(2) *Isis* 323, 327 at note 13 (citing J. M. Thorburn & Co. Catalogue [1908]).

breeders, therefore, could not guarantee the quality of every crop grown from their plant varieties. Writing about the difficulty of defining plant varieties, Robert Cook, in 1932, explained:

> ...plants are different from anything heretofore patentable, and different in a number of ways. Perhaps the most striking difference lies in their great variability. A lever is always a lever, a cam is always a cam, and even a complex chemical compound stays the same in molecular structure. But not so with plants. Change the conditions and the plant changes. The Washington navel orange, which is the basis of the California orange industry, is practically worthless in Florida. The conditions are different, and the plants are different too; and other varieties have to be used. [83]

Yet another obstacle to the protection of plant varieties under patent law was the nature of protection granted under patent law. Generally speaking, while a patent is a personal property right that can be assigned and licensed to others, it was seen to be inappropriate and ineffective for plant varieties. While a patent provides the patentee with an exclusive right to stop others from making, using and selling the patented invention, there was limited benefit in protecting individual plants in this way. The dual nature of plants – as both the product and the means of production – means that there was little benefit in protecting individual plants.[84] Unlike patent law where the scope of protection is generally limited by the valid claims of the granted patent, there was a need to specifically delineate the nature and scope of protection that is appropriate in the context of plant breeding.[85] On the one hand, this was needed to ensure a level of protection that would provide an incentive to breed new plant varieties. On the other hand, the delineation of the scope of protection of plant varieties was necessary so as not to unreasonably restrict the uses of protected plant varieties.

While the way that patent law was characterised in the debates during the late nineteenth and early twentieth centuries may be open to criticism – particularly in light of the experience of the United States in relation to the Plant Patents Act 1930 and the apparent increase in patent applications for plant varieties throughout the world since the late 1900s – the arguments against including plant varieties under patent law were widely accepted and, consequently, those people wanting to protect plant varieties had to look to a new form of intellectual property distinct from

[83] R. Cook, 'Patents for new plants' (1932) 27 *American Mercury* 66, 66. Also cited in Heitz, 'The history of plant variety protection', p. 65.

[84] A plant is 'not only a product . . . but carries within itself the ability to produce more of the same': Kingsbury, *Hybrid: The History and Science of Plant Breeding*, p. 381.

[85] See M. Janis and J. Kesan, 'Designing an optimal intellectual property system for plants: A US Supreme Court debate' (2001) 19 *Nature/Biotechnology* 981.

patent law.[86] The idea of a separate law designed to protect new plant varieties was taken up at the ASSINSEL Congress of 1956.[87] ASSINSEL approached the French government, which was largely supportive of stronger protection for the work of plant breeders.[88] The support of the French government opened the way for a Diplomatic Conference on the Protection of Plant Varieties, and in 1957, the French government extended an invitation to twelve countries – including Austria, Belgium, Denmark, France, Holland, Italy, Norway, Spain, Sweden, Switzerland, Britain and West Germany – to consider a separate international regime for the protection of new varieties of plants that could deal specifically with the issues and needs of plant breeders.[89] Significantly, countries that already provided limited protection over plant varieties, or may have opposed a new treaty for plant varieties, were excluded from the Diplomatic Conference. The United States, for example, was not invited to the Diplomatic Conference because it had 'confined itself to plant patents for vegetatively reproduced varieties, with at best only a minor part to play as foods' and may have been antagonistic towards a broad and inclusive regime of intellectual property for plant varieties.[90]

2.4.2 A New Regime of Intellectual Property for Plant Varieties

Developing a new international treaty for plant varieties was a challenging process. Although the Diplomatic Conference began in 1957, it took a number of years to formalise and adopt an international treaty for plant varieties. Between the first session (held from 7 to 11 May 1957) and the second session (held from 21 November to 2 December 1961) of the Diplomatic Conference, a number of meetings were held.[91] The move

[86] While the specifics of the new scheme were not discussed, the 1952 AIPPI Congress has been described as a 'welcome invitation for establishing an entirely new and exclusive form of plant variety protection': S. Bent, R. Schwaab, D. Conlin and D. Jeffrey, *Intellectual Property Rights in Biotechnology Worldwide* (Basingstoke, Macmillan, 1987), p. 51.

[87] A separate scheme for the protection of plant varieties had also been proposed at the 1952 AIPPI Congress by delegates from Luxemburg, the Netherlands, Switzerland and the United Kingdom.

[88] ASSINSEL acknowledged its own influence on the UPOV Convention, at the Diplomatic Conference of 1978 ASSINSEL, stating that: '[a]s indicated in the official records of the first Conference (Actes des Conférences Internationales pour la Protection des Obtentions Végétales, 1957–1967, 1972; page 14) this first Conference, ultimately leading to the 1961 Convention, was convened by the French Government at the proposal formulated in 1956 by ASSINSEL': UPOV, *Records on the Geneva Diplomatic Conference on the Revision of the International Convention for the Protection of New Varieties of Plants 1978*, Publication No. 337(E) (UPOV, 1981), p. 8.

[89] Heitz, 'The history of plant variety protection', p. 82. [90] Ibid., p. 82.

[91] These meetings are discussed chronologically in Heitz, 'The history of plant variety protection', pp. 53–96.

towards an international treaty for the protection of new plant varieties was characterised by a series of determined and progressive efforts. First, the first session of the Diplomatic Conference of 1957–1961 is notable for the fact that delegates recognised the legitimacy of plant breeders' rights. Second, the requirements of distinctness, homogeneity and stability were agreed upon. Third, a Committee of Experts was established to study the legal problems arising out of the protection of the breeders' rights. The Committee of Experts also made changes to the basic technical and economic principles presented at the conference, and prepared the first draft of an international convention for submission to a later session of the conference. Fourth, a Drafting Committee and various Groups of Legal Experts were established to examine the protection of plant breeders' rights in new plant varieties.[92]

Delegates at the second session of the Diplomatic Conference, held between 21 November and 2 December 1961, were presented with a draft international treaty on the protection of plant varieties. In preparing and considering the draft text, conference delegates, the Committee of Experts, the Drafting Committee and the various Groups of Legal Experts had to answer many of the questions discussed earlier in relation to patent law. For example: How can the work of plant breeders be recognised? How can plant varieties be described? What was the appropriate scope of protection for plant varieties? After some debate, the UPOV Convention was signed on 2 December 1961 by delegates from Belgium, France, the Federal Republic of Germany, Italy and the Netherlands. In 1962, Denmark and the United Kingdom signed the UPOV Convention. Then, on 10 August 1968, with ratification by the Netherlands, the Federal Republic of Germany and the United Kingdom, the UPOV Convention entered into force. The key to the success of the Diplomatic Conference 1957–1961 was the ability to be able to address the particular issues relevant to plant breeding, which was reflected in the preamble to UPOV 1961 which states that UPOV members were 'conscious of the special problems arising from the recognition and protection of the right of the creator in this field'.[93]

2.5 Conclusion

What does the emergence of the UPOV Convention tell us about the nature of the Convention? The emergence of the UPOV Convention as

[92] Other groups to be established included the Group of Legal Experts on the Relations Between the Protection of the Names of New Plant Varieties and Trademark Protection and the Group of Legal Experts on the Relations Between the Paris Convention for the Protection of Industrial Property and the Preliminary Draft of the Convention for the Protection of New Varieties of Plants: Ibid., pp. 82–88.

[93] UPOV 1961, Preamble.

the most significant international treaty that regulates plant development was not an inevitable response to advances in science and technology. Rather, the development of the UPOV Convention was the result of an ongoing dialectic of resistance and accommodation, involving a large number of influences and actors. The growing economic importance of plants, the intentions and will of breeder and intellectual property organisations, the nature and characteristics of plants as biological entities and the (un)suitability of existing protection mechanisms – including hybridisation, patent, trade mark and certification laws – played a part in the emergence of the UPOV Convention.

Importantly, though, despite the fact that the emergence of a separate legal regime to regulate plant varieties marked a radical change in the international intellectual property landscape, the UPOV Convention was not the creation of a completely new legal regime, with new concepts and legal rights. The emergence of UPOV-engaged actors from many different legal, scientific, technical, commercial and political perspectives. In this way, the UPOV Convention and its concepts and practices are embedded with and have links to past practices, people and performances. Perhaps the interwoven, interconnected and contingent nature of the UPOV Convention's emergence and development can be best summed up by the sentiments of Andrew Pickering, a scientist, sociologist and philosopher who suggests that in the same way in which 'new machines are modelled on old ones, so are new conceptual structures modelled on their forebears'.[94] Acknowledging that the UPOV Convention is not isolated but is modelled on its 'forebears' is not only important to situate the UPOV Convention within broader legal, political and scientific frameworks but also for the way in which specific principles and concepts – such as plant breeder, plant variety, the scope of protection, farm-saved seed and EDVs – are considered. Indeed, as we will see in the remaining chapters of this book, the key concepts and principles of the UPOV Convention continue to be a reworking of historical, botanical, legal, scientific and commercial particulars. Before turning to these concepts and practices in Part II, it is worth examining the intergovernmental organisation that oversees the UPOV Convention: the International Union for the Protection of New Varieties of Plants (UPOV).

[94] A. Pickering, 'Concepts and the mangle of practice: Constructing quaternions' in R. Hersh (ed.) *18 Unconventional Essays in the Nature of Mathematics* (Springer Science and Business Media Inc, 2006), pp. 250–288, 253.

3 UPOV's Legitimacy

From Members and Trade to Objectives, Structure and Norms

3.1 Introduction

The International Union for the Protection of New Varieties of Plants ('UPOV') has been, and continues to be, under scrutiny. Established in 1961, UPOV is the central body dealing with the protection of new plant varieties and plays a crucial role in international policy aimed at promoting the development of new plants.[1] However, there continues to be a contest for legitimacy over plant variety protection, and some commentators, policy makers and advocates have accused UPOV of lacking transparency, being unaccountable and, in its relationships with developing countries, misusing its influence and power.[2] UPOV has also been criticised for being self-serving and biased, with some civil society organisations and governments urging developing and least developed countries to avoid UPOV.[3] During the late 1990s, for example, GRAIN argued that UPOV was biased towards industrial agriculture and had pressured many developing countries into joining UPOV.[4] More recently, UPOV has been accused of ignoring its own Convention by approving the African Regional Intellectual Property Organisation's ('ARIPO') *Draft Protocol for the Protection of New Varieties of Plants*

[1] UPOV is an intergovernmental organisation based in Geneva, Switzerland: UPOV 1991, Article 24.

[2] In 2011, for example, Graham Dutfield likened UPOV to an exclusive, selective 'club' and emphasised concerns over transparency, democratic accountability and lack of public debate: G. Dutfield, *Food, Biological Diversity and Intellectual Property: The Role of the International Union for the Protection of New Varieties of Plants (UPOV)* (Quakers United Nations Office, 2011), pp. 12–14.

[3] See C. Correa, *Plant Variety Protection in Developing Countries: A Tool for Designing a Sui Generis Plant Variety Protection System: An Alternative to UPOV 1991* (APBREBES, 2015), p. 4. GRAIN/GAIA, 'Ten reasons not to join UPOV', *Global Trade and Biodiversity in Conflict*, Issue No. 2 (1988), www.grain.org/article/entries/1-ten-reasons-not-to-join-upov; J. Ekpere, *TRIPs, Biodiversity and Traditional Knowledge: OAU Model Law on Community Rights and Access to Genetic Resources* (UNCTAD/ICTSD, 2000).

[4] See GRAIN, 'UPOV on the warpath', *Seedling* (June 1999), www.grain.org/es/article/entries/257-upov-on-the-war-path; GRAIN/GAIA, 'Ten reasons not to join UPOV'.

('*Draft Protocol* ') on 11 April 2014.[5] At times these criticisms of and challenges to UPOV have manifested in public protest. For example, there have been public protests by farmers, indigenous communities and civil society groups against UPOV and the UPOV Convention in Thailand (2013), Costa Rica (2014), Ghana (2014) and Chile (2014).[6]

Many of the criticisms of UPOV, and the UPOV Convention, have been informed and sustained by interests and values that are different to, and sometimes compete with, those championed by UPOV. The proliferation of forums and institutions dealing with plants and plant genetic resources has meant that there are a range of differing interests related to plants, plant genetic resources and people. For example, the CBD and the Nagoya Protocol impose an obligation to equitably share the benefits of genetic resources.[7] And the Plant Treaty aims to guarantee food security through the conservation, exchange and sustainable use of the world's plant genetic resources for food and agriculture and has in so doing recognised farmers' rights.[8] The differing and sometimes competing interests over plants and plant breeding include the conservation of biodiversity; farmers' rights and practices such as saving, exchanging and selling saved seed; and food security and sovereignty. As a consequence some countries – such as India, Thailand, Malaysia and the Philippines – have opted not to join UPOV and have instead implemented *sui generis* national plant variety protection laws.[9] In large

[5] UPOV, 'UPOV Council Holds its Thirty-First Extraordinary Session' (Press Release, 96, 11 April 2014), www.upov.int/export/sites/upov/news/en/pressroom/pdf/pr96.pdf. Specifically, it has been suggested that in ratifying ARIPO's Draft Protocol UPOV ignored Articles 34 and 30(2) of UPOV 1991: See e.g., S. Gura, *UPOV Breaking Its Own Rules to Tie-In African Countries* (11 April 2014) APBREBES, www.apbrebes.org/press-release/upov-breaking-its-own-rules-tie-african-countries; Alliance for Food Sovereignty in Africa, *ARIPO sells out African Farmers, seals Secret Deal on Plant Variety Protection* (8 July 2015), http://afsafrica.org/aripo-sells-out-african-farmers-seals-secret-deal-on-plant-variety-protection/. The Draft Protocol was replaced with the *Arusha Protocol for the Protection of New Varieties of Plants* on 6 July 2015: see H. Haugen, 'Inappropriate processes and unbalanced outcomes: Plant variety protection in Africa goes beyond UPOV 1991 requirements'(2015) 18(5) *The Journal of World Intellectual Property* 196.

[6] See D. Jefferson, 'Development, farmers' rights, and the Ley Monsanto: the struggle over the ratification of UPOV 91 in Chile' (2014) 55(1) *IDEA* 31; GRAIN, *Thai Farmers and Civic Groups Protest UPOV Lobby* (19 November 2013), www.grain.org/bulletin_board/entries/4833-thai-farmers-and-civic-groups-protest-upov-lobby; APBREBES, *Massive Protests in Ghana over UPOV Style Plant Breeders' Rights Bill* (11 February 2014), www.apbrebes.org/news/massive-protests-ghana-over-upov-style-plant-breeders-bill.

[7] *Convention on Biological Diversity* (1992) 1760 UNTS 79 ('CBD').

[8] *International Treaty on Plant Genetic Resources for Food and Agriculture* (2001) 2400 UNTS 303 ('Plant Treaty').

[9] In 2002, India attempted to join UPOV, although the proposed national laws were deemed to be incompatible with the UPOV Convention by the UPOV Council,

part the governments of these countries believe that their own approaches are better able to meet the needs and objectives of plant breeders and farmers and at the same time achieve the goals of conservation, biodiversity and food security.

Unfortunately, while these issues are important, some of the discussions and criticisms of UPOV, and the UPOV Convention, are underpinned by these competing interests. As a consequence much of the criticism of UPOV is rich with rhetoric, dichotomy and irreconcilable difference. In this chapter I examine UPOV (rather than the UPOV Convention) and highlight and sketch some of the sources of UPOV's legitimacy over plant varieties. But first, what do I mean by legitimacy? While there are numerous theories and definitions of legitimacy,[10] in its simplest form, legitimacy means that an institution, such as UPOV, is broadly recognised and accepted as a representative institution. Weber, for example, defined legitimacy in terms of both recognition and belief: so that an institution is legitimate if it is representative, recognised as legitimate, and other actors believe that it is legitimate.[11] So, in this chapter I will examine whether UPOV is broadly recognised and accepted as a representation institution. Of course not everyone accepts UPOV nor its actions. However, legitimacy transcends specific acts (e.g. problems with ARIPO), and despite the reservations that some governments and civil society groups have about UPOV, it is necessary to consider whether the 'international community' as a whole supports UPOV. Importantly, too, sources of legitimacy are not always commensurate; they are not mutually exclusive; and they are contingent. Nonetheless, by identifying some of the sources of UPOV's legitimacy, I am able to show that despite accusations of bias, unaccountability and so on, UPOV is recognised and accepted as a representative institution. Importantly, though, this chapter is not intended as a definitive list or treatment of all of the sources of UPOV's legitimacy, but a condensed and partial one. As we will see in this chapter, UPOV has legitimacy based on a mix of factors including expanding membership,[12] and clearly defined

particularly in relation to farmers' rights: see Matthews, *Intellectual Property, Human Rights and Development: The Role of NGOs and Social Movements* (Edward Elgar, 2011), pp. 196–198.

[10] For an analysis and discussion of organisation legitimacy, see M. Suchman, 'Managing legitimacy: Strategic and institutional approaches' (1995) 20(3) *Academy of Management Review* 571.

[11] M. Weber, *The Theory of Social and Economic Organization* (The Free Press, 1947), translated by A.M. Henderson and Talcott Parsons, p. 213.

[12] As of 15 April 2016, UPOV had seventy-four members, and a further sixteen states and one intergovernmental organisation have initiated the procedure for acceding to the UPOV Convention, and a further twenty-four states and one intergovernmental

objectives and hierarchical structure (which sets out clearly delineated functions and spheres of competence and technical abilities). Perhaps most importantly, UPOV has played and continues to play a crucial role in the standardisation and normalisation of concepts and practices in and around plant variety protection. It is therefore clear that despite some misgivings, UPOV has legal, technical, social and normative legitimacy over plant variety protection.

3.2 Expanding Membership

The legal basis for UPOV's existence can be found in the UPOV Convention. Article 1(2) of UPOV 1961 established that the Member States to the UPOV Convention 'constitute a Union for the Protection of New Varieties of Plants'. The existence of UPOV was subsequently reiterated and reinforced by UPOV 1978 and UPOV 1991.[13] The UPOV Convention clearly establishes that it is the Members that constitute UPOV and that, therefore, UPOV is a member-driven institution. And as a member-driven institution, UPOV relies on its members for finances, structures, decision making and ultimately its legitimacy over plant variety protection. Further, despite a close and peculiar relationship with WIPO,[14] UPOV has legal personality[15] and enjoys the legal capacity 'necessary for the fulfilment of the obligations of the Union and for the exercise of its functions' in each Member State.[16] This is a prerequisite for UPOV signing the UPOV Convention in its own name and means that UPOV can make and amend rights and obligations around plant variety protection. UPOV also has the capacity to exercise its functions in each of its Member States.

organisation have sought assistance with the development of laws based on the UPOV Convention: UPOV, *Members of the International Union for the Protection of New Varieties of Plants*, (15 April 2016), www.upov.int/export/sites/upov/members/en/pdf/pub423.pdf.

[13] UPOV 1978, Articles 15–28; UPOV 1991, Articles 23–29.

[14] Although UPOV is formally independent of WIPO, there is substantial overlap of some of the administrative functions and resources of UPOV and WIPO. For example, the UPOV Office and headquarters are located in the same building as WIPO; the Secretary-General of UPOV is the Director-General of WIPO (although the current Secretary-General has not taken a salary from UPOV); WIPO is involved in the appointment of the Vice Secretary-General of UPOV; and UPOV receives a number of administrative support services from WIPO. See UPOV, *Agreement between the World Intellectual Property Organisation and the International Union for the Protection of New Varieties of Plants*, UPOV Document No. UPOV/INF/8 (26 November 1982), www.upov.int/edocs/infdocs/en/upov_inf_8.pdf.

[15] UPOV 1991, Article 24(1). [16] UPOV 1991, Article 24(2).

Because UPOV is a member-driven institution and it draws much legitimacy from its members, it is necessary to trace the expansion of UPOV's membership.

The reasons for UPOV's expanding membership are complex, varied and contingent. In her book, *The Implementation Game: The TRIPS Agreement and the Global Politics of Intellectual Property Reform in Developing Countries*, Carolyn Deere calls joining and implementing international intellectual property treaties an 'implementation game': a process in which there is a mix of complex politics, diverse needs of countries, capacity building and expertise and pressure (particularly on developing countries).[17] As we saw in Chapter 2, UPOV was founded by a handful of European countries so that the early iterations of it consisted of, and were limited by, a small and homogenised membership.[18] By the mid-1970s, there were still only eight members of UPOV, none of which from developing countries. For much of the early to mid-twentieth century, then, UPOV was largely a specialist agency that had a homogenous and narrowly focused membership. In this sense UPOV was, and to some extent still is, seen as proxy for the European plant breeding organisations that had been instrumental to its formation. Importantly, UPOV acknowledged that its small, homogenous membership was not ideal and from as early as the 1970s actively sought to increase its membership. For example, in 1974, a meeting was held to 'discuss the conditions which might need to be fulfilled to make UPOV attractive to states which did not yet belong to it'.[19] In the end, UPOV oversaw various amendments to the UPOV Convention, with key revisions of the Convention occurring in 1978 and 1991, which were in part motivated by a desire to attract new members. While these revisions were a response to various factors – including changes in technology, advances in plant breeding and lobbying from various plant breeding organisations – there was also the realisation by UPOV that the substantive provisions of the UPOV Convention needed to be amended to attract new Members. Heitz states that by the mid-1970s:

> member States of UPOV had already realized that there was a need to revise the substantive provisions of the Convention and to ascertain whether they were not too stringent for States that otherwise were favourably disposed towards

[17] C. Deere, *The Implementation Game: The TRIPS Agreement and the Global Politics of Intellectual Property Reform in Developing Countries* (Oxford University Press, 2009).

[18] The first signatories were Belgium, France, the Federal Republic of Germany, Holland and Italy. Followed, in 1962, by Denmark, the United Kingdom and Sweden.

[19] UPOV, *Summary of the Main Amendments to Convention in the Revised Text of 1978*, Document DC/PCD/I (referring to UPOV Publication No. 330 (1975)) in UPOV, UPOV Publication No. 337(E).

introducing a system of plant variety protection and acceding to UPOV, but had difficulty in conforming to the Convention.[20]

In order to attract new members, a number of concessions were made leading up to the Diplomatic Conference of 1978. Most importantly perhaps there was a pragmatic, if not flexible, mind-set among delegates that made compromise possible. It was explicitly acknowledged that concessions were necessary and that 'it was unavoidably necessary to sacrifice somewhat the ideal of the founders of the 1961 Convention'.[21] One of the key concessions made was to allow non-members and observer organisations to participate in the Diplomatic Conference of 1978 and contribute to the revisions of the UPOV 1961. As a result, twenty-seven non-members (called 'observer delegates') – including Argentina, Australia, Bangladesh, Brazil, Canada, Mexico, Thailand and the United States – participated in the Diplomatic Conference of 1978. In addition, three international organisations (i.e. the FAO, European Economic Community and International Seed Testing Association) and six international non-governmental organisations (e.g. AIPPI, ASSINSEL and the International Commission for the Nomenclature of Cultivated Plants of the International Union of Biological Sciences) participated in the Diplomatic Conference of 1978.[22] Non-members were also permitted to submit proposals to the Diplomatic Conference of 1978 and in so doing were able to comment on a range of issues important or relevant to them. Bangladesh and India, for example, submitted official documents on variety denomination, and Mexico and Peru argued for the inclusion of Spanish as an official language of UPOV.[23] Non-members also expressed the relevance of broader social and political considerations on their decision to join, or not join, UPOV. For example, the delegate for Libyan Arab Jamahiriya, Dr. A. Ben Saad, stated that his country lamented:

> the fact that the Republic of South Africa, which practised racial discrimination was a member of the Union, and moreover that the Republic of South Africa had been elected to serve in the Credentials Committee. This would seriously affect the desire of many countries, including the Socialist People's Libyan Arab Jamahiriya, which would like to join the Union but which could not do so under those circumstances.[24]

[20] A. Heitz, 'The history of plant variety protection' in *The First Twenty-Five years of the International Convention for the Protection of New Varieties of Plants* (UPOV, 1987), p. 91.

[21] Heitz, 'The history of plant variety protection', p. 92.

[22] For a full list of participants, see UPOV, Publication No. 337(E), pp. 290–296.

[23] Ibid., Documents 8, 65 and 66, respectively. [24] Ibid., p. 131 [53].

To facilitate countries joining UPOV, a number of changes were eventually made to the UPOV Convention by UPOV 1978. A summary of the changes made in aid of membership are summarised in UPOV document, *Amendments to Facilitate the Joining of the Union by Further States*,[25] and include changes to the preamble, the botanical genius and species protected, novelty and variety denomination. Further, UPOV 1978 removed the supervision of the Swiss Confederation and gave UPOV legal personality and capacity to carry out its functions in Member States.[26] While there were a number of changes made to the UPOV Convention explicitly to encourage membership, nonetheless, UPOV remained small and homogenous. There were still only eighteen members by 1 January 1990, all of which were developed countries, most of them European.[27]

Further attempts to make UPOV membership attractive were made leading up to UPOV 1991. The Diplomatic Conference of 1991 was attended by twenty-five 'observer delegations' including Argentina, Austria, Bolivia, Brazil, Cameroon, Chile and Ecuador[28] and twenty-five 'observer organisations' including WIPO, GATT, FAO, AIPPI and CIOPORA. These observer delegations and organisations made numerous submissions and proposals on topics such as farm-saved seed, the breeder's exemption, dependency and the extension of protection beyond propagating material. The opportunity to contribute to these discussions was valued and acknowledged by the observer delegations and organisations: for example, Mr. Clucas from ASSINSEL said that inclusion 'had created a fertile debate and a creative momentum, and ASSINSEL hoped that the same climate of creativity would continue'.[29] The Diplomatic Conference of 1991 was, however, a difficult negotiation as there were somewhat contradictory goals: on the one hand, a number of participants were strongly advocating for strengthening the UPOV Convention, so that it more closely resembled patent law. For example, in comparing the UPOV system of plant variety protection with patents, the International Chamber of Commerce argued that the UPOV Convention was 'inherently less suitable to stimulate the desired research and progress' because it 'provides neither the necessary degree of exclusivity

[25] Ibid., p. 281 (Document DC/PCD/1). [26] Ibid.

[27] The members were Australia, Belgium, Denmark, France, Germany, Ireland, Israel, Italy, Japan, the Netherlands, New Zealand, Poland, South Africa, Spain, Sweden, Switzerland, the United Kingdom and the United States.

[28] Although many more were invited: UPOV, *Records on the Diplomatic Conference for the Revision of the International Convention for the Protection of New Varieties of Plants 1991*, UPOV Publication No. 346(E) (UPOV, 1992).

[29] Ibid., p. 183 [80.1].

to stimulate the heavy research investment required, nor the necessary element of early public description and disclosure to aid further research'.[30] On the other hand, developing countries were hoping to make the UPOV Convention more flexible and amenable to their diverse national interests, particularly around farmers' rights, biodiversity and the fair and equitable sharing of benefits arising from the use of plant varieties.

In the end, the push for stronger plant variety protection, and the desire to have the United States join UPOV, won out. And the changes made by UPOV 1991 increased the strength of plant variety protection including the removal of the ban on dual protection, making the farm-saved seed exception optional, extending protection to harvested material and products derived from the harvested material in certain circumstances and introducing the concept of essential derivation.[31] While these changes strengthened the UPOV Convention, and went some way to attracting membership of developed countries, it was acknowledged that UPOV 1991 was likely to be less attractive to developing countries. As a consequence, transitional arrangements were put in place so that, in some ways at least, developing countries were not completely overlooked by the UPOV Council: UPOV 1978 was left open for ratification by developing countries until UPOV 1991 came into force.[32]

Yet it was not until the mid-1990s that developing countries began joining UPOV and implementing national UPOV-based plant variety protection. And international treaties and trade agreements have played a key role in expanding UPOV's membership, with more than fifty countries – including Argentina (1994), Brazil (1999) and Mexico (1997) – joining UPOV from the mid-1990s. One of the key reasons for the expanding UPOV membership was the inclusion of intellectual property and plant variety protection in broader international regimes related to trade, particularly through the World Trade Organization's (WTO) *General Agreement on Tariffs and Trade 1994* (GATT) and bilateral and regional trade agreements.[33] TRIPS, which formed part of the broader WTO-GATT agreement, imposed a mandatory obligation on

[30] (1978) 18(2) *IIC* 223, 226.

[31] For a summary of the changes made of UPOV 1991, see, e.g., N. Byrne, *Commentary on the Substantive Law of the 1991 UPOV Convention for the Protection of Plant Varieties* (University of London, 1996).

[32] UPOV 1991, Article 37(1). UPOV 1991 came into force on 24 April 1998 which meant that developing countries initially had until 24 April 1999 to sign up to UPOV 1978.

[33] The *General Agreement on Tariffs and Trade 1994* (GATT) entered into force on 1 January 1995.

WTO members to provide some form of protection for plant varieties in their national laws.[34] Specifically, Article 27(3)(b) of TRIPS states that '[m]embers shall provide for the protection of plant varieties either by patents or by an effective sui generis system or by any combination thereof'.[35] It was, therefore, up to WTO member states to determine whether to protect plant varieties using patents, an effective *sui generis* system, or a combination of both.

There was no assistance given to countries as to what an 'effective sui generis system' might be. And while explicit reference to the UPOV Convention in Article 27(3)(b) was contemplated, it was not included in TRIPS because some countries believed that the UPOV Convention was not suitable for them and consensus about the role of the UPOV Convention could not be reached.[36] Another reason for the omission of specific reference to the UPOV Convention from TRIPS was more pragmatic and related to timing: UPOV was 'in-between' Conventions at the time TRIPS was negotiated. As Watal notes, 'a reference to UPOV 1978 was considered inadequate, while a reference to UPOV 1991 was considered premature'.[37]

Despite TRIPS being silent on the UPOV Convention, it was, and still is, generally accepted that national laws based on the UPOV Convention will be recognised as an 'effective sui generis' system of plant variety protection. Despite the ambiguity of Article 27(3)(b), TRIPS has nonetheless been good for UPOV membership. Le Buanec, for example, points out that 'one of the primary effects of the adoption of the TRIPS Agreement is the strong increase in the numbers of members of UPOV, from 24 States in 1994 to 58 States and one intergovernmental organization (the European Community) in July 2004'.[38]

More recently UPOV membership has benefitted from trade and economic partnership agreements particularly those involving the United States and Europe. Many of these trade and economic agreements became known as 'TRIPS-Plus' because the overall effect of them was to strengthen national intellectual property laws beyond those set out in

[34] For a discussion of the negotiation and provisions of TRIPS, see D. Gervais, *The TRIPS Agreement: Drafting History and Analysis* (Sweet & Maxwell, 2012). For consideration of interpretative possibilities for the TRIPS' text, see J. Malbon, C. Lawson and M. Davison, *The WTO Agreement on Trade-Related Aspects of Intellectual Property Rights: A Commentary* (Edward Elgar, 2014).

[35] TRIPS, Article 27(3)(b). Least developed countries do not currently need to provide intellectual property protection for plant varieties: TRIPS, Article 66(1).

[36] Gervais, *The TRIPS Agreement*.

[37] J. Watal, *Intellectual Property Rights in the WTO and Developing Countries* (Springer, 2001), p. 14.

[38] B. Le Buanec, 'Protection of plant-related innovation: Evolution and current discussion' (2006) 28 *World Patent Information* 50, 53.

TRIPS.[39] While some trade agreements contemplate and allow non-UPOV protection of plant variety rights,[40] the majority of trade and economic agreements are not as flexible as TRIPS and oblige partner countries to join or reaffirm their commitment to UPOV and the UPOV Convention. Numerous countries – including Bangladesh, Chile, Costa Rica, Jordan, Malaysia, Morocco and Singapore – have been obliged to join UPOV 1991 as a consequence of various trade agreements they have entered into. Some notable trade agreements and their provisions on plant variety protection include the:

- North American Free Trade Agreement (NAFTA):[41] Article 1701 of chapter 17, *Intellectual Property*, requires signatories to 'at a minimum, give effect to' UPOV 1978 or UPOV 1991.
- United States-Central America Free Trade Agreement (CAFTA):[42] Article 5(a) of chapter 15, *Intellectual Property Rights*, requires signatories to 'ratify or accede to [UPOV 1991]'.[43]
- Economic Partnership Agreement between West African and the European Union (EPA):[44] The specific provisions on intellectual property are to be completed; Article 106 states that parties would continue to negotiate on 'intellectual property and innovation'.
- Trans Pacific Partnership ('TPP'):[45] Article 18.7.2(d) and Annex 18-A state that if parties are not already a member of UPOV, they will 'ratify or accede' to UPOV 1991 'by the date of entry into force of this Agreement'.[46]

The specific impact of various trade agreements on UPOV's legitimacy is important but varied. Most broadly, the requirement to accede to, or ratify, UPOV indicates that UPOV is an accepted way of providing

[39] See M. El-Said, 'The road from TRIPS-Minus, to TRIPS-Plus' (2005) 8(1) *The Journal of World Intellectual Property* 53; P. Drahos, *Information Feudalism: Who Owns the Knowledge Economy* (Earthscan, 2002).

[40] The *Contour Agreement*, for example, requires signatories to implement some form of plant variety protection but not necessarily a scheme that is UPOV-based: Contour Agreement, Article 46.1.

[41] Signed, in 1993, by Canada, the United States and Mexico, and taking effect from 1994.

[42] Signed, in 2004, by Costa Rica, El Salvador, Guatemala, Nicaragua, the Dominican Republic and the United States, and taking effect between 2006 and 2009.

[43] The CAFTA also specified the dates by which signatories have to have met this obligation: Nicaragua (1 January 2010), Costa Rica (1 June 2007) and all other parties (1 January 2006).

[44] Signed, in 2014, by the European Union and sixteen West African states.

[45] Signed on 5 October 2015 by Australia, Brunei, Canada, Chile, Japan, Malaysia, Mexico, Peru, the United States and Vietnam. Although, the future of the TPP is uncertain: on 24 January 2017, United States President, Donald Trump, signed an Executive Order withdrawing the U.S. from the TPP.

[46] Article 18.3.7(4) of the TPP also requires parties to ensure 'patents are available at least for inventions that are derived from plants'.

Table 3.1 *Members of UPOV (as of 22 October 2015)*[47]

State/organisation	Date state/organisation became a UPOV Member	Convention to which state/ organisation is party
African Intellectual Property Organisation	10 July 2014	UPOV 1991
Albania	15 October 2005	UPOV 1991
Argentina	25 December 1994	UPOV 1978
Australia	1 March 1989	UPOV 1991
Austria	14 July 1994	UPOV 1991
Azerbaijan	9 December 2004	UPOV 1991
Belarus	5 January 2003	UPOV 1991
Belgium	5 December 1976	UPOV 196/1972
Bolivia	21 May 1999	UPOV 1978
Brazil	23 May 1999	UPOV 1978
Bulgaria	24 April 1998	UPOV 1991
Canada	4 March 1991	UPOV 1991
Chile	5 January 1996	UPOV 1978
China	23 April 1999	UPOV 1978
Colombia	13 September 1996	UPOV 1978
Costa Rica	12 January 2009	UPOV 1991
Croatia	1 September 2001	UPOV 1991
Czech Republic	1 January 1993	UPOV 1991
Denmark	6 October 1968	UPOV 1991
Dominican Republic	16 June 2007	UPOV 1991
Ecuador	8 August 2007	UPOV 1978
Estonia	24 September 2000	UPOV 1991
European Union	29 July 2005	UPOV 1991
Finland	16 April 1993	UPOV 1991
France	3 October 1971	UPOV 1991
Georgia	29 November 2008	UPOV 1991
Germany	10 August 1968	UPOV 1991
Hungary	16 April 1983	UPOV 1991
Iceland	3 May 2006	UPOV 1991
Ireland	8 November 1981	UPOV 1991
Israel	12 December 1979	UPOV 1991
Italy	1 July 1977	UPOV 1978
Japan	3 September 1982	UPOV 1991
Jordan	24 October 2004	UPOV 1991
Kenya	13 May 1999	UPOV 1978
Kyrgyzstan	26 June 2000	UPOV 1991
Latvia	30 August 2002	UPOV 1991
Lithuania	10 December 2003	UPOV 1991
Mexico	9 August 1997	UPOV 1978
Montenegro	24 September 2015	UPOV 1991

[47] UPOV, *Members of the International Union for the Protection of New Varieties of Plants.*

Table 3.1 (*cont.*)

State/organisation	Date state/organisation became a UPOV Member	Convention to which state/ organisation is party
Morocco	8 October 2006	UPOV 1991
Netherlands	10 August 1968	UPOV 1991
New Zealand	8 November 1981	UPOV 1978
Nicaragua	6 September 2001	UPOV 1978
Norway	13 September 1993	UPOV 1978
Oman	22 November 2009	UPOV 1991
Panama	23 May 1999	UPOV 1991
Paraguay	8 February 1997	UPOV 1978
Peru	8 August 2011	UPOV 1991
Poland	11 November 1989	UPOV 1991
Portugal	14 October 1995	UPOV 1978
Republic of Korea	7 January 2002	UPOV 1991
Republic of Moldova	28 October 1998	UPOV 1991
Romania	16 March 1991	UPOV 1991
Russian Federation	24 April 1998	UPOV 1991
Serbia	5 January 2013	UPOV 1991
Singapore	30 July 2004	UPOV 1991
Slovakia	1 January 1993	UPOV 1991
Slovenia	29 July 1999	UPOV 1991
South Africa	6 November 1977	UPOV 1978
Spain	18 May 1980	UPOV 1991
Sweden	17 December 1971	UPOV 1991
Switzerland	10 July 1977	UPOV 1991
The former Yugoslav Republic of Macedonia	4 May 2011	UPOV 1991
Trinidad and Tobago	30 January 1998	UPOV 1978
Tunisia	31 August 2003	UPOV 1991
Turkey	18 November 2007	UPOV 1991
Ukraine	3 November 1995	UPOV 1991
United Kingdom	10 August 1968	UPOV 1991
United Republic of Tanzania	22 November 2015	UPOV 1991
United States of America	8 November 1981	UPOV 1991
Uruguay	13 November 1994	UPOV 1978
Uzbekistan	14 November 2004	UPOV 1991
Vietnam	24 December 2006	UPOV 1991

intellectual property over plant varieties. Further, trade agreements have consequences for countries who are not yet obliged to comply with TRIPS. First, signatories that are not UPOV Members may be required to join. This is particularly important for developing and least developed countries. The TPP, for example, if implemented will require Brunei and Malaysia to join UPOV 1991. And, while the intellectual provisions of the

Economic Partnership Agreement between West African and the European Union are yet to be finalised, if these require accession or ratification to UPOV, this will impact the national plant variety protection laws of West African countries: currently all sixteen West African countries party to the trade agreement are *not* UPOV Members. Secondly, trade agreements may oblige countries to sign up to the stricter provisions of UPOV 1991. For example, while the NAFTA gave Mexico the choice between UPOV 1978 and UPOV 1991, by signing the TPP, in 2015, Mexico agreed to accede to UPOV 1991 (within 3 years of the TPP coming into force) and will need to implement stronger national plant variety protection laws that are UPOV 1991 compliant, as opposed to UPOV 1978.

It is clear that international treaties and trade agreements have played an important role in expanding UPOV's membership. Yet to suggest that expanding UPOV membership is merely due to international pressure is overly simplistic and disregards other influences. There is no doubt that decisions to join UPOV and implement UPOV-based plant variety protection are imbricated in local and regional issues, and there are often national economic and political pressures to implement international treaties, including UPOV.[48] For example, UPOV membership is seen as a way of positively affecting interstate relations. Kenya and Colombia, for example, joined UPOV in 1999 and 1996, respectively, and subsequently amended or implemented national plant variety protection schemes.[49] While these countries are members of the WTO and have to comply with TRIPS, much of the pressure to join UPOV and implement UPOV-based plant variety protection came from domestic and foreign horticulture and floriculture industries, many of which were looking to cultivate and grow flowers in Kenya and then export them to the European Union.[50] By joining UPOV, and becoming part of the 'UPOV community', countries such as Kenya and Columbia assuaged some of the concerns of local and foreign breeding industries about protecting their investments and the profitability of introducing new plant varieties. In a paper published in 2014, Rangnekar highlights that domestic pressures played a crucial role in Kenya's accession to UPOV in 1999 and that joining UPOV was seen as a

[48] Deere, *The Implementation Game.*

[49] See, e.g., P. Munyi, B. De Jonge, and B. Visser, 'Opportunities and threats to harmonisation of plant breeders' rights in Africa: ARIPO and SADC' (2016) 24(1) *African Journal of International and Comparative Law*, 86; W. Jaffe, et al., *The Impact Of Plant Breeder Rights In Developing Countries: Debate And Experience in Argentina, Chile, Colombia, México and Uruguay* (University of Amsterdam, 1995).

[50] See, D. Rangnekar, 'Geneva rhetoric, national reality: The political economy of introducing plant breeders' rights in Kenya' (2014) 19(3) *New Political Economy* 359; N. Louwaars et al., *Impacts of Strengthened Intellectual Property Rights Regimes on the Plant Breeding Industry in Developing Countries*, Report commissioned by the World Bank (Wageningen, 2005).

way to enhance relations within the region and globally.[51] In so doing Rangnekar suggests that there is often a difference between the rhetoric of countries at international forums such as the WTO (e.g. arguing for a relaxation of TRIPS obligations for developing countries and drawing attention to issues such as biodiversity, farmers' rights and food security) and national reality (e.g. the need to protect the rights of particular groups such as plant breeders in a manner consistent with international norms). In addition, Rangnekar suggests that national governments who joined UPOV have ready-made discourses about their efforts to safeguard investment, promote plant breeding and encourage development. So, even if there is no direct evidence of UPOV's benefits, the discursive and persuasive force of UPOV membership is powerful and may facilitate countries' participation in regional and international agriculture, horticulture and floriculture markets. Going even further, perhaps joining UPOV is a way for countries to assert their sovereignty. Raustiala, for example, has argued that international institutions 'are now the primary means by which states may prosper and achieve social objectives' and that through these institutions 'states may reassert or express their sovereignty'.[52]

In the preceding paragraphs, I have shown that some countries undoubtedly felt pressure to join UPOV and implement UPOV-based plant variety protection. Perhaps these countries feared being excluded from the trade or economic benefits of plant variety protection or of being left behind in development, growth or other areas. Yet there are other sources of UPOV's legitimacy including the benefits of being part of the 'UPOV community', the realisation that there are regional and global benefits of joining UPOV and even as a way to reassert or express sovereignty. In the next part of this Chapter I show that UPOV is broadly accepted and legitimate because it is a goal-oriented institution and has targeted activities, clear objectives and a defined hierarchy of offices which set out delineated spheres of competence, functions and technical abilities.

3.3 Objectives, Hierarchy and People

In this part I argue that UPOV's clearly defined objectives and hierarchical structure – which sets out clearly delineated functions and spheres of competence and technical abilities – contribute significantly to UPOV's legitimacy over plant varieties. Since its inception in 1961, UPOV has

[51] Rangnekar, 'Geneva rhetoric, national reality: The political economy of introducing plant breeders'.

[52] K. Raustiala, 'Rethinking the sovereignty debate in international economic law' (2003) 8 *Journal of International Economic Law* 841, 860.

worked passionately, uniformly and effectively towards providing and promoting plant breeding through UPOV-based plant variety protection and 'an effective system of plant variety protection, with the aim of encouraging the development of new varieties of plants, for the benefit of society'.[53] Even the last part of UPOV's mission statement – 'for the benefit of society' – which appears to be broad in intent and scope is interpreted consistently with the aim of encouraging plant breeding and the development of new plant varieties. As Peter Button, Vice Secretary-General, elaborates:

> In the context of UPOV, "for the benefit of society" is encouraging the development of new varieties of plants. New varieties of plants are a crucial means of delivering new technologies to farmers and growers and, ultimately, of course, delivering benefits through to consumers. It is virtually impossible to list all the benefits that new plant varieties offer to farmers and consumers, but they can include higher yield, resistance to pests and diseases, tolerance to stresses (eg drought, heat), greater efficiency in the use of inputs, improved harvestability and crop quality. New plant varieties also offer diversity of choice to farmers that can improve their access to national and international markets.[54]

Having clear and narrowly defined objectives is important to UPOV's legitimacy because it informs and underpins UPOV's decisions and activities as well as its discursive and persuasive authority over plant variety protection. Indeed, the majority, if not all, of UPOV's activities are directed towards promoting plant breeding and encouraging the development of new varieties of plants. In this way UPOV's clear objectives and targeted activities can be compared (favourably) to the objectives of the WIPO and the WTO: broadly stated, WIPO's objectives are to promote the protection of intellectual property throughout the world through cooperation among States,[55] and the WTO has a number of goals centred on trade liberalisation.[56] The objectives of WIPO and the WTO have been underpinned by broad sentiments, aspirations or agendas: most notably, development or sustainable development. One of the problems for WIPO and the WTO achieving these goals has been the differing views about what constitutes development or sustainable development, let alone how to achieve either of these things. And while there is no disputing the importance of such broadly stated objectives, they are difficult if not impossible to realise: promoting plant variety rights protection is straightforward by comparison.

[53] UPOV, *Mission Statement*, www.upov.int/about/en/mission.html.

[54] J. Sanderson, 'Why UPOV is relevant, transparent and looking to the future: a conversation with Peter Button' (2013) 8(8) *Journal of Intellectual Property Law & Practice* 615, 620.

[55] *Convention Establishing the World Intellectual Property Organization* (Signed at Stockholm on 14 July, 1967 and as amended on 28 September, 1979), Articles 3 and 4.

[56] *Marrakesh Agreement Establishing the World Trade Organization*, 1867 U.N.T.S. 154, 33 I.L.M. 1144 (1994).

Efforts to realise UPOV's objectives are facilitated by UPOV's structure and people. As mentioned earlier, UPOV has a clearly defined hierarchy of offices and functions (see Figure 3.1) and clearly delineated spheres of competence and technical abilities.[57] While the full details of the various UPOV bodies can be found on the UPOV website, it is useful to briefly outline some of the key bodies and their functions here.[58] The UPOV Council and the Office of the Union are the permanent organs of UPOV,[59] with the UPOV Council being the highest body, having responsibility for 'safeguarding the interests and encouraging the development of the Union and for adopting its programme and budget'.[60] The Council is comprised of an elected President (See Table 3.2), Vice-President and one representative from each of the Member States of the Union. The tasks of the Council are established in Article 26(5) of UPOV 1991 and include establishing UPOV's rules of procedure, giving the Secretary-General 'all necessary directions' to accomplish UPOV's tasks and to 'take all necessary decisions to ensure that the efficient functioning' of UPOV.[61] Significantly, decisions of the UPOV Council must be consistent with the objectives and aims of the UPOV Convention. Because the UPOV Council is constituted by one representative from each Member State, it is largely a representative body. In recent years some important measures have been adopted by the Council. For example, it is the members who make decisions on the circumstances in which observers, such as NGOs, can participate in UPOV.[62] And, in 2011, the Council launched a redesign of the UPOV website with the aim of improving the availability of information including access to UPOV collection material and a database of

[57] According to Weber's classifications, UPOV is a 'rational-legal' authority and is, therefore, an effective authority structure: see Weber, *The Theory of Social and Economic Organization*, pp. 328, 330.

[58] UPOV, *Terms of Reference and Composition of the UPOV Bodies and a Brief History of their Development*, www.upov.int/about/en/organigram.html.

[59] UPOV 1991, Article 26(5).

[60] UPOV, *Rules of Procedure of the Council*, UPOV Document No. UPOV/INF/7 (15 October 1982) International Union for the Protection of New Varieties of Plants, www.upov.int/edocs/infdocs/en/upov_inf_7.pdf. Typically, the Council is convened once a year; however, it has recently met twice yearly in either March or April and then in October. The tasks of the Council are set out in Article 26(5) of UPOV 1991. Also see UPOV 1961, Article 20; UPOV 1978, Article 21.

[61] Also see UPOV 1978, Article 21; UPOV 1961, Article 20.

[62] For example, in 2010, the UPOV Council initiated a process to review the rules concerning observers and to recommend appropriate changes. These rules were adopted by the UPOV Council on 1 November 2012 and make it clear that intergovernmental and international non-governmental organisations will only be granted observers status if they have 'competence in areas of direct relevance in respect of matters governed by the UPOV Convention. UPOV, *Rules Governing the Granting of Observer Status to States, Intergovernmental Organizations and International Non-Governmental Organizations in UPOV Bodies*, Document UPOV/INF/19/1 (1 November 2012), www.upov.int/edocs/infdocs/en/upov_inf_19_1.pdf.

Table 3.2 *Presidents of the Council*

Name (term)	UPOV Member
Luis Salaices (2016–2018)	Spain
Kitisri Sukhapinda (2013–2015)	United States of America
Keun-Jin Choi (2010–2012)	Republic of Korea
Doug Waterhouse (2007–2009)	Australia
Enriqueta Molina Macias (2004–2006)	Mexico
Karl Olov Öster (2001–2003)	Sweden
Ryusuke Yoshimura (1998–2000)	Japan
Bill Whitmore (1995–1997)	New Zealand
Ricardo López de Haro y Wood (1992–1994)	Spain
Wilhelmus F.S. Duffhues (1988–1991)	Netherlands
S.D. Schlosser (1987–1987)	United States of America
J. Rigot (1984–1986)	Belgium
Dr. W. Gfeller (1981–1983)	Switzerland
Mr. Halvor Skov (1978–1980)	Denmark
Mr. B. Laclavière (1975–1977)	France
Professor L. Pielen (1972–1974)	Germany
L.J. Smith (1968–1971)	United Kingdom

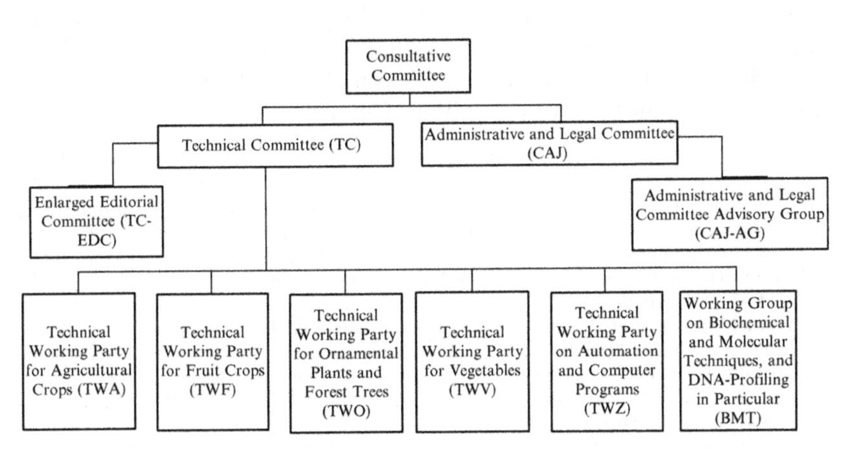

Figure 3.1 UPOV Organigram[63]

[63] UPOV, *Organigram International Union for the Protection of New Varieties of Plants*, www.upov.int/export/sites/upov/about/en/pdf/upov_bodies.pdf.

UPOV Members' laws, UPOV Lex.[64] In 2011, the Council also increased access to documents of the Administrative and Legal Committee (CAJ), Technical Committee (TC) and Technical Working Parties (TMPs), which were previously only available to Members and Observers.

The Office of UPOV carries out the duties and tasks assigned to it by the Council and is under the direction of the Secretary-General.[65] The current Secretary-General of UPOV is Mr. Francis Gurry: a lawyer and former academic who has served as Director-General of WIPO since 2008.[66] However, the day-to-day running of UPOV is carried out by the Vice Secretary-General, currently Mr. Peter Button.[67] Mr. Peter Button has a background in science and plant breeding, having worked in plant breeding since 1981 and previously held the role of Technical Director at UPOV. The Vice Secretary-General is responsible 'for the staff of the Office of UPOV and for the delivery of the results indicated in the programme and budget approved by the UPOV Council'.[68] In addition to the Vice Secretary-General, there are only a small number of full-time staff in the Office of the Union: two Directors, four professionals and five general service staff.[69] These staff are largely responsible for the technical work of UPOV, including answering questions related to the substantive provisions of the UPOV Convention. The staff are also responsible for managing all of the activities relating to the international cooperation in the protection of new plant varieties.[70] In this way the UPOV Office plays a key role in guiding potential members through the procedure of assessing whether their proposed national laws are UPOV compliant, before making a recommendation on the matter to the UPOV Council.[71]

64 'UPOV Lex contains the legislation of members of the Union that has been notified in accordance with the UPOV Convention, the UPOV Convention notifications concerning individual members of the Union (e.g. accessions, ratifications) and the text of the UPOV Convention and its Acts': UPOV, *UPOV Lex*, www.upov.int/upovlex/en/.

65 UPOV 1991, Article 23(1).

66 Dutfield has suggested it is necessary to raise debate 'as to the appropriateness of the head of a United Nations Agency at the same time also leading a non-UN agency operating in such a commercially sensitive and strategic area': Dutfield, *Food, Biological Diversity and Intellectual Property: The Role of the International Union for the Protection of New Varieties of Plants (UPOV)*, p. 142.

67 A full list of Vice Secretaries-General are Peter Button (United Kingdom), 2010–current; Rolf Jolf Jördens (German), 2000–2010; Barry Greengrass (United Kingdom), 1988–2000; Walter Gfeller (Switzerland), 1986–1988; Heribert Mast (German), 1974–1985; and Halvor Skov (Denmark), 1970–1973.

68 Sanderson, 'Why UPOV is relevant, transparent and looking to the future: a conversation with Peter Button'.

69 Directors include the post of Secretary-General: Personal communication with Mr. Peter Button, Vice Secretary-General of UPOV, 11 March 2016.

70 UPOV 1991, Article 6.

71 UPOV will carry out all the duties and tasks assigned to it by the UPOV Council: UPOV 1991, Article 27.

The specific tasks of the UPOV Council are performed by a number of Committees and Working Parties (See Figure 3.1).[72] These Committees and Working Parties have played a key role in standardising and normalising concepts and practices in and around plant variety protection. The Technical Committee ('TC') and its Working Parties are responsible for advising the Council on matters related to the 'practical implementation and international harmonization of the technical approach to plant variety protection'.[73] The TC is assisted by a number of Working Parties including the Technical Working Party for Agricultural Crops ('TWA'), Technical Working Party for Fruit Crops ('TWF'), Technical Working Party for Ornamental Plants and Forest Trees ('TWO'), Technical Working Party for Vegetables ('TWV'), Technical Working Party on Automation and Computer Programs ('TWC') and a Working Group on Biochemical and Molecular Techniques and DNA-Profiling in Particular ('BMT').[74] Each of these Working Parties is responsible for specific issues including the drafting and production of *Test Guidelines*, as well as issues that are less species specific (e.g. the BMT considers developments and applications of biochemical and molecular techniques).[75] The Technical Working Parties also provide a forum in which experts and non-experts from UPOV Members, particularly those that have recently adopted the UPOV Convention, can receive specialised training.[76] Significantly, the rate of attendance at the meetings of the Working Parties by UPOV Members and Observers has steadily increased indicating growing interest in, and importance of, UPOV's technical approach to promoting plant variety protection in a harmonised manner.[77]

Other important aspects of UPOV's legitimacy are the people imbricated in UPOV and its Committees and Working Parties. While there are a range of backgrounds in UPOV – including in plant breeding, agronomy, economics and law – many of the people at UPOV have strong science or plant breeding backgrounds. In many ways the scientific and plant breeding backgrounds of these people reflect and reinforce the objectives of UPOV and UPOV's principle functions, particularly the technical assistance and support roles played by UPOV. However, there

[72] The Consultative Committee sits below the UPOV Council and is split into two committees, the Technical Committee (TC) and the Administrative and Legal Committee (CAJ): See UPOV, Document No. UPOV/INF/19/1.
[73] UPOV, *Report of the sixteenth session of the Consultative Committee*, Document CC/XVI/5 (Geneva, 6–9 December 1977).
[74] These Technical Working Parties typically convene once a year.
[75] UPOV, *Report of the Thirty-Eighth Session of the Technical Committee*, TC/38/16 (Geneva, 7–15 April 2002) [204].
[76] Ibid. [77] Ibid.

is an increasing role to be played by people other than breeders or scientists in UPOV. Lawyers, for example, are increasingly crucial in UPOV's activities, particularly as membership has expanded, and the likelihood of legal issues and disputes, reliance on legal structures and the need to provide guidance and information increased. The key body dealing with legal issues is the Administrative and Legal Committee ('CAJ'). Established in 1977 to 'study the relations between competition rules and plant variety protection',[78] CAJ is responsible for 'matters, mainly of an administrative and legal nature'.[79] The current Chair of CAJ is Mr. Martin Ekvad, of the European Union. Mr. Ekvad is also President of the Community Plant Variety Office ('CPVO') and has extensive knowledge of intellectual property rights and in particular plant variety rights.[80] Further, in 2005, the CAJ established an advisory group – the Administrative and Legal Committee and a Technical Committee ('CAJ-AG') – to assist in the preparation of information documents and explanatory materials.[81] The agreed approach between the Office of the Union, CAJ and CAJ-AG was explained by UPOV's Vice Secretary-General, Peter Button, as:

> the Office of the Union would develop certain draft materials which it considered covered aspects of a straightforward nature and would circulate those to the CAJ for comments within a specified time. In other cases, where it was considered that there were difficult issues, where discussions at a CAJ session would be important for the development of suitable information materials, and also in cases where the drafts on seemingly straightforward materials provoked unexpected concerns when circulated for comments, it was

[78] UPOV, Document No. CC/XVI/5.

[79] UPOV, *Report of the sixteenth session of the Consultative Committee*, Document CC/XVI/5 (Geneva, 6–9 December 1977) [58]; UPOV, *Report of the Sixteenth Session of the Consultative Committee*, UPOV Document No. CC/XVI/5 (5 December 1977) International Union for the Protection of New Varieties of Plants, [19](iii) and [20].

[80] Prior to his appointment as President in 2011, Mr. Ekvad had previously held the position of Head of the Legal Unit since 2003. He is a trained lawyer, holding both a Masters of Law and a Masters in European Law, who prior to commencing at the CPVO worked both in the public and private sector. See FloraCulture International, *Martin Ekvad joins CPVO as new President*, 20 July 2011, www.floraculture.eu/?p=6314; UPOV, *Martin Ekvad, Head of the Legal Affairs Service, Community Plant Variety Office (CPVO)*, UPOV Document No. UPOV/SYM/GE/08/4A (23 October 2008), www.upov.int/edocs/mdocs/upov/en/upov_sym_ge_08/upov_sym_ge_08_4a.pdf.

[81] However, in many ways the CAJ-AG was a victim of its own success and its life was short-lived. In 2014, the CAJ decided that the CAJ-AG should only be convened on an *ad hoc* basis. The rationale for suspending the work of the CAJ-AG was that they had been so effective that there was 'relatively little discussion in the CAJ' and that 'participation in the CAJ-AG had increased substantially and regularly comprised almost all the members of the Union that attended the CAJ-AG.': UPOV, *Report of the Seventieth Session of the Administrative and Legal Committee (CAJ)*, CAJ/70/11, 13 October 2014, [12].

agreed that the assistance of the CAJ-AG would be sought prior to the CAJ being invited to discuss those matters at its sessions.[82]

While lawyers play a necessary role in the drafting and developing of legal texts, it is largely the contributions, and passion, of people with backgrounds in science and plant breeding in this process that ensure that UPOV can adapt to changes and progressions in plant breeding practices. In *The Making of Law: An Ethnography of the Conseil d'Etat*, Bruno Latour examined the daily practice of the Conseil d'Etat ('the Conseil'): one of the highest authorities in France, specialised in administrative law. In so doing, Latour points to some differences between law and other fields such as science. While Latour describes both the law and science as being similar in their pursuit of 'enunciating truth', he observes that differences lie insofar as 'lawyers appear to be much less passionate about their files and cases than scientists about their experiments; lawyers are more concerned with affirming the timeless stability of law than with generating new information and bringing about paradigm change'.[83]

Having clear objectives, defined hierarchy and passionate people has benefitted UPOV and its members in numerous ways. Most notably it has enabled UPOV to target its activities and play a significant educative, training and capacity building role in and around plant variety protection. UPOV has not passively sat while its membership expanded. While I cannot highlight all of UPOV's educative, training and capacity-building activities here, it is worth mentioning some notable features. The main purpose of many of UPOV's activities is to promote UPOV membership and the UPOV Convention and to make it as easy as possible for States and intergovernmental organisations to join UPOV and implement UPOV-based national plant variety protection laws.[84] In order to achieve this, UPOV engages in a broad range of activities. In the 1990s, the purpose of many of these activities was to convince countries that UPOV membership and UPOV-based national plant variety protection satisfied the requirements of 'effective sui generis system' under Article 27(3)(b) of TRIPS.[85] In addition to various seminars and workshops, UPOV has provided technical

[82] Personal communication with Mr. Peter Button, Vice Secretary-General of UPOV, 11 March 2016.

[83] B. Latour, *The Making of Law: An Ethnography of the Conseil d'État* (Polity, 2010), pp. 219–220.

[84] See UPOV, *Guidance for the Preparation of Laws Based on the 1991 Act of the UPOV Convention*, UPOV/INF/6/3 (UPOV, 2013), p. 5.

[85] Some of these seminars were held in conjunction with WIPO and the WTO and coincided with the introduction to the TRIPS Agreement: UPOV and WIPO had,

assistance by way of supplying countries with model laws that are consistent with the UPOV Convention. Most recently UPOV has designed and published numerous explanatory notes ('UPOV/EXN' Series), information documents ('UPOV/INF' Series) and standards on a range of topics for those countries looking to become members and those that are already UPOV Members. Many of these documents clarify aspects of the Convention and include the topics of the definition of plant variety under the Convention, EDVs, and the notion of the plant breeder, enforcement and exceptions.[86] Having published a range of explanatory and information materials, UPOV believes that its focus can shift and 'be placed on other important activities to maintain and improve the effectiveness of the UPOV system'.[87] One of these activities is assisting existing Members and non-Members, particularly those looking to develop national laws consistent with UPOV 1991.[88] Some examples of the assistance that UPOV provides in this context include *Guidance for the Preparation of Laws Based on the 1991 Act of the UPOV Convention* (which provides example text and explanations for articles based on UPOV 1991),[89] *Guidance on How to Become a Member of UPOV* (which outlines the eligibility requirements and procedures to become a member of UPOV)[90] and *Guidance for Members or UPOV on How to Ratify, or Accede to the 1991 Act of the UPOV Convention* (which provides guidance on how to develop national laws compliant with UPOV 1991 and deposit an instrument of ratification or accession).[91]

UPOV facilitates and promotes its style of plant variety protection in other ways. Since 2000, UPOV has gone on a publicity campaign to help to improve its image.[92] In so doing UPOV has made a large

from 1996–1999, an agreement with the WTO to help countries to comply with their TRIPS Agreement commitments: WIPO, *WIPO's Legal and Technical Assistance to Developing Countries for Implementation of the TRIPS Agreement: From 1 January 1996, to 31 March 1999*, WIPO Document No. PCIPD/1/3 (28 April 1999), www.wipo.int/edocs/mdocs/mdocs/en/pcipd_1/pcipd_1_3.pdf at [60].

[86] See UPOV, *Explanatory Notes on the UPOV Convention*, www.upov.int/explanatory_notes/en/.

[87] UPOV, *Program and Budget for the 2016–2017 Biennium*, C/49/4 Rev, [2.2.2.5].

[88] Ibid., [2.3.2.2]. [89] UPOV, UPOV/INF/6/3. [90] UPOV, UPOV/INF/13/1.

[91] UPOV, *Guidance for Members of UPOV on How to Ratify, or Accede to the 1991 Act of the UPOV Convention*, UPOV/INF/14/1 (UPOV, 2009).

[92] UPOV began to publish a number of documents on its website in 2000. The documents of the ordinary sessions of the UPOV Council have been freely available on the UPOV website since that time. Although, the UPOV Council decided that not all documents should be made publicly available. In particular, documents concerning the sessions of Administrative and Legal Committee (CAJ), the Technical Committee (TC) and the Technical Working Parties (TWPs) were placed in a 'first restricted area' of the UPOV website, with access for UPOV Members and for States or organisations with observer

portion of its discussions and decisions publicly available, published a raft of information documents and explanatory notes and administered various databases (such as PLUTO and GENIE) and other resources. They have also simplified and clarified the rules around observer status[93] and have attempted to personalise the discourse around plant variety protection. Personalising the discourse and narrative around plant variety protection is important because, as we saw in Chapter 1, evidence to support the benefits of all intellectual property, including UPOV-based plant variety protection, is notoriously difficult to measure, distil and substantiate. In many ways, then, UPOV has adopted a strategy similar to that adopted by civil society organisations criticising UPOV and the UPOV Convention, a strategy that relies on rhetoric, anecdote and speculation. Perhaps the best example of UPOV personalising its message is the case study of Japanese plant variety Ashiro Rindo, where, according to UPOV, plant variety rights protection has led to benefits to individuals, business and the region as a whole. UPOV asserts, for example, that '[a]s of 2012, the Ashiro region takes in US$13 million annually through sales of Ashiro Rindo cut stems, and the brand has become one of the world's most desired gentian varieties'. In so doing, UPOV highlights a specific, local example of the benefits of UPOV-based plant variety protection such as licensing, trademarks and commercialisation. As the following extract illustrates, the story of Ashiro Rindo provides little specific information about plant variety protection:

> In the late 1960s, rice farmer Hideo Kudo found it difficult to support his family in Ashiro solely with his yearly harvest. To supplement his income, every winter he left his family and went to Tokyo to work in the construction industry. However, Mr. Kudo found this took an emotional toll on him and his family. "It was really hard to leave my wife alone," he explained. Resolving to stay close to his family in Ashiro during the cold winter months, in 1971 Mr. Kudo and a few other Ashiro farmers decided to pool their resources together and form a collective – Ashiro Gentian Growers (AGG) – and take the risky move of farming flowers.[94]

status at the Council, CAJ, TC or TWPs. An important reason for restricting access to these documents was to minimise the risk of confusion concerning the status of documents and, in particular, to avoid the impression that documents prepared for discussion by the CAJ, TC and TWPs represented the views of UPOV. See Sanderson, 'Why UPOV transparent and looking to the future: a conversation with Peter Button'.

[93] UPOV, *Rules Governing the Granting of Observer Status to States, Intergovernmental Organizations and International Non-Governmental Organizations in UPOV Bodies*, Document UPOV/INF/19/1.

[94] See UPOV, *Ashiro Rindo Story*, www.upov.int/multimedia/en/2011/ashiro_rindo.html.

Before moving to a discussion of the role of UPOV's concepts and practices in its legitimacy, it is worth summarising UPOV's legitimacy so far. First, UPOV is a member-driven, goal-oriented institution. Second, UPOV has a clearly defined hierarchy and functions and clearly delineated spheres of competence and technical abilities. Thirdly, UPOV and its various Working Parties are filled with dedicated and passionate people skilled in and passionate about plant breeding and plant variety protection. In many ways these three characteristics of UPOV have contributed to and reinforced its legitimacy around plant variety protection. In pursuing the goal of promoting plant variety protection, UPOV (and its people) firmly believes that '[f]or smooth operation, international trade requires uniform, or at least mutually compatible, rules'[95] and that expanding UPOV membership and implementing UPOV-based plant variety protection are the best ways of achieving this. Taken together the clear objectives, hierarchy and people have benefitted UPOV and its members in numerous ways.[96] UPOV has not been impassive, while its membership expanded due to obligations under TRIPS or trade agreements. In making information and materials more available, it has shown itself to be a dynamic institution. And while some civil society organisations and commentators have criticised UPOV for its lobbying and advocacy activities,[97] there needs to be some credit given for the way in which UPOV has promoted UPOV-styled plant variety protection. Many, if not all, of UPOV's activities go directly to building the expertise and capacity of its members and those countries looking to become members. The importance of UPOV's educative, training and capacity-building activities cannot be underestimated, particularly as many developing countries do not have the financial, technical and legal capacity to devise, implement and administer their own unique plant variety rights system. As Deere argues:

In the area of plant variety protection, many developing countries expressed interest in adopting sui generis approaches. In practice, however, governments were constrained by limited expertise, institutional capacity, and experience in this area. In lieu of devising completely new laws, most countries were persuaded

[95] UPOV, *International Union for the Protection of New Varieties of Plants: What It Is, What It Does*, UPOV Publication No. 437(E) (8 July 2011), www.upov.int/export/sites/upov/en/about/pdf/pub437.pdf.

[96] R. Jördens, 'Progress of plant variety protection based on the International Convention for the protection of new varieties of plants (UPOV convention)' (2005) 27(3) *World Patent Information* 232, 235.

[97] See, for example, GRAIN, 'UPOV on the warpath'; Dutfield, *Food, Biological Diversity and Intellectual Property: The Role of the International Union for the Protection of New Varieties of Plants*, p. 14.

by technical assistance providers to adopt or adapt the "off-the-shelf" solutions supplied by the Union for the Protection of Plant Varieties (UPOV).[98]

3.4 Concepts and Practices

Institutions often create and formalise their own standards, concepts and practices. In so doing they set the normative framework around their subject matter. UPOV is no exception. In the following paragraphs, I argue that by standardising and normalising concepts and practices around plant variety protection, UPOV has been increasingly accepted by many actors. This is, perhaps, the most crucial and substantial element in UPOV's legitimacy.

The UPOV Convention stipulates that Member States must implement certain defined criteria for plant variety protection into their domestic laws. A crucial aspect of this is the standardisation and normalisation of concepts and practices around plant variety protection. This norm-setting function is not unique to UPOV and can be found in other international intellectual property organisations. In relation to WIPO, for example, Arpad Bogsch argued that one of the key tasks of WIPO is to establish 'norms that oblige Member States to grant a certain level of protection' and that these norms 'require constant revision because the social, cultural, technical and economic conditions of mankind are constantly evolving and the institutions or our civilization – including the institution of intellectual property – must evolve with them to remain useful'.[99] Specifically, then, UPOV can be said to perform three broad roles in relation to the concepts and practices in and around plant variety protection: (1) establishing the appropriate concepts and practices, (2) socialising and normalising the concepts and practices and (3) revising the concepts and practices.

Other than the UPOV Convention itself, UPOV has produced numerous texts and documents. These have gone a long way to establish a 'precise and internationally harmonised understanding and application of the technical criteria for protection'.[100] Notably UPOV has developed technical texts, standards, guidelines, information and explanatory notes and practices to enable Member States to 'provide and promote an effective system of plant variety protection, with the aim of encouraging

[98] Deere, *The Implementation Game*, p. 315.

[99] A. Bogsch, *Brief History of the First Twenty-Five Years of the World Intellectual Property Organization*, WIPO Publication No 881 [E], (Geneva, 1992).

[100] Jördens, 'Progress of plant variety protection based on the International Convention for the Protection of New Varieties of Plants (UPOV Convention)', 238.

the development of new varieties of plants, for the benefit of society'.[101] One of the notable achievements of UPOV is that its standards, guidelines and practices have helped to reinforce and stabilise the way in which plant varieties are identified, described and distinguished. In this way UPOV has reinforced the central role of *characteristics* in plant breeding and the importance of the 'end result' in plant breeding. As we will see in Chapter 7, reinforcing phenotypic *characteristics* for plant variety protection is particularly important given the push by some scientists, policy makers and academics to (re)focus the UPOV Convention to genotype and genetic differences. There are two key texts that relate to the examination of plant varieties. The first of these is the *General Introduction to the Examination of Distinctness, Uniformity and Stability* ('*General Introduction*') which provides guidance on the development and observation of characteristics of plant varieties by setting out detailed principles for the conduct of the examination of new plant varieties.[102] The *General Introduction* outlines general aspects of examination that apply across all plant species or plant varieties. The second key text, or more precisely set of texts, are the specific *Test Guidelines* for particular plant varieties or species ('*Test Guidelines*').[103] The specialised *Test Guidelines* extrapolate the principles in the *General Introduction* into detailed practical guidance for the harmonised examination of distinctness, uniformity and stability. In 2016, there were more than 300 specific *Test Guidelines* including for maize, wheat, rice, barley, oats, chrysanthemum, persimmon and Echinacea.[104] The *Test Guidelines* contain highly technical data relevant to each species against which any new plant variety can be compared and may include written technical information, drawings and photographs. Generally speaking the *Test Guidelines* set out the relevant *characteristics* and their states, numerical values for electronic processing and example varieties. For example, the *Guidelines* for Brussel Sprouts set out the relevant *characteristics* and their states including plant height, leaf blade/size/colour/

[101] UPOV, *Mission Statement*. Bourdieu talks about texts in law as 'sanctifying a correct or legitimized vision of the social world': P. Bourdieu, 'Force of law: Toward a sociology of the juridical field' (1986) 38 *The Hastings Law Journal* 805, 817.

[102] UPOV, *General Introduction to the Examination of Distinctness, Uniformity and Stability and the Development of Harmonized Descriptions*, UPOV Publication No. TG/1/3 (19 April 2002), www.upov.int/en/publications/tg-rom/tg001/tg_1_3.pdf.

[103] UPOV, *Development of Test Guidelines*, Document TGP/7 (2004); UPOV, *List all Test Guidelines by TG Reference* International Union for the Protection of New Varieties of Plants, www.upov.int/test_guidelines/en/list.jsp. Currently there are approximately 300 *Test Guidelines*, see www.upov.int/en/publications/tg_rom/tg_index.html.

[104] See UPOV, *List all Test Guidelines by TG Reference* International Union for the Protection of New Varieties of Plants, www.upov.int/test_guidelines/en/list.jsp.

length/waxiness/cupping and sprout shape/colour/spacing;[105] and the *Test Guidelines* for Chrysanthemum set out the relevant characteristics and their states including for plant height, stem colour and flower bud type/colour.[106]

Taken together, UPOV's *General Introduction* and *Test Guidelines* have helped to standardise and normalise the way in which plant varieties are described, tested and examined. Significantly, UPOV Members cooperate with one another in examining plant varieties: often either one Member accepts the test results produced by others as the basis for granting protection or another UPOV Member conducts testing on behalf of others. So ingrained in plant breeding, and plant variety protection, are *characteristics* that even those countries opting out of UPOV (e.g. India and Thailand) make use of similar provisions for the assessment of distinctness, uniformity and stability. In certain circumstances these concepts and practices are used by plant breeders and scientists, with plant breeders taking into account UPOV's *Technical Guidelines* – and using the same *characteristics* outlined therein – when selecting plants for breeding programmes and trials.[107] Further, UPOV's concepts and practices on *characteristics* are often used in the registration of plants under national plant registration schemes that regulate the marketing and selling of new plant varieties. For example, to be eligible for the United Kingdom's national list, a variety must be a new plant variety that is distinct, uniform and stable. The national list is maintained by the Plant Variety Rights and Seeds Office (PVS), and the same characteristics are used in determining eligibility for plant variety protection and registration on the national list.[108]

Another way in which UPOV's legitimacy has been strengthened and reinforced is through the stabilisation of variety denominations.[109] One of the conditions for grant of plant variety rights protection is that the application must include a denomination for the new variety.[110] As we

[105] UPOV, *Brussel Sprouts: Guidelines for the Conduct of Tests for Distinctness, Uniformity and Stability*, TG/54/7 (31 March 2004).

[106] UPOV, *Chrysanthemum: Guidelines for the Conduct of Tests for Distinctness, Uniformity and Stability*, TG/26/5Corr.2 (16 March 2010).

[107] F. Schneider, 'The concept of distinctness in plant breeders' rights control of Plant Variety Rights' in B.T. Styles (ed.), *Infraspecific Classification of Wild and Cultivated Plants* (Oxford, 1986), p. 395.

[108] See UK Government, *Add a New Plant Variety to the National List*, www.gov.uk/guidance/national-lists-of-agricultural-and-vegetable-crops#check-its-a-new-variety.

[109] The argument that plant intellectual property, including patent and plant variety protection, plays an important role in stabilising plant names has been made by Brad Sherman: see B. Sherman, 'Taxonomic property' (2008) 67(3) *The Cambridge Law Journal* 560.

[110] UPOV (1991), Article 20(7).

will see more fully in Chapter 6, to attain order and stability in the naming of plants, botanists and taxonomists have developed and refined a set of rules and procedures that govern the naming of plants. These rules and procedures include the *International Code of Botanical Nomenclature* ('*Botanical Code*') which was adopted in 1867 and the *International Code of Nomenclature for Cultivated Plants* ('*Cultivated Plant Code*') which was first published in 1953.[111] There is also a system for the registration of plant names that includes numerous International Cultivar Registration Authorities ('ICRAs') which tries to ensure that plant names are applied consistently, with limited duplication.[112] Although the *Botanical Code*, *Cultivated Plant Code* and ICRAs have helped 'to transform local knowledge of plants, critical to the survival of indigenous people anywhere, into a comprehensive system of naming, of ordering and classifying, which now embraces every known plant in the world',[113] nonetheless aspects of naming plants remain unsettled and problematic.[114] The relevance and effectiveness of the *Codes* and ICRAs have been constrained because they are voluntary and provide no legal protection of plants or plant names.[115] The *Codes* and ICRAs are, therefore, open to disregard and misuse by plant breeders, traders and marketers.

Since its introduction in 1961, the UPOV Convention has played an increasingly significant role in the naming of plants and, to some extent at least, has ameliorated some of the concerns and confusion over plant names stemming from the *Plant Codes*. By developing rules and practices on variety denomination, UPOV established a legal framework that facilitates the consistent and effective naming of plants: anyone that commercialises a plant variety must use the variety denomination, even after the plant variety right expires. It is conceivable that UPOV and the UPOV Convention have supplanted the *Plant Codes* and ICRAs in importance in relation to ordering and stabilising variety denominations. UPOV and the UPOV Convention have had a number of effects on the use of plant names. When a name is registered under

[111] The most recent versions of the *Botanical Code* and *Cultivated Plant Code* are 2011 and 2009, respectively.
[112] For more details see Chapter 6.
[113] D. Gledhill, *The Names of Plants* (Cambridge, 2002), p. 4.
[114] As early as the nineteenth century, botanists and taxonomists lamented the confusion of Latin plant names: see W.T. Stearn, 'Historical survey of the naming of cultivated plants' (1986) *Acta Horticulturae* 19.
[115] See, e.g., E. Scott, 'Plant breeders rights trials for ornamentals: The international testing system and its interaction with the naming process for new cultivars' in S. Andrews, A. Leslie and C. Alexander, *Taxonomy of Cultivated Plants: Third International Symposium* (Royal Botanical Gardens, 1999), pp. 89–94.

national plant variety rights laws, that name must generally be used as the generic name of the plant for the purposes of legal protection. Further, while the rights granted to breeders are of limited duration, the name that is given to a plant is perpetual. Both of these factors have helped to normalise the use of plant names throughout the world. The fact that UPOV's practices on variety denominations have been taken up by so many countries means that in certain situations UPOV practices have been embraced by scientists, plant breeders and lawyers and accepted by taxonomists and horticultural bodies. Indeed, UPOV's requirements and rules for naming plants are recognised under the *Cultivated Plant Code* as providing de facto registration of a plant name by horticultural authorities; once a plant variety is registered for plant variety rights protection under national UPOV-based schemes, the designated name is recognised in the relevant ICRA as the final name of the plant, and often the examination of variety denomination under UPOV is carried out in conjunction with the lists of the International Registration Authority.[116] In some situations, the relationship between UPOV and plant taxonomy is also 'reinforced by the fact that the work of Registration Authorities and that of national plant variety rights offices have been fused into what is effectively a single process'.[117] For example, in Australia, the Australian Cultivar Registration Authority (ACRA) registers all Australian varieties accepted by the Plant Breeder's Rights Office;[118] and in Poland, all the 'activities connected with plant variety testing, the maintenance of the Register of Cultivars and the Register of Plant Breeder's Rights are provided by the Research Centre for Cultivar Testing'.[119]

UPOV has also helped to order and stabilise variety denominations in more practical ways. By developing and maintaining the UPOV Code System (UPOV Code) and various databases (e.g. GENIE and PLUTO), UPOV has established a repository of the UPOV Code and information

[116] U. Loscher, 'Variety denomination according to plant breeders' rights' (1986) 182 *Acta Horticulturae* 59, 61.

[117] Sherman, 'Taxonomic property', 582. In many countries, the checking of variety denomination under UPOV is carried out in conjunction with the lists of the ICRAs. U. Loscher, 'Variety denomination according to plant breeders' rights', 61.

[118] See Australian Cultivar Registration Authority (ACRA), www.anbg.gov.au/acra/.

[119] See J. Borys, 'DUS testing of cultivars in Poland' in *Taxonomy of Cultivated Plants: Third International Symposium of Cultivated Plants* (Kew 1999), p. 199. And also J. Sadie, 'Cultivar registration for statutory and non-statutory pin South Africa', in *Taxonomy of Cultivated Plants: Third International Symposium of Cultivated Plants* (Kew, 1999), p. 101. Julia Borys, 'DUS testing of cultivars in Poland' in *Taxonomy of Cultivated Plants: Third International Symposium of Cultivated Plants* (Kew, 1999), p. 199; C. Thomson, 'Classification of Brussels sprout cultivars in the UK' in *Taxonomy of Cultivated Plants: Third International Symposium of Cultivated Plants* (Kew, 1999), p. 439.

about variety denominations including botanical and common names.[120] This provides easily accessible information about variety denominations to the authorities of UPOV Members and in so doing has helped to resolve some of the concerns around synonyms by enhancing the usefulness of UPOV's databases. Specifically, the UPOV Code requires that variety denominations have an alphabetical element of five letters indicating genus and where necessary three characters to indicate species and a further three characters to indicate subspecies.[121] This ensures that plant varieties are given a denomination consistently with other members and that synonyms for the same plant taxa are given the same Code. While the UPOV Code and the information contained therein are not necessarily new – it was informed largely by the International Seed Testing Association's list of stabilised plant names and the Germplasm Information Network databases – the fact that it is adopted by UPOV Members, is easily accessible and is supported by a legal framework is crucial to its success.

In addition to standardising and normalising various concepts and practices around plant varieties, UPOV has created or reimagined concepts related to plant varieties. One of the best examples of this is the concept of EDVs, which was introduced by UPOV 1991 but is still very much a work in progress. While EDVs are the focus of Chapter 9, it is worth highlighting some of the notable features of the concept here and how they have contributed to UPOV's legitimacy. Leading in to the Diplomatic Conference of 1991, there were concerns among UPOV Members that the combination of Article 6(1)(a) of UPOV 1978 (in which any variety was protectable that is clearly distinguishable) and Article 5(3) (protected variety can be used as an initial source of variation for creating new plant varieties) allowed, or even encouraged, copying and plagiarism,[122] the combination of which resulted in an unfair advantage to second and subsequent plant breeders and weakened plant variety protection.[123] One of the specific reasons given for introducing the EDVs concept was to prevent the exploitation of mutations of protected varieties and varieties that had undergone a minor change in relation to the initial variety, for example, by using biotechnology, without the holder of the initial variety right being able to share in the revenues.[124]

[120] See, UPOV, *Guide to the UPOV Code System* (22 February 2013), ww.upov.int/genie/en/pdf/upov_code_system.pdf.

[121] Ibid.

[122] For example, the ISF define plagiarism in plant breeding as 'any act or use of material/technology in a breeding process that purposely makes a close imitation of an existing plant variety': ISF, *ISF View on Intellectual Property* (ISF, 2012).

[123] UPOV, Publication No. 346(E), 338–334.

[124] International Union for the Protection of New Varieties of Plants, Fourth Meeting with International Organisations: Revision of the Convention, IOM/IV/2 (1989). See also

The introduction of EDVs in UPOV 1991 created a new scientific and legal concept to be qualified and quantified. Briefly stated, EDVs are varieties that are predominantly derived from an existing, protected plant variety. As well as introducing the concept into UPOV 1991, UPOV resolved to 'start work immediately…on the establishment of draft standard guidelines for adoption by the Council of UPOV, on essentially derived varieties'.[125] UPOV has contributed to the understanding and implementation of the concept.[126] In 2013, UPOV held a *Seminar on Essentially Derived Varieties* in Geneva that canvassed some of the technical and legal aspects of EDVs.[127] However, a major part of the difficulty for UPOV is that the introduction of EDVs created a scientific, legal and pragmatic problem. And while UPOV has mobilised plant breeders, scientists, policy makers and lawyers attempting to examine and identify EDVs, general agreement about how to examine and identify EDVs has not been reached. As we will see in Chapter 9, scientists have studied essential derivation in various plant species using molecular techniques directed to plant DNA, including non-expressed molecular markers and statistical models to provide useful measures of similarity and difference.[128] Breeder organisations have also contributed to the concept of essential derivation. Most notably, the International Seed Federation (ISF) has developed position papers,[129] guidelines[130] and lists of arbitrators for disputes[131] that combines science, law and pragmatism. The concept of essential derivation has also been the subject of legal dispute and litigation that has attempted to reconcile scientific (quantitative) and legal (qualitative) questions around the concept, as

World Intellectual Property Organisation, *Introduction to Plant Variety Protection under the UPOV Convention* (2003), WIPO/IP/BIS/GE/03/00, [53]–[57].

125 UPOV, Publication No. 346(E), 349.

126 For example, Working Group on Biochemical and Molecular Techniques, and DNA-Profiling in Particular, Concepts of Dependence and Essential Derivation. The Possible Use of DNA Markers, BMT/11/24 (2008); Working Group on Biochemical and Molecular Techniques, and DNA-Profiling in Particular, Essentially Derived Varieties (EDV) in the Area of Asexually Reproduced Ornamental and Fruit Varieties, BMT/11/22 (2008).

127 UPOV, *Seminar on Essentially Derived Varieties*, Publication 358 (Geneva, 2013), www.upov.int/edocs/pubdocs/en/upov_pub_358.pdf.

128 See, for examples, E. Noli, M. Teriaca and S. Conti, 'Identification of a threshold level to assess essential derivation in Durum Wheat' (2012) 29(3) *Molecular Breeding* 687; A. Kahler, et al., 'North American study on essential derivation in Maize: II. Selection and evaluation of a panel of simple sequence repeat loci' (2010) 50(2) *Crop Science* 486.

129 See, e.g., ISF, *View on Intellectual Property*.

130 See ISF, *Guidelines for Handling a Dispute on Essential Derivation in Ryegrass* (ISF, 2009); ISF, *ISF Guidelines for the Handling of a Dispute on Essential Derivation of Maize Lines* (ISF, 2008); ISF, *Guidelines for the Handling of a Dispute on Essential Derivation in Oilseed Rape* (ISF, 2007); ISF, *Issues to be Addressed by Technical Experts to Define Molecular Marker Sets for Establishing Thresholds for ISF EDV Arbitration* (ISF, 2010).

131 ISF, *List of International Arbitrators for Essential Derivation* (ISF, 2010).

well as distinctly legal issues such as the standard of proof required and whether the plaintiff or defendant has the burden of proof.[132] As I argue in more detail in Chapter 9, as the concept of EDVs takes shape and is itself standardised and normalised, it renders UPOV more legal.

3.5 Conclusion

In this Chapter I have argued that UPOV has technical, social and normative legitimacy over plant variety protection. As UPOV's membership has expanded, so too has its legitimacy around plant variety protection. This means that the UPOV Convention is generally accepted as delimiting an effective way of protecting plant varieties. Generally, UPOV is accepted by Members, and the international community more broadly, as necessary and inevitable. A large part of UPOV's legitimacy and success is based on clear congruence between its objectives, hierarchy, activities and practices. Of course, not all UPOV Members want the same thing. And not all plant breeders, civil society organisations or governments will agree with UPOV's direction and approaches. Therefore, from time to time, UPOV will make decisions against the interests of some of its members, non-members and other groups and communities such as farmers and indigenous peoples. One thing is clear, however: UPOV makes decisions and directs its activities towards the promotion of plant breeding. This should not be a surprise to anyone.

As a membership-driven and goal-oriented institution, UPOV has achieved a great deal for plant variety protection. UPOV represents the interests of its member and plant breeders with passion, commitment and a clear sense of purpose. Importantly, it is representative of all of its members, not one member or interest group. UPOV is unified and stable, and as much as a member-based international organisation such as UPOV can, it has shown itself to be dynamic and increasingly democratic and transparent. Since its formation, UPOV has actively sought new members and has made itself more open to observers. Perhaps the key to understanding the nature of UPOV, therefore, is to accept that – while it is not entirely indifferent to the broader social issues related to famers' rights, conservation and biodiversity – it has its own clearly articulated objectives, hierarchy of offices, delineated spheres of competence, functions and technical abilities. Importantly, too, the sources of legitimacy will vary between UPOV Members. For some, UPOV membership brings technical assistance and other resources. For others,

[132] *Danziger* v. *Astée* 105.003.932/01, Court of Appeal, The Hague (2009); *Danziger* v. *Azolay* 1228/03, District Court, Tel-Aviv-Jaffa (2009).

membership and being part of a 'UPOV family' brings regional or global acceptance and benefits. Perhaps, most importantly though, UPOV's legitimacy comes largely from standardising and normalising concepts and practices around plant variety rights. In so doing UPOV informs and underpins the work of many actors including plant breeders, taxonomists, researchers, lawyers, administrators and advocates.

Part II

4 Recognising Plant Breeders, Protecting Discoveries

4.1 Introduction

A conceptual problem faced by those looking to protect the work of plant breeders was that intellectual property had been reluctant to protect the kind of work that plant breeders did.[1] As we saw in Chapter 2, despite the fact that during the late nineteenth and early twentieth centuries plant breeders had numerous legal and non-legal ways of protecting their innovations, these mechanisms were largely inadequate or were restricted to a complex and disparate set of national laws and regulations that could not be repeated on an international level. One of the problems was that intellectual property law valorised a particular kind of human agency – whether the inventor in patent law, the author in copyright law or the designer in design law – which was largely based on the idea of the human as creative agent or the producer of an output deserving protection.[2] Plant breeding – defined crudely as the identification of variation in plants, the selection of a particular plant with a desired characteristic or trait and the propagation of the selected plant – utilised existing materials (germplasm) and, therefore, tended to be seen as technical rather than creative, derivative rather than innovative and a process of discovery rather than invention. The distinction between discoveries and inventions was based on the view that inventions were creations, while

[1] Other difficulties for the inclusion of plans in patent law – including nonobviousness, description and exhaustion – are discussed in more detail in Chapter 2. Also see the 2013 decision of the United States Supreme court in *Bowman* v. *Monsanto Co et el* 569 U.S. (2013) in which the Court unanimously affirmed the Federal Circuit and held that the patent exhaustion doctrine does not permit a farmer to plant and grow saved patented seeds without the patent owner's permission.

[2] While the trade mark owner seems anomalous, it has been argued that the trade mark owner is a '"quasi-author" who "creates" a particular set of meanings attached to a mark by investing time, labor and money': K. Aoki, 'Authors, inventors and trademark owners: private intellectual property and the public domain Part I' (1993–1994) 18 (1–2) *Columbia-VLA Journal of Law and the Arts* 1; K. Aoki, 'Authors, inventors and trademark owners: private intellectual property and the public domain Part II' (1993–1994) 18 (3–4) *Columbia-VLA Journal of Law and the Arts* 191.

discoveries were simply findings of something not created by 'man'.[3] In the United States, for example, the 1862 decision of *Morton* v. *New York Eye Infirmary* provided:

> In its naked ordinary sense, a discovery is not patentable. A discovery of a new principle, force or law operating, or which can be made to operate, on matter, will not entitle the discoverer to a patent. It is only when the explorer has gone beyond the mere domain of discovery, and has laid hold of the new principle, force or law, and connected it with some particular medium or mechanical contrivance by which, or through which, it acts on the material world, that he can secure the exclusive control of it under the patent laws ... A discovery may be brilliant and useful, and not patentable. No matter though what long, solitary vigils, or by what importunate efforts, the secret may have been wrung from the bosom of Nature, or to what useful purpose it may be applied. Something more is necessary.[4]

In many ways the distinction between an inventor and plant breeder was based on the view that new plants originated from plants that already existed or were merely mixtures of plants that already existed.[5] This sentiment was consistent with the way some breeders described and advertised their new plant varieties. For example, a 1966 advertisement of one of the United Kingdom's leading plant breeding companies, Gartons, characterised a new grass variety as a mixture and claimed '[e]ight distinct grass strains will be included in our mixture for higher yield per acre'.[6] The characterisation of plant breeding as discovery (or mixture) rather than invention presented a conceptual, if not a legal, obstacle to the inclusion of plant varieties under patent law because it was generally felt that it was difficult to satisfy the

[3] B. Sherman and L. Bently, *The Making of Modern Intellectual Property Law* (Cambridge University Press, 1999), pp. 44–47. The distinction between invention and discovery has also been a theoretical concern for those involved in the fields of science, literature and philosophy. For example, in Aristotle's *Physics* the Stagirite argues that a bed cannot be natural, since if a planted bed could grow and bloom, it would not sprout beds but trees. Also, in the seventeenth century, the object of science began to be demarcated along two lines: experiments of light (that sought to discover the cause of things) and experiments of fruit (that applied knowledge to practical ends): see H. Brown, *The Wisdom of Science: Its Relevance to Culture and Religion* (Cambridge University Press, 1986). Although not everyone views discovery as a mysterious process: see P. Feyerabend, *Against Method* (New Left Books, 1975), who argues that intellectual progress depends on the creativity of scientists; A. Brannigan, *The Social Basis of Scientific Discovery* (Cambridge University Press, 1981), who recasts the question of scientific discovery as social construction or attribution; I. Stengers, *The Invention of Modern Science* (University of Minnesota Press, 2000), who argues that science is neither 'objective' or 'socially constructed'.

[4] Quoted in J. Wigmore and F. Ruffini, 'Scientific Property' (1927) xxii *Illinois Law Review* 355, 356.

[5] See, e.g., Sherman and Bently, *The Making of Modern Intellectual Property Law*, pp. 44–47.

[6] *Farmer and Stockbreeder*, 11 November 1966, p. 100.

requirements needed to obtain patent protection.[7] New plants were not seen as inventions (but as discoveries) and plant breeders were not seen as inventors (but as discoverers).[8] Significantly, then, it was generally assumed that plant breeding developments did not meet the threshold requirement of patent law, one that excluded discoveries from the realm of patent law.[9]

In order to more fully recognise the work of plant breeders, it was therefore necessary to adapt or modify intellectual property's traditional position of humans as creative agents. This chapter examines how the plant breeder was conceived and defined under the UPOV Convention.[10] The chapter begins by showing that although it was clear that plant breeders needed protection, it was not immediately clear how the notion of plant breeder could be articulated in the UPOV Convention. Part of the difficulty was that on the one hand, those 'within' UPOV felt that they knew what a plant breeder was and that it was unnecessary to explicitly define the concept and on the other hand, the UPOV Convention was increasing interacting with other laws and regulation including patent and biodiversity. As a result, UPOV 1961 did not include a definition of a plant breeder but instead focused on protecting 'plant improvements'. In so doing, the UPOV Convention endowed a discoverer of a new plant variety with a legal right as the 'obtainer' of that variety. However, by the 1980s, the changing nature of plant breeding and concerns over the exploitation of plant genetic resources meant that the absence of a definition was seen as untenable. As a result, Article 1(iv) of UPOV 1991 was introduced and defines a plant breeder to include the person who 'bred or discovered and developed' a new plant variety. Clarifying that discoverers and developers were plant breeders acknowledged that a plant breeder

[7] See UPOV, *Industrial Patents and Plant Breeders' Rights – Their Proper Fields and Possibilities for Their Demarcation* (UPOV, 1984), pp. 73–95.

[8] See C. MacLeod, 'Concepts of invention and the patent controversy in victorian Britain' in Robert Fox (ed.), *Technological Change: Methods and Themes in the History of Technology* (Harwood Academic Publishers, 1998), pp. 137–154; R. Magnuson, 'A short discussion on various aspects of plant patents' (1948) 30 *Journal of the Patent Office Society* 493, 496; A. Pottage and B. Sherman, 'Organisms and manufactures: On the history of plant inventions' (2007) 31 *Melbourne University Law Review* 539.

[9] In addition, it was argued that patent law could not adequately protect plant breeding because of the difficulty of reporting unlawful propagation. See A. Caudron, 'Plant breeding: A common understanding for public laboratories, breeding firms and users of varieties' in UPOV, *The First Twenty-Five Years of the International Convention for the Protection of New Varieties of Plants* (UPOV, 1987), p. 43.

[10] The breeder is entitled to file an application for plant variety rights (UPOV 1991, Article 10); a variety is not new if the breeder has sold or otherwise disposed, or given consent for this to happen (UPOV 1991, Article 6(1)); and the breeder can authorise certain acts in relation to the protected material (UPOV 1991, Article 14).

was an improver of plant varieties: through, for example, the discovery of mutations or variants and acknowledged that discoveries and nature played a key role in the creation of new plant varieties but that plant breeders were required to complete the process. Indeed, UPOV believes that '[t]he UPOV Convention would have failed in its mission if it had … withheld from discoverers the incentive to preserve and propagate useful discoveries for the benefit of the world at large'.[11]

Finally, the chapter considers concerns over farmers' rights and asks whether traditional farmers can be plant breeders under the UPOV Convention. While the legal answer to this question is likely to be yes (and farmers can be plant breeders under the UPOV Convention), there are a number of challenges for farmers wanting to protect new or modified plant varieties under UPOV-based schemes. In response to concerns over farmers' rights, some countries such as India, Thailand and the Philippines have introduced national plant variety rights laws outside of the UPOV system. These national laws change the concept of plant breeder even further and introduce more diverse and contingent plant breeders into national plant variety rights schemes. Yet by trying to achieve so many things, these *sui generis* national schemes might have pushed the concept of breeder too far and may, in fact, be ameliorating the protection given to both breeders and farmers.

4.2 Focusing on Plant Improvements and Protecting Discoveries

Despite a desire to protect the work of plant breeders, it was not immediately clear how the notion of the plant breeder could be articulated in an international treaty on plant varieties. In considering how to protect the work of plant breeders, delegates of the Diplomatic Conference of 1957–1961 considered existing laws to see how these laws dealt with plant breeders. Because most countries ruled out patent protection for plant breeders, in part because plant breeders were not considered to be inventors, considering why (and how) existing laws excluded plant breeders was an important step in elevating plant breeders to the position of creator in intellectual property. As we have already seen, because plant development had traditionally been characterised as the result of perseverance or chance, plant varieties were not considered to be inventions, nor were plant breeders considered to be inventors.[12]

[11] UPOV, *The Notion of Breeder and Common Knowledge*, C (Extr.)/19/2 Rev (UPOV, 2002), p. 4.

[12] During the 1920s there was an (unsuccessful) attempt to gain agreement on an international scheme extending intellectual property to scientific discoveries. The

When thinking about how to accommodate the work of plant breeders, the United States *Plant Patent Act 1930* was of particular interest to delegates at the Diplomatic Conference of 1957–1961. The United States government recognised the difficulty of protecting the work of plant breeders under patent law and enacted the *Plant Patent Act 1930*.[13] The *Plant Patent Act 1930* explicitly recognised the work of plant breeders by engrafting plant patent provisions onto the general patent law. In so doing the United States was one of the first jurisdictions to legally acknowledge the place of the plant breeder in intellectual property law. In explaining the role of the breeder in this context, the United States Congress stated that the work of a plant breeder was an aid to nature that was patentable and that providing protection to plant varieties would ensure that 'young agriculturalists of America will be enlisted in a profitable work of invention and discovery of new plants that will revolutionize agriculture as inventions in steam, electricity, and chemistry have revolutionized those fields and advanced our civilization'.[14] In accommodating the work of plant breeders, the *Plant Patent Act 1930* introduced a liberal policy regarding discoveries.[15] More specifically the *Plant Patent Act 1930* authorised the grant of a patent to 'any person who has invented or discovered any new and useful art, machine, manufacture or composition of matter, or asexually reproduced any distinct and new variety of plant'.[16] A complication for delegates of the Diplomatic Conference of 1957–1961, however, was that despite the explicit inclusion of discoveries by the *Plant Patent Act 1930*, delineating a boundary between an inventor and a discoverer was a difficult task. Indeed, the United States legislature and judiciary appeared to treat the inventor and discover as one and the same person, and the requirement that a plant breeder 'invents or discovers' was often conflated so that it was the invent aspect of the definition that was paramount. In so doing, the United States law appeared to emphasise invention over discovery.[17] As Fay said

argument was that an international convention on scientific property would grant scientists individual rights in their discoveries so that 'if and when inventors and industry applied those discoveries, the scientist would be entitled to a royalty': D. Miller, 'Intellectual property and narratives of discovery/invention: The League of Nations' Draft Convention on "scientific property" and its fate' (2008) 46 *History of Science* 299, 306.

[13] *Plant Patent Act 1930* 35 U.S.C. § 161–164.

[14] United States Congress, *Hearings on H.R. 11372 Before the House, Committee on Patent*, 71st Congress (1930), pp. 3, 2.

[15] C. Fowler, *Unnatural Selection: Technology, Politics and Plant Evolution* (Gordon and Breach, 1994), p. 93.

[16] *Plant Patent Act 1930* 35 U.S.C. § 161.

[17] Later, in 1976, the invention requirement was considered in *Yoder Brothers Inc* v. *California-Florida Plant Corporation* 193 U.S.P.W.Q. 264 (5th Circuit, 1976) 264.

in 1937, '[t]he law, both before and after the Plant Amendment, authorized the patenting of any invention or discovery, etc. Are these two things or one? The courts seem to have treated them as one, requiring a discovery to be an invention also'.[18]

In addition to considering the place of plant breeders under existing laws, delegates of the Diplomatic Conference of 1957–1961 were also given an *aid-mémoire* prepared by the State Secretariat for Agriculture of France.[19] The *aid-mémoire* included a range of questions to be considered including whether it was appropriate to grant a right to 'every person who is able to prove that he is the first to bring a new variety of plant into cultivation' or should 'only those *obtentions* which result immediately and directly from a process of acting on the genetic structure of the plant' be considered 'true creations'. The *aid-mémoire* referred to creating a new plant variety by means of various plant breeding techniques. The list included techniques such as bulk or pedigree selection within an existing population, the discovery of a natural mutation, as well as chance and deliberate cross-pollination. Despite the *aid-mémoire's* focus on breeding techniques, delegates at the Diplomatic Conference of 1957–1961 decided not to define plant breeding through a list of techniques. Part of the difficulty in trying to define the work of the plant breeder was, and continues to be, that plant breeding is not reducible to a single definition nor can plant breeding be explained by a series of common techniques or practices. It is the resultant plant variety that is important.

So instead of listing all the breeding techniques used to produce new plant varieties, delegates agreed that it was more appropriate to focus on the result of plant breeding. In many ways a plant breeder was, and commonly still is, considered a person who arrives at a result: a plant improvement. Thus, the focus of the UPOV Convention became the

While the United States Court of Appeal found that to develop or discover a new plant variety that retains the desirable qualities of the parent stock, and adds significant improvements, is a challenging proposition, the Court generally associated the discovery requirement under the *Plant Patent Act 1930* to an improvement, so that to judge a new plant variety, all the plant's characteristics had to be taken into account. For example, if the plant is used as food, then a discovery might relate to nutritive content or prolificacy; if the plant is medicinal, a discovery might relate to changed therapeutic values; or if the plant is ornamental, a discovery might relate to increased beauty, usefulness in the industry and how much of an improvement over other ornamental plants: See D. Jeffery, 'The patentability and infringement of sport varieties: Chaos or clarity' (1977) 59 *Journal of the Patent Office Society* 645.

[18] A. Fay, 'Are plant patents "inventions"?' (1937) 28 *The Journal of Hereditary* 261, 261. See also R. Cook, 'Plant patent 110 declared invalid' (1936) 27 *The Journal of Hereditary* 394.

[19] UPOV, *The Notion of Breeder and Common Knowledge.*

output of the plant breeder – an improved plant variety – rather than the skills or techniques used by the breeder in creating the new plant. The question then became: what sort of plant improvements should be protected? One of the proposals presented at the Diplomatic Conference of 1957–1961 was to protect either 'creative' or 'effective' improvements.[20] While the issue of requiring qualitative changes in new plant varieties was discussed at the Diplomatic Conference in relation to distinctness,[21] this was also an important issue in the rationale for rejecting the need for qualitative intervention on the part of the plant breeder. Ultimately, however, the need for a plant breeder to engage in 'creative' or 'effective' work and thus intervene qualitatively was rejected. There were a number of reasons for this including the fact that the results of plant breeding are not easily measured. So, requiring 'effective' or 'creative' plant breeding would have placed an unworkable qualitative judgment on the work of plant breeders.[22] By only protecting 'effective' or 'creative' plant breeding, the proposed Convention would have bound the protection of plant varieties to improvements such as increased yield, resistance to pest or tolerance to drought in new plant varieties. Linking plant breeding with 'effective' or 'creative' work would have meant that a plant breeder was synonymous with an improver or modifier of plant varieties. Rather than being creative, plant breeding was often a practical and pragmatic exercise where a plant breeder identified problems and then, often through hard work and perseverance, set about overcoming them. Luther Burbank, for instance, was largely seen to rely on intuition, perseverance and a process of trial and error with a large number of plants.[23] Indeed, Burbank was well known for 'crossing and re-crossing until something worthwhile came up'[24] and even referred to himself as an 'evoluter' of new plants.[25] Even given all of these ingredients, there was the real possibility that the new plant variety might not be an improvement on existing plant varieties. Furthermore, predicting the outcome of plant breeding is often complicated because plant breeding is generally contingent on environmental factors such as

[20] UPOV, *Actes des Conférence Internationales pour la Protection des Obtentions Végétales 1957–1961, 1972*, UPOV Publication No. 316 (UPOV, 1972); UPOV, *The Notion of Breeder and Common Knowledge.*

[21] The requirement of distinctness is discussed in more detail in Chapter 7.

[22] UPOV, *The Notion of Breeder and Common Knowledge*, p. 3.

[23] There is an element of hard work and perseverance for research in most disciplines: in 1854, Pasteur said, chance only favours the mind which is prepared.

[24] N. Kingsbury, *Hybrid: The History and Science of Plant Breeding* (The University of Chicago Press, 2009), pp. 189, 191.

[25] J. Smith, *The Garden of Invention: Luther Burbank and the Business of Plant Breeding* (Penguin, 2009), p. 2.

soil quality and climatic conditions. In other words, 'effective' plant breeding in one country or region was not necessarily 'effective' in another. Another concern about requiring plant breeders to carry out 'effective' or 'creative' plant breeding was that it would have placed a heavy burden on independent authorities such as national Departments of Agriculture and Intellectual Property Offices. The reluctance to place a qualitative criterion on new plant varieties was summed up by the British Committee on Transactions in Seeds which stated: '[t]he authority's decision in any particular case would not only determine whether the breeder could claim royalties, but might in practice also settle the question for the breeder whether it might be worthwhile marketing the variety or abandoning work on it'.[26]

In the end it was agreed that defining the plant breeder in relation to 'creative' or 'effective' work, or by reference to various plant breeding techniques, posed a number of difficulties and was not tenable. Because the intention of those present at the Diplomatic Conference of 1957–1961 was to fill the gap left by patent law by articulating what was meant by a plant breeder, the delegates of the Diplomatic Conference of 1957–1961 contrasted the patent system (which protected inventions but not discoveries) with the proposed UPOV scheme (in which discoveries would be protected) and decided the best course of action was to not define a plant breeder in UPOV 1961. Put simply, a breeder was deemed to be someone distinct from the inventor in patent law. This was important because it allowed the UPOV Convention to accommodate and provide protection to plant breeders as discoverer and practitioner. This was significant for intellectual property law as it endowed the plant breeder with a legal right as the 'obtainer' of new plant varieties.

4.2.1 Plant Breeder as Discoverer

Although a plant breeder was not defined in UPOV 1961, the view that a plant breeder under UPOV 1961 encapsulates discoverers comes from a number of sources. One of these is the name of the Convention. The acronym UPOV is derived from the French name of the organisation, *Union Internationale pour la Protection des Obtentions Végétales*. In the official (French) language of the UPOV, the word *obtention* translates to

[26] Committee on Transactions in Seeds, *Plant Breeders' Rights: Report of the Committee on Transactions in Seeds*, Cmnd. 1092 (London, July 1960), p. 56. Even though this statement was made in relation to using qualitative criteria to assess distinctness, it can be applied more broadly.

'getting' or 'obtaining'.[27] This can be taken to mean that a plant breeder includes a person who 'gets' or 'obtains' a new variety of plant by way of discovery. However, simply substituting the English word 'breeder' for the French *obtention* or sélectionneur is problematic. As UPOV puts it:

Obtenteur in French means a person who achieves a result particularly as a result of trials or research. It is usually translated into English as "breeder". "Breeding" in its strict sense connotes a process involving sexual reproduction as a source of variability but in practical usage the activity of plant breeding is much wider and includes, in particular, selection within pre-existing sources of variation. "Obtenteur" might be better translated into English as "plant improver" rather than "breeder."[28]

Another reason why UPOV 1961 is taken to encapsulate the discoverer of new plant varieties is because the opening words of Article 6(1)(a) of UPOV 1961 states that a new plant variety, '[w]hatever may be the origin, artificial or natural, of the initial variation', must be clearly distinguishable. As discoverers make selections based on 'natural' sources of variation, the implication of this wording is that a discoverer can be a plant breeder under UPOV 1961.[29] Further support for including discoverers under UPOV 1961 comes from the Office of UPOV, which states that:

The UPOV Convention differs from the patent system in its treatment of discoveries. Discoveries are not patentable. However, the "discovery" of mutations or variants in a population of cultivated plants is indeed potentially a source of new improved varieties. The UPOV Convention would have failed in its mission if it had excluded such varieties from protection and withheld from discoverers the incentive to preserve and propagate useful discoveries for the benefit of the world at large. The United States of America adopted the same approach in 1930 when it made the plant patent available to "whoever invents or discovers and asexually reproduces any distinct and new variety[30]

By including discoverers in the UPOV Convention, the importance of discovery and collection to plant development was acknowledged. It also reflected and accommodated centuries of plant development and improvement in which botanists had collected plants in the name of science for centuries, and plant collectors collected thousands of seeds, cuttings and whole plants from throughout the world with, for example,

[27] The official languages of UPOV 1961 and UPOV 1978 are French, English and German: UPOV 1961, Article 28, UPOV 1978, Article 28. The official languages of UPOV 1991 are French, English, German and Spanish: UPOV 1991, Article 28.

[28] UPOV, *The Notion of the Plant Breeder,* p. 2. In Europe, the definition of 'plant varieties' was also considered in the case of *Tomatoes II/State of Israel,* T 1242/06 (31 May 2012).

[29] UPOV 1961, Article 6(1)(a).

[30] UPOV, *The Notion of Breeder and Common Knowledge,* p. 4.

cinchona collected from South America, tea from China and eucalypts from Australia.[31] While early plant collectors were primarily concerned with scientific endeavours (and the pursuit of herbarium specimens), plants were also collected so that nurseries could propagate and sell them. Individuals such as Joseph Banks, who collected more than 1,300 new plant species for the Royal Botanic Garden established at Kew, and Francis Masson, who embarked on a number of plant collecting expeditions and sent back more than 1,700 unknown plant species, typified plant collection in the name of science.[32] That said, even from these scientific specimens, plants were propagated and sometimes improved by hybridisation in the Kew greenhouses and then sent out to the colonial gardens.[33] The economic potential of plant discoveries and improvements played a role in plant collection, particularly as plants came to be viewed as valuable commodities.[34]

One of the clearest examples of the importance of plant collection, plant introduction and economics involved *The Horticulture Society* in London. The *Horticulture Society* sent plant collectors to 'exotic' places in order to obtain living specimens and seeds that nurseries could then propagate, promote and sell. Gardeners and scientists like David Douglas and brothers William and Thomas Lobb sent plant specimens and seeds back to the United Kingdom which were then propagated and sold.[35] The English brothers, William and Thomas Lobb were part of the new breed of collectors who emerged in the mid-eighteenth century, who trekked around the world and sent back sacks full of seeds, instead of the herbarium specimens, and who were motivated by 'what to collect for a nurseryman rather than one who only appraised plants with a botanist's ego'.[36] This aspect of plant development continued to be important, and in the early twentieth century, plant introduction was considered to be a valuable method of plant improvement, where '[t]he acquisition of superior varieties by

[31] C. Heiser, *Of Plants and People* (University of Oklahoma Press, 1985); T. Whittle, *The Plant Hunters: Tales of the Botanist-Explorers Who Enriched Our Gardens* (The Lyons Press, 1997).

[32] Most of these specimens were brought back to England as dried herbia, not as seeds or living plants: see, for example, M. Gribbin and J. Gribbin, *Flower Hunters* (Oxford University Press, 2008)

[33] Packets of seeds were sent out by mail from the Royal Botanic Gardens: L. Brockway, *Science and Colonial Expansion: The Role of the British Royal Botanic Gardens* (Yale University Press, 2002), pp. 83–86.

[34] N. Scourse, *The Victorians and their Flowers* (Croom Helm, 1983).

[35] See Gribbin and Gribbin, *Flower Hunters*, pp. 163–188. E. Killip, *The Botanical Collections of William Lobb in Columbia* (Smithsonian Institute, 1932); J. Ewan, *William Lobb, Plant Hunter for Veitch and Messenger of the Big Tree* (University of California Press, 1973).

[36] J. Veitch, cited in Gribbin and Gribbin, *Flower Hunters*, pp. 171.

importing them from other areas accomplishes the same purpose as developing superior varieties in deliberate breeding programs'.[37] In many ways, then, including the discoverer of new plant varieties, the notion of the plant breeder in UPOV 1961 provided an important concession to the long-standing practices of plant collection and plant introduction. And, in so doing UPOV 1961 endowed the discoverer of a plant variety with a legal right as the 'obtainer' of new plant varieties.

As well as providing protection to a discoverer of new plant varieties, the notion of the plant breeder under UPOV 1961 encapsulated the breeder as practitioner and technician. Accepting that the practical and technical nature of plant breeding was worth protecting was important because this aspect of plant breeding had been one of the reasons for excluding plant breeding from patent law. Because plant development had traditionally been characterised as the result of technical skill and chance, an obstacle to including plant varieties under patent law was that it was difficult to satisfy the requirements needed to obtain patent protection.[38] This created a number of problems for breeders obtaining intellectual property protection. One of which was the concern that plant breeding, because the work of a plant breeder was generally considered to be obvious or non-inventive, would not meet the requirements of patent law.[39] The problem of obviousness and non-inventiveness is exemplified by breeding techniques that were prevalent during the 1950s. At the time UPOV 1961 was being negotiated, one of the most commonly used plant breeding methods was hybridisation, a technique by which different plants are crossed with the intention of combining some of the best qualities of each parent.[40] Hybridisation involves the

[37] R. Allard, *Principles of Plant Breeding* (John Wiley & Sons, 1960), p. 19.

[38] See UPOV, *Industrial Patents and Plant Breeders' Rights – Their Proper Fields and Possibilities for Their Demarcation* (UPOV, 1984), pp. 73–95.

[39] It was also argued that patent law could not adequately protect plant breeding because of the difficulty of reporting unlawful propagation. See A. Caudron, 'Plant breeding: A common understanding for public laboratories, breeding firms and users of varieties' in UPOV, *The First Twenty-Five Years of the International Convention for the Protection of New Varieties of Plants* (UPOV, 1987), pp. 43–49.

[40] While the technology has changed, hybridisation remains a widely used technique. For a century the creation of a hybrid plant was accomplished by mechanically grafting limbs of plants together or obtaining pollen from a plant and using it to fertilise another plant. The development of tissue culture changed the method and speed at which hybridisation is accomplished and a more recent method of introducing new genetic material into plant cells is known as somatic cell hybridisation. This technique involves taking cultures of the so-called somatic cells (cells concerned with structure or functions other than reproduction) of two different plants and trying to make the cells fuse with one another. There are also molecular biological techniques such as DNA hybridisation, where the DNA encoding of the desired characteristic is inserted into the plant genome. See N. Kingsbury, *Hybrid: The History and Science of Plant Breeding*.

processes of selecting the parents with the desired characteristics, crossing the parents selected, selecting the progeny that develop the desired characteristics and recrossing the progeny with the desired characteristics.[41] It was unlikely that such selection techniques constituted an invention under patent law because of a lack of inventiveness. The kind of rationale that excluded plant breeding from patent protection can be seen in the United States case of *Cole Nursery Co.* v. *Youdath Perennial Gardens Inc:*[42] while not contributing to the invalidity of the patent, in *obiter dicta* the court stated that while the inventor claimed that through five generations, he had practised a process of selection and genetics that 'the use of nature and knowledge of propagation of plant life seem to me to have been the forces behind the development of the upright variety of barberry. I am not prepared to accord invention to the result produced by such uses in respect of the upright barberry'.[43]

In addition to including selectors of new plant varieties, UPOV 1961 also encompassed plant breeders who had obtained a new plant variety through changing the deoxyribonucleic acid ('DNA') of a plant via mutations or sports.[44] While some mutations are harmful, a minority provide variation in plant material that provides a new range of material for selection.[45] It was felt that a new plant variety (notably a mutant) deserved to be rewarded according to the same criteria as the results of hybridisation.[46] More specially, in 2002, UPOV reiterated that 'mutations or variants in a population of cultivated plants is indeed potentially a source of new improved varieties'.[47]

4.2.2 *Plant Breeder as Discoverer and Developer*

While the absence of a definition of plant breeders remained unchanged in UPOV 1978 and was largely not an issue for the plant breeding community, during the last half of the twentieth century, the absence of

[41] Allard, *Principles of Plant Breeding.*
[42] *Cole Nursery Co* v. *Youdath Perennial Garden Inc* 17 F. Supp. (1951) 159.
[43] Ibid., 160. Although, the patent was held invalid on the ground that the plant was in public use.
[44] R. Cook, 'Plant patent 110 declared invalid'.
[45] Committee on Transactions in Seeds, *Plant Breeders' Rights: Report of the Committee on Transactions in Seeds*, p. 62. Mutation is a rare process taking place with a frequency of about 1 in 100,000 and may result in a change in part of the plant (e.g. mutation in a leaf bud during mitose) or change in the whole plant (e.g. mutation in the flower during meiose). While mutations can be a natural phenomenon, they are also induced by a range of techniques including x-rays, gamma rays and chemical reagents. See N. Kingsbury, *Hybrid: The History and Science of Plant Breeding*, pp. 266–72.
[46] R. Cook, 'Patents for new plants' (1932) 27 *American Mercury* 66.
[47] UPOV, *The Notion of Breeder and Common Knowledge*, p. 4.

a definition of plant breeder became, politically at least, problematic. However, before I consider some of the issues around the concept of plant breeder in UPOV 1961 and UPOV 1978, two related points are worth making. First, much of the concern over the concept of plant breeder came from outside the UPOV community, and the absence of a definition of a breeder was largely a non-issue for plant breeder organisation such as ASSINSEL. As said during the discussion by Nicole Bustin from France, in 1991, the concept of breeder had 'enjoyed over 25 years of interpretation' and was largely unproblematic.[48] Second, there is overlap between the concept of plant breeder and the requirement that a plant variety be distinct. Specifically, it was felt that many of the issues around the concept of plant breeder could be solved through the correct implementation of the principle of distinctness.[49] Despite this there was criticism of the way in which the absence of a definition of plant breeder in the UPOV Convention resulted in such an inclusive concept of breeding.

One reason why the undefined, broad concept of plant breeder was considered problematic was because of the changing identity of plant breeders. More precisely, as it became increasingly possible for many scientists – using knowledge from cytogenetics, molecular genetics, biochemistry, statistics and so on – to be involved in the development of new plant varieties, the idea of a 'traditional' plant breeder was changed during the second half of the twentieth century.[50] The fragmentation of plant breeders is perhaps best illustrated by the example of hybrid corn in the United States during the early to mid-1900s. The development of hybrid corn coincided with advances in herbicide technology, where plant pathologists and entomologists helped plant breeders to add disease and insect resistance to their varieties. And while questions over the actual impact of biotechnology on plant breeding remain, during the 1970s and 1980s, the context of plant variety production became more fragmented, as did the identity of the plant breeder.[51] This means that

[48] Nicole Bustin, delegate for France: UPOV, UPOV, *Records on the Diplomatic Conference for the Revision of the International Convention for the Protection of New Varieties of Plants 1991*, UPOV Publication No. 346(E) (UPOV, 1992), p. 191 [118]

[49] See E. Massoud, 'Call for moratorium on seedbank patents' (1998) 391 *Nature* 728. Noting that the Genetic Resources Policy Committee of the CG Centres under the chairmanship of M.S. Swaminathan agreed that this had nothing to do with the definition of the breeder but a wrong application of the principle of distinctness and difficulties of the reference collections: Personal correspondence with Mr. Peter Button (Vice Secretary-General of UPOV) and Bernard Le Buanec, 7 August 2013.

[50] See J. Brown and P. Caligari, *An Introduction to Plant Breeding* (Blackwell Publishing, 2008); R. Schlegel, *Dictionary of Plant Breeding* (CRC Press, 2003).

[51] See, for example, D. Freebairn, 'Did the green revolution concentrate incomes? A qualitative study of research papers' (1995) 23(2) *World Development* 265; V. Shiva, *The Violence of the Green Revolution: Third World Agricutlure, Ecology and Politics* (Zed

the concept of a plant breeder as discoverer and practitioner was increasingly complicated and fragmented. The consequence of this shift was that it was difficult to clearly distinguish or demarcate the work of a plant breeder. The identity of the plant breeder was shifting from a person in the field selecting and propagating plants to include a range of scientists: people in the laboratory wearing lab coats and using chemistry, radiation and chromosomes to produce new varieties of plants.

Another reason why the absence of a definition of plant breeder in UPOV 1978 was perceived to be problematic was because of the apparent controversy over the development, use and exploitation of plant genetic resources that occurred in the 1980s.[52] This issue of 'plant piracy' was particularly relevant in Australia as it was alleged that the Australian plant variety rights scheme was confronted with the 'biggest scandal ever faced concerning intellectual property over plant varieties'.[53] The Australian plant variety 'scandal' was revealed in the 1990s when the Rural Advancement Foundation International ('RAFI')[54] and Heritage Seeds Curators Australia ('HSCA') argued that at least 118 plant breeders' rights claims in Australia were potentially invalid.[55] RAFI and HSCA found 'significant irregularities' in plant

Books, 1992); K. Mechlem, 'Agricultural biotechnologies, transgenic crops and the poor: Opportunities and challenges' (2010) 10(4) *Human Rights Law Review* 749.

52 See G. Dutfield, *Intellectual Property Rights, Trade and Biodiversity: Seeds and Plant Varieties* (Earthscan, 2000); V. Shiva, *Biopiracy: The Plunder of Nature and Knowledge* (South End Press, 1997); F. Burhenne-Guilmin and S. Casey-Lefkowitz, 'The convention on biological diversity: A hard won global achievement' (1992) 3 *Yearbook of International Environmental Law* 43; C. Juma, *The Gene Hunters: Biotechnology and the Scramble for Seeds* (Princeton University Press, 1989).

53 Rural Advancement Foundation International, *Plant Breeders Wrongs* (1998), www.etcgroup.org/en/node/399; see also Commonwealth, *Parliamentary Debates*, Senate, 21 October 2002, 5553 (John Cherry); Rural Advancement Foundation International, *Australia's Unresolved Plant Piracy Problems* (1999), www.etcgroup.org/en/node/373.

54 The Rural Advancement Foundation International changed its name in 2001 to 'Action Group on Erosion, Technology and Concentration' ('ETC Group'), see ETC Group, *'RAFI' Becomes 'ETC Group': International Advocacy Group Changes Name, President, and Widens Agenda* (2001), www.etcgroup.org/upload/publication/248/01/news_rafietc.pdf.

55 RAFI and HSCA, *Plant Breeders Wrongs: An Inquiry into the Potential for Plant Piracy Through International Intellectual Property Conventions*, A Report by the Rural Advancement Foundation International (RAFI) in Partnership with Heritage Seed Curators Australia (HSCA). The controversy was (partly) resolved when the Australian Plant Breeders' Rights Office revoked the offending applications. However, further claims were made by the Rural Advancement Foundation International that Australian breeders held plant breeders' rights certificates on plant varieties held in trust by CGIAR. See also B. Hankin, 'Australia bungles plant breeders' rights' (1998) 19(2) *Australasian Research (Incorporating Search)* 43. Others have argued that these problems were not problems of definition but rather there was a lack of communication between researchers and seed banks, see M. Rimmer, 'Blame it on Rio: Biodiscovery, native title, and traditional knowledge' (2003) 1 *Southern Cross University Law Review*, 40.

variety certificates issued from 1987: allegedly invalid claims were granted largely under national laws that were based on UPOV 1978 and were mostly related to plant breeders' right claims on FAO Trust plant germplasm (which are held 'in trust' by CG Centres on behalf of the international community).[56] More specifically RAFI and HSCA claimed that 'plant piracy' was rife because of the way in which the UPOV Convention was interpreted and implemented by national plant variety rights authorities. And, while this was largely a problem of distinctness and implementation, the absence of a definition of a plant breeder in UPOV 1961 and 1978 was also seen as a contributing factor. It was thought that the UPOV Convention overemphasised the role that discoverers played in plant breeding and made it possible for plant collectors and those who merely 'introduced' new plants to Australia to gain plant variety protection. In this way it was argued the concept of plant breeder under UPOV 1961 and UPOV 1978 failed to adequately account for the contribution of indigenous peoples and farmers to the traditional breeding of plants. Bill Hankin, President of the HSCA, argued that the problem was that:

> under Australian law, the national Plant Breeders' Rights Office had no authority to grant certificates for plant varieties that had clearly not been bred by the Australian applicants. In many cases, the varieties were bred by Third World farmers. In other cases, the varieties were obviously just pulled from the ground in the Mediterranean region, for example, and claimed in Australia.[57]

Another reason why the absence of a definition of plant breeder in UPOV 1978 was seen to be problematic was because the relationship between the UPOV Convention and other legal and non-legal instruments and forums was becoming more prevalent. Most notably there was growing concern about biological diversity and genetic resources. Further, during the 1980s a number of suggestions were made on how best to deal with the issue of the misappropriation of biological materials including traditional and indigenous biological materials, known as biopiracy.[58]

[56] Ibid.

[57] ETC Group, *Toward a Global Moratorium on Plant Monopolies* (1998), www.etcgroup.org/en/node/421.

[58] A possible 'discoverer's right' was mooted to protect the rights of individuals or communities who, through their traditional knowledge, discover new plant properties: see M. Gollin, 'An intellectual property rights framework for biodiversity prospecting' in Laird, et al. (eds.), *Biodiversity Prospecting: Using Genetic Resources for Sustainable Development* (World Resources Institute, 1993), p. 180. Also, see R. Sedjo, 'Property rights and the protection of plant genetic resources' in Kloppenburg (ed.), *Seeds and Sovereignty* (Duke University Press, 1988), p. 293; S. Brush and D. Stabinsky (eds.), *Valuing Local Knowledge: Indigenous People and Intellectual Property Rights* (Island Press, 1996).

Ultimately, the concerns over biological resources found a home in two international conventions: the CBD and Plant Treaty. The CBD, which was concluded at the Earth Summit in 1992 and which entered into force for Contracting Parties in 1993, was introduced because of concerns over biopiracy and traditional knowledge.[59] The CBD addresses issues of sovereignty, the distribution of benefits of biological organisms, the appropriateness of granting intellectual property over living organisms and access to technology and benefit sharing.[60] It has three objectives: (a) the conservation of biological diversity(b) the sustainable use of its components and (c) the fair and equitable sharing of the benefits arising out of the utilization of genetic resources, including by appropriate access to genetic resources and by appropriate transfer of relevant technologies, taking into account all rights over those resources and to technologies, and by appropriate funding.[61]

The second convention dealing with biological resources is the Plant Treaty. The Plant Treaty was the culmination of many years of negotiation over food security and access to plant germplasm and was adopted in 2001 and entered into force on 29 June 2004.[62] Under Article 12.3(d) of the Plant Treaty, recipients of germplasm 'shall not claim any intellectual property or other rights that limit the facilitated access to the plant genetic resources for food and agriculture, or their genetic parts or components, in the form received from the Multilateral System'. Importantly, the Plant Treaty recognised the role of intellectual property for plant genetic resources. Article 13.2(d)(ii), for example, states:

> [t]he Contracting Parties agree that the standard Material Transfer Agreement referred to in Article 12.4 shall include a requirement that a recipient who commercializes a product that is a plant genetic resource for food and agriculture and that incorporates material accessed from the Multilateral System, shall pay to the mechanism referred to in Article 19.3f, an equitable share of the benefits arising from the commercialization of that product, except whenever such a product is available without restriction to others for further

[59] *United Nations Conference on Environment and Development: Convention of Biological Diversity*, 5 June 1992, UNEP/Bio.Div/N7-INC 5/4, 31 I.L.M.818 (1993), which entered in force December 1993. Although, the CBD expressly does not deal with agricultural plants.

[60] K. Aoki, *Seed Wars: Controversies and Cases on Plant Genetic Resources and Intellectual Property* (Carolina Academic Press, 2008), pp. 79–80.

[61] CBD, Article 1.

[62] Despite the fact that the Plant Treaty was adopted on 3 November 2001, the Plant Treaty was negotiated and developed throughout the 1980s and included the 1983 *International Undertaking on Plant Genetic Resources*.

research and breeding, in which case the recipient who commercializes shall be encouraged to make such payment.[63]

Concerns over genetic resources can be seen in the comments and proposals made by some of the delegates at the Diplomatic Conference of 1991. For example, the representative of Turkey, who at the time was an observer delegation, stated that 'Turkey was a country endowed by nature with a rich variety of flora and fauna'. [64] At the time many developing countries, and Turkey was in that category more than 20 years ago, had the feeling that they were sitting on a gold mine and that breeders were pirating their genetic resources. Faced with largely political concerns over the identity of the plant breeder, the place of the discoverer in intellectual property law and UPOV's role in wider legal and non-legal frameworks, it was felt that an express definition of plant breeding was needed in UPOV 1991.[65] More precisely it was felt that the place of discoverers within the UPOV Convention, as well as intellectual property law more generally, needed to be clarified.

As we saw earlier, because the notion of the plant breeder in UPOV 1961 and UPOV 1978 emerged primarily as a response to the (un) suitability of patent law to protect the work of plant breeders, the plant breeder was configured in a broad and inclusive way as a plant improver to include a person who had discovered a plant variety. However, because of the continued economic significance of discovering plant varieties and issues around genetic resources and biodiversity, there was some support for explicit reference to the discoverer of new plant varieties to be included in UPOV 1991.[66] Indeed, the definition of plant breeder included in the proposed 'Basic Text' was 'the person who bred or discovered a variety'.[67]

Explicit reference to a person who discovered a variety was ardently debated at the Diplomatic Conference of 1991. Support for the 'Basic Text' came from a number of delegates. Mr. Harvey, from the United Kingdom, argued that the use of the word discovered in the 'Basic Text' was deliberate and that there were occasions (e.g. in the case of

[63] This article is in fact based on the work of Bernard Le Buanec: Personal correspondence with Mr. Peter Button (Vice Secretary-General of UPOV) and Bernard Le Buanec, 7 August 2013. See also Aoki, *Seed Wars: Controversies and Cases on Plant Genetic Resources and Intellectual Property*.

[64] UPOV, Publication 346(E), p. 179 [72.2].

[65] The separate definition of plant breeding in UPOV 1991 also simplified the requirement of distinctness.

[66] UPOV, *Records on the Diplomatic Conference for the Revision of the International Convention for the Protection of New Varieties of Plants 1991*, Publication No. 346(E) (UPOV, 1992), pp. 190–193 [109–127], 198–200 [148–160].

[67] Ibid., p. 17.

mutations) when a new plant variety was discovered.[68] While acknowledging that there were concerns over the way in which the UPOV Convention treated discoverers, the New Zealand delegate, Mr. Whitmore, did not oppose the notion of granting a breeder's right to a new variety that had been discovered.[69] Further support for the inclusion of 'discovered' came from Denmark,[70] Poland[71] and France.[72] However, despite the support for the 'Basic Text', and the express inclusion of discoveries in the definition of plant breeder, the strongest arguments came in opposition to the reference to discoveries in the UPOV Convention.[73] The delegate for Turkey, Mr. Demir, argued that the word 'discovered' should be excluded so as to avoid 'that old landraces could be protected' and the possibility of 'conflicting with the notion of farmers' rights' under the Plant Treaty.[74] The Australian delegate, Mick Lloyd, argued that there were concerns about the place of the discoverer in UPOV 1961 and that there was considerable opposition to protecting discoveries because conservation groups held concerns over 'vast and undiscovered' indigenous species and that discoveries held an 'emotive connotation for conservation groups in relation to the vast and as yet undiscovered array of indigenous species in Australia and other countries'.[75] It was also argued that the act of discovery was a chance process that fell outside the scope of systematic intellectual endeavours and 'should therefore not be subject to intellectual property rights'.[76] Finally, Mr. Lloyd suggested that in reality a discoverer contributed little to plant breeding,[77] and on this basis, an alternative definition of the plant breeder (that removed any explicit reference to a discovery) should be included in UPOV 1991.[78] It was generally agreed, therefore, that unless they had contributed something more, a mere discoverer of a plant variety should not be entitled to protection under UPOV 1991. While the requirement of 'something more' than a mere discovery may be implied from the distinctness requirement, the need for 'something more' in relation to the input of the plant breeder was reflected in the suggestions of a number of delegates, including Denmark who stated that the discovery had to be 'finalised' by the person who discovered the plant variety[79]

[68] Ibid., p. 191 [115]. [69] Ibid., p. 190 [113]. [70] Ibid., p. 116 [191].
[71] Ibid., p. 117 [191]. [72] Ibid., p. 118 [191].
[73] The proposed definition of plant breeder presented to the Diplomatic Conference of 1991 was 'the person who bred or discovered a variety': Ibid., p. 16.
[74] Ibid., pp. 190 [111]. [75] Ibid., pp. 190 [110.1]. [76] Ibid., p. 190 [110.2].
[77] Ibid.
[78] Initially, the Australian proposal was verbally presented; however, so that the proposal could be officially discussed, it was later set out in DC/91/27 where the plant breeder was defined as 'the person who bred or developed' a new plant variety: Ibid., 106.
[79] Ibid., p. 191 [116].

and the United States who argued that 'there had to be the additional act of reproduction' and proposed the definition of 'bred or discovered and reproduced a variety'.[80] The German delegate, Mr. Burr, argued that the definition of plant breeder should include 'discovered and developed'.[81] Subsequently, it was the proposal of Mr. Burr that was adopted, and plant breeders were defined in Article 1(iv) of UPOV 1991 as 'the person who bred, or discovered and developed, a variety'.[82]

Importantly, though, the introduction of an express definition of a plant breeder in UPOV 1991 has clarified, rather than modified, the concept of plant breeder.[83] Defining a plant breeder as 'the person who bred, or discovered and developed, a variety' made a number of important accommodations. Most broadly, the definition of a breeder in Article 1(iv) of UPOV 1991 recognised the growing concerns over biological materials and the questions over the place of discoveries in intellectual property law. While the relationship between the UPOV Convention, the CBD and the Plant Treaty continues to be a political issue, the need to reconcile the concept of the plant breeder with concerns over the appropriation of plant genetic resources was important for UPOV 1991.[84] The introduction of Article 1(iv) of UPOV 1991 and the phrase 'discovered and developed' mediated the issue of biopiracy as it relates to the UPOV Convention. Article 1(iv) of UPOV 1991 makes it clear that the discoverer of a new plant variety growing in the wild does not qualify the person as the 'creator' of new plant varieties, which reduces plagiarism and piracy. Furthermore, the Australian Expert Panel on Plant Breeding has suggested that it is not a discovery if the variety is commonly known or if someone else provides the particulars of its existence to that person.[85] By reformulating the concept of the plant breeder to be a person who 'discovered and developed' a plant variety, a person who merely discovers a plant variety is not considered a plant breeder under UPOV 1991.

Most notably the definition of plant breeder in UPOV 1991 makes it clear that discovery alone is not enough to warrant protection under

[80] Ibid., p. 192 [120]. [81] Ibid., p. 198 [150].

[82] 'Breeder' also means the person who is the employer of the aforementioned person or who has commissioned the latter's work, where the laws of the relevant Contracting Party so provide, or the successor in title of the first or second aforementioned person, as the case may be: UPOV 1991, Article 1(iv).

[83] UPOV, Publication No. 346(E), p. 191 [116] (Mr. Espenhain, Denmark).

[84] UPOV, *Letter to the Executive Secretary of the Secretariat of the Convention on Biological Diversity* (2008), www.upov.int/export/sites/upov/en/about/pdf/upov_cbd_17_04_2008.pdf.

[85] See, for example, Expert Panel on Plant Breeding, *Clarification of Plant Breeding Issues under the Plant Breeders' Rights Act 1994: Report of the Expert Panel on Breeding* (2002), www.anbg.giv.au/breeder/index.html#link02.

UPOV 1991. There must also be 'development' of the new plant variety. Thus, the addition of the term 'developed' clarified that the UPOV scheme does not reward plagiarism or piracy and that the discovery of a plant variety (e.g. by finding and/or importing it) by itself does not make a person a plant breeder.[86] It must be said, however, that the definition in UPOV 1991 was more of a theoretical (or political) clarification than a practical one, as the risk of a person merely discovering a new field crop or vegetable is very unlikely or even impossible. Further, as we will see in the next part, perhaps the principles of distinctness and common knowledge play a bigger role in mitigating the chance of mere discoverers obtaining plant variety protection.

The definition of a plant breeder as 'a person who bred, or discovered and developed, a variety' raises a number of questions. Most notably does a plant breeder need to change the plant variety in some way? On one reading of the definition of a plant breeder, it is possible that mutations could be denied protection under UPOV 1991, as most mutations are propagated unchanged. However, the process of developing new plant varieties through mutation conforms to the first part of the definition of Article 1(iv) that defines a plant breeder as the person who 'bred' a new plant variety. According to UPOV, the inclusion of the term 'developed' does not require that the plant variety be changed in some way, rather it merely reflects the need for 'propagation and evaluation' in plant breeding.[87] In response to the suggestion that the criterion of 'development' is only satisfied if the discovered plant itself is subsequently changed in some way and that the propagation of the plant unchanged would not constitute 'development', an Australian Expert Panel on Plant Breeding concluded that:

> This approach would require the discovered plant to be propagated sexually and for a selection to be made in the progeny in order to demonstrate development. It is suggested that this approach cannot be correct since selection in the progeny would constitute "breeding." This approach would also deny protection to most mutations, since the mutation is usually propagated unchanged.[88]

Finally, it is important to note that the UPOV Convention also contains nullity and cancellation provisions. Pursuant to Article 21(iii) of UPOV 1991, a breeder's right must be declared null and void if it is established that the breeder's right has been granted to a person who is not entitled to it, unless it is transferred to the person who is so entitled. If it is shown, for example, that the person has not 'bred or discovered and developed'

[86] Ibid. [87] UPOV, *The Notion of Breeder and Common Knowledge*, p. 4.
[88] Ibid., p. 4.

the plant variety, then there are grounds under the UPOV Convention to make any plant variety certificate that has already been granted null and void. In addition to nullifying a plant variety right because there is not a plant breeder, under Article 22 of UPOV 1991 states that a breeder's right may be cancelled if, among other things, the plant breeder does not provide the relevant authority with the information, documents or material deemed necessary for verifying the maintenance of the variety.

4.3 Farmers as Breeders or Custodians

There are important questions to be asked about the relationship between farmers' rights and the concept of plant breeder under the UPOV Convention. One of the recurring criticisms of the UPOV Convention is that it does not adequately balance the needs of breeders and farmers. Indeed, over the last twenty-five years, some of the most animated and ardent debates over plant genetic resources, particularly in developing countries, have centred on farmers' rights. A complete history and analysis of farmers' rights is not within the scope of this chapter and can be found elsewhere:[89] nonetheless, it is worth outlining what is meant by farmers' rights.

In 1978, Pat Mooney and Cary Fowler joined RAFI, a non-profit organisation that focuses on the socio-economic impact of new technologies on rural societies.[90] In order to highlight the valuable, but often exploited and unrewarded involvement of farmers to plant genetic resources for food and agriculture, Mooney and Fowler coined the term 'farmers' rights' in the early 1980s.[91] Subsequent to this, farmers' rights had its legal origins in the international legal frameworks surrounding the Food and Agriculture Organization of the United Nations (FAO) in the 1980s. Known as the 'Resolution on Farmers' Rights' paragraph 108 of Resolution 5/89 of the *International Undertaking of Plant Genetic Resources* outlines the substance of the farmers' rights concept as 'rights originating from past, present and future contributions of farmers to conservation, development and availability of plant genetic resources, particularly those from centers of origin/diversity'.[92] Though important, the

[89] For a discussion of farmers' rights, see R. Anderson, *The History of Farmers' Rights: A Guide to Central Documents and Literature*, Conference on Plant Genetic Resources (Leipzig, 1996), p. 50.

[90] In 2001, RAFI changed its name to ETC group.

[91] C. Fowler and P. Mooney, *Shattering: Food, Politics, and the Loss of Genetic Diversity* (University of Arizona Press, 1990); Fowler, *Unnatural Selection: Technology, Politics and Plant Evolution*, p. 192.

[92] For a discussion of the development and emergence of farmers' rights, see, e.g., J. Santilli, *Agrobiodiversity and the Law: Regulating Genetic Resources, Food Security and*

International Undertaking of Plant Genetic Resources was light on detail, particularly on how the farmers' rights concept should be implemented. More significant though was the fact that the *International Undertaking* was not legally binding.

The Plant Treaty, which was adopted in 2001 and entered into force on 29 June 2004, was the first legally binding agreement dealing exclusively with the management of plant genetic resources. The Plant Treaty has the objectives of conserving and sustaining plant genetic resources, as well as the fair and equitable sharing of the benefits arising from their use.[93] It also includes provisions on farmers' rights. Unsurprisingly, farmers' rights were one of most controversial topics in the negotiations for the Plant Treaty. On the one hand, the drive to include farmers' rights in the Plant Treaty was headed by the African group who argued 'simultaneously for fair access, for Farmers' Rights and for a consistent scope',[94] and threatened to withdraw from the Treaty negotiations if any of these matters were not included in the final text of the Plant Treaty. On the other hand, developed countries such as the United States wanted farmers' rights to be left out of the Treaty because they feared the weakening of intellectual property rights and thought that the appropriate scope of the Plant Treaty was individual (not collective) rights. Eventually a compromise was reached, and it was agreed that the Plant Treaty would acknowledge the contributions that indigenous or small-scale farmers had on the preservation and conservation of plant genetic resources. Reference to farmers' rights is found in a number of places in the Plant Treaty. First, the Preamble to the Plant Treaty recognises farmers' rights and affirms 'the past, present and future contributions of farmers in all regions of the world'.[95] Second, Article 9 of the Plant Treaty sets out the substance of farmers' rights by recognising the 'enormous contribution that the local and indigenous communities and farmers of all regions of the world' have made to the conservation and development of plant genetic resources (Article 9.1); listing some general, non-specific steps that can be taken to protect farmers' rights such as the protection of traditional knowledge and the right to participate in decision making (Article 9.2);

Cultural Diversity (Routledge, 2012); S. Brush, 'Farmer's rights and genetic conservation in traditional farming systems' (1992) 20(11) *World Development* 1617.

[93] Plant Treaty, Article 1.

[94] T. Egziabher, G. Egziabher, E. Matos and G. Mwila, 'The African regional group: Creating fair play between north and south' in C. Frison, F. Lopéz and J. Equinas-Alcázar (eds.), *Plant Genetic Resources and Food Security: Stakeholder Perspectives on the International Treaty on Plant Genetic Resources for Food and Agriculture* (Earthscan, 2011), pp. 41–56, 48.

[95] Plant Treaty, Preamble.

and affirming farmers' right to 'save, use, exchange and sell farm-saved seed/propagating material, subject to national laws and as appropriate' (Article 9.3). In many ways, however, the concept of farmers' rights was enfeebled in the negotiation of the FAO, and farmers' rights as articulated in the Plant Treaty remain largely aspirational. Countries are required to do no more than 'acknowledge' and 'recognise' farmers' rights.[96] Perhaps most significantly, Article 9(2) of the Plant Treaty makes it clear that farmers' rights are optional, not legally binding and it is up to national governments to recognise farmers' rights according to 'their needs and priorities'. Furthermore, while Article 9(3) of the Plant Treaty affirms the right of farmers to 'save, use, exchange and sell farm-saved seed/propagating material', it makes it clear that such rights are optional and 'subject to national laws and as appropriate'.

While discussion of farmers' rights and the UPOV Convention tends to focus on farm-saved seed,[97] there is also a question of whether farmers have the 'right' to be considered plant breeders. Although there is no universally accepted definition of farmers' rights, there have been numerous attempts to define the nature and scope of farmers' rights. These attempts have often included the need to recognise and reward farmers' contribution to new plant varieties. One of the most active organisations in the area of farmers' rights is the Fridtjof Nansen Institute of Norway which, in order to facilitate a common understanding of farmers' rights, started the *Farmers' Rights Project* in 2005 to support the 'implementation of Farmers' Rights as they are recognized in the International Treaty on Plant Genetic Resources for Food and Agriculture'.[98] The Fridtjof Nansen Institute define farmers' rights as:

> the customary rights of farmers to save, use, exchange and sell farm-saved seed and propagating material, their rights to be recognized, rewarded and supported for their contribution to the global pool of genetic resources as well as to the development of commercial varieties of plants, and to participate in decision making on issues related to crop genetic resources.[99]

Some commentators argue that farmers' rights were introduced to mitigate the effects of the UPOV Convention on traditional farmers. One of

[96] See M. Lightbourne, *Food Security, Biological Diversity and Intellectual Property Rights* (Ashgate, 2009).

[97] See Chapter 10 for more details.

[98] Fridtjof Nansen Institute, *Farmers' Rights: Resource Pages for Decision-Makers and Practitioners*, www.farmersrights.org/index.html. Also see Regine Anderson, *The Farmers' Rights Project – Background Study 2 – Results from an International Stakeholder Survey on Farmers' Rights* (Fridtjof Nansen Institute of Norway, 2005).

[99] Fridtjof Nansen Institute, *About Farmers' Rights?*, www.farmersrights.org/about/index.html.

the specific issues was that traditional farmers were not treated as breeders by the UPOV Convention. Correa, for example, suggests that the introduction of plant variety rights caused 'a serious asymmetry' in the legal protection of plant genetic resources because plant breeders are able to gain rights over their plant varieties but traditional farmers, who had been improving plant varieties for centuries, are not.[100] According to Mooney, international non-government organisations argued that farmers should be treated as plant breeders in part because 'farmers' varieties were the product of farmer genius and should not be treated in any way as being less than varieties produced by the public or private sector'.[101] Other commentators suggest that the UPOV Convention favours commercial plant breeders and in doing so marginalises farmers and their contribution to improved and new plant varieties. Borowaik, for example, has argued that '[o]ne of the most notable features of the successive institutionalizations of plant variety protection is the way they have systematically elided farmers' role as breeders'.[102] Clearly, if farmers are not plant breeders under UPOV 1991, they do not have the right to seek protection under UPOV-based schemes and subsequently will not have the means to protect and control newly developed plant varieties.

When thinking about whether farmers are plant breeders under UPOV 1991, it is necessary to examine both the nature of farming and the definition of plant breeder in UPOV 1991. For centuries farmers have been cultivating crops by carefully and deliberately selecting, saving and modifying. Even today, in many developing countries, farmers are involved in seed improvement and production, and plant breeding is merged with the growing of crops and the work of farmers.[103] When farmers choose plants from which to collect seed, they are selecting (sometimes subconsciously and other times consciously) a particular plant or crop that has desirable characteristics. In this way it can be argued that farmers have 'bred, or discovered and developed, a variety'. If farmers do more than merely select for desirable characteristics, they may be breeders under Article 1(vi) of UPOV 1991. The fact that farmers

[100] C. Correa, *Options for the Implementation of Farmers' Rights at the National Level* (2000, South Center,) 4. See also J. Kloppenburg and D. Kleinman, 'Seed wars: Common heritage, private property and political strategy' (1987) 95 *Socialist Review* 32.

[101] P. Mooney, 'International Non-governmental Organizations' in *Plant Genetic Resources and Food Security: Stakeholder Perspectives on the International Treaty on Plant Genetic Resources for Food and Agriculture* (2012) 135, p. 177.

[102] C. Borowiak, 'Farmers' rights: Intellectual property regimes and the struggle over seeds' (2004) 32(4) *Politics & Society*, 511, 517–518.

[103] Kingsbury, *Hybrid: The History and Science of Plant Breeding*, p. 39.

may be breeders under the UPOV Convention has been confirmed by UPOV, and a number of presenters at the 2012 *Symposium on the Benefits of Plant Variety Protection for Farmers and Growers* demonstrated that farmers can and do use UPOV-based plant variety protection.[104] Further support for the recognition of farmers as plant breeders comes from the UPOV's Vice Secretary-General, Mr. Peter Button, who states that:

> The definition of breeder in the 1991 Act of the UPOV Convention gives no consideration to the process used for obtaining the plant variety. This means that the process for the granting of the plant breeders' rights will not take into account whether the plant variety was bred using the most modern techniques or whether it was the outcome of a simple selection process. What is relevant is the outcome itself. That is, it is the attainment of a new plant variety that entitles the breeder to apply for a plant breeder's right.[105]

There are, however, a number of obstacles for farmers claiming to be plant breeders under the UPOV Convention. One of these challenges goes to the heart of intellectual property law: the need to identify a creator. On the one hand, where a farmer can be identified as a plant breeder (and they can show that they have 'bred, or discovered and developed, a variety'), there would be no problem in seeking protection under UPOV 1991. On the other hand, if a plant breeder cannot be identified, then plant variety protection will not be granted. The identification of a plant breeder may be an obstacle to farmers obtaining plant variety protection because a significant portion of plant breeding is carried out at a community or collective level. While a community can be the title holder of plant variety rights – in much the same way that plant breeders assign their plant variety rights to their employer or research institution – a community cannot be the breeder of a new plant variety.[106] In many developing countries, plant breeders are associated with the community and is a collective endeavour. In regard to traditional farmers, collaborative and participatory varietal selection is the most familiar form of farmer participation in plant breeding,[107] with

[104] UPOV, *Symposium on the Benefits of Plant Variety Protection for Farmers and Growers* (Geneva, 2012), www.upov.int/meetings/en/details.jsp?meeting_id=26104.

[105] J. Sanderson, 'Why UPOV is relevant, transparent and looking to the future: a conversation with Peter Button' (2013) 8(8) *Journal of Intellectual Property Law & Practice* 615, 618.

[106] The question of collective ownership is not new to intellectual property: see S. Scafidi, 'IP and cultural products' (2001) 81 *Boston University Law Review* 793.

[107] See, e.g., M. Ruiz and R. Vernooy (eds.), *The Custodians of Biodiversity: Sharing Access to and Benefits of Genetic Resources* (Routledge, 2012); M. Morris and M. Bellon, 'Participatory plant breeding research: opportunities and challenges for the international crop improvement system' (2004) 136(1) *Euphytica* 21.

farmers involved in collaborative and collective plant breeding using various methods to create new plant varieties. Salazar, Louwaars and Viser, for example, outline collaborative farming techniques that range from the testing of plant varieties produced at breeding stations to publicly supported farmer breeding.[108] Going into more detail about collective plant breeding, Salazar, Louwaars and Viser discuss a semi-dwarf variety of rice, IR36, which is resistant to pests and diseases. Since its release in 1976 by the International Rice Research Institute (IRRI), IR36 has been grown by farmers throughout the developing world so that there are possibly more than fifty phenotypically different IR36 rice types based on farmers' experimentation. The collective character of traditional farming is further encapsulated in the following statement, made by Brush:

> The lack of possessive individualism among peasant farmers regarding seeds and genetic resources might be seen as an adaptive cultural trait in the face of the risks in agriculture and the importance of diversity in meeting those risks. The efficacy of peasant seed systems is the fact that particularly good cultivars spread rapidly and over a wide area.[109]

In recent times collaborative plant breeding such as participatory plant breeding (PPP) has held the attention of farmers, policy makers and scholars. PPP is an approach to breeding new varieties of plants that 'pools the knowledge, labour, equipment, seeds and other resources of farmers and plant breeders to improve crops and contribute to better rural livelihoods'.[110] PPP includes programmes in which farmers test plant varieties (often produced at breeding stations) and programmes in which farmers are supported by public institutions and non-government organisations to conduct their own plant breeding. It is argued that PPP has the potential of exponentially increasing the role of farmers in producing new cultivars, of substantially increasing the number of new farmers' varieties being developed and contributing to biodiversity and conservation.[111]

Another obstacle for traditional farmers relates to the requirements of grant. The challenge for traditional farmers (as plant breeders) of satisfying the requirements of grant manifests in two ways. First, to be

[108] R. Salazar, N. Louwaars and B. Visser, '"Protecting farmers" new varieties: New approaches to rights on collective innovation in plant genetic resources' (2007) 35(9) *World Development* 1515.

[109] S. Brush, 'Bio-Cooperation and the benefits of crop genetic resources: the case of Mexican maize' (1998) 26(5) *World Development* 755, 761.

[110] See Ruiz and Vernooy (eds.), *The Custodians of Biodiversity: Sharing Access to and Benefits of Genetic Resources*.

[111] Ibid.

registered a plant variety must satisfy the level of distinctiveness, uniformity and stability (DUS). The requirements of distinctness (i.e. 'clearly distinguishable from any other variety whose existence is a matter of common knowledge at the time of the filing of the application'),[112] uniformity (i.e. 'if, subject to the variation that may be expected from the particular features of its propagation, it is sufficiently uniform in its relevant characteristics')[113] and stability (i.e. 'if its relevant characteristics remain unchanged after repeated propagation or, in the case of a particular cycle of propagation, at the end of each such cycle')[114] can be difficult for traditional farmers to satisfy. Most notably farmers' varieties are often locally adapted and heterogeneous which means that they can fall outside the requirements of uniformity and stability. Second, even if a plant breeder has a plant variety that is distinct, uniform and stable, they must show this to be the case. This may be difficult for farmers because the examination of DUS is based mainly on growing tests. These growing tests are carried out by the national authority charged with granting plant variety rights, a separate institution such as public research institutes or by the plant breeder. The various UPOV *Test Guidelines* set out the principles used in the examination of distinctness, uniformity and stability.[115] Generally speaking, the DUS examination provides a description of the plant variety based on its relevant characteristics (e.g. plant height, leaf shape) by which the requirements of grant can be assessed. For traditional farmers, conducting and administering the DUS examination proves difficult.[116] It takes time, money and other resources that many traditional farmers just do not have.

Finally, when considering the role of farmers in plant breeding and plant variety protection, it is also necessary to consider non-UPOV schemes. As we have already seen, pursuant to Article 27(3)(b) of TRIPS, members of the WTO have an obligation to 'provide for the protection of plant varieties either by patents or by an effective sui generis system or by any combination thereof'.[117] In addition to the obligation to protect plant varieties, under the Plant Treaty, national legislators have the opportunity of implementing farmers' rights based on their national 'needs and priorities'. In meeting their obligations under TRIPS and in their desire to safeguard farmers' rights, a number of countries have chosen to protect plant variety rights outside of the UPOV system. In

[112] UPOV 1991, Article 7. [113] UPOV 1991, Article 8. [114] UPOV 1991, Article 9.

[115] UPOV, *General Introduction to the Examination of Distinctness, Uniformity and Stability and the Development of Harmonized Descriptions of New Varieties*, TG/1/3 (Geneva, 2002).

[116] H. Ghijsen, 'Plant variety protection in a developing and demanding world' (1998) 36 (1) *Biotechnology and Development Monitor* 2.

[117] TRIPS, Article 27(3)(b).

the following paragraphs, I will briefly outline how India and Thailand have accommodated farmers, as plant breeders, under their respective plant variety protection laws.

India is not a member of UPOV and has chosen to meet its obligation to provide protection of plant varieties with a *sui generis* system – the *Protection of Plant Varieties and Farmers' Rights Act 2001* (PPVFR Act). And as such India was the first country to implement a plant variety rights scheme that encompasses aspects of the UPOV Convention, the Plant Treaty and the CBD. The PPVFR Act was the result of thorough consideration of the needs of Indian plant breeders, farmers and consumers. It was also the result of the ongoing and passionate involvement of small-scale farmers and non-government organisations. Gene Campaign, for example, argued that it was not enough to merely permit farmers to save seed as provided for in the UPOV Convention; rather, if India was to grant plant variety rights, it would have to balance the rights of farmers and commercial plant breeders, taking into account Indian conditions and practices. More specifically Gene Campaign insisted that the farming community had to retain control over seed production and use and that this could not be achieved by merely allowing farmers to save seed from the harvest to sow for the next crop.[118] Farmers must also be entitled to obtain plant variety protection.

While the regime for plant variety protection set out in the PPVFR Act is similar to the UPOV Convention – and the requirements for protection are novelty, distinctness, uniformity and stability – the PPVFR Act explicitly deals with the question of whether Indian farmers are plant breeders. In so doing the PPVFR recognises farmers as plant breeders in a number of ways.[119] First, and perhaps most generally and symbolically, both the title of the PPVFR Act and the name of the administrative authority reflect the central place of farmers in the scheme. The title of the Act makes reference to the fact that the law is about the protection of farmers' rights. Furthermore, the Indian plant variety authority is known as the Protection of Plant Varieties and Farmers' Rights Authority.[120] Second, Section 2(c) of India's PPVFR Act defines a 'breeder' as a 'person or group of persons or a farmer or a group of farmers or any institution which has bred, evolved, or developed any plant variety'. This

[118] Gene Campaign, *Advocacy to Protect Farmers' Rights*, www.genecampaign.org/farmers-rights/.

[119] For a discussion of India's plant variety protection regime, as it was being debated and introduced, see S. Ragavan and J. O'shields, 'Has India addressed its farmers' woes-A story of plant protection issues' (2007) 97 *Georgetown International Environmental Law Review* 97.

[120] PPVFR Act, s 3.

definition of breeder makes it clear that farmers can be plant breeders by stating that 'farmers or a group of farmers' are capable of breeding, evolving or developing a plant variety. In doing so the PPVFR Act, theoretically at least, provides farmers with the same level of protection as other plant breeders. Third, Section 39(1) of the PPVFR Act unequivocally provides that Indian farmers, as individuals or a as a community of farmers, have a right to register a plant variety with the National Register of Plant Varieties. The PPVFR Act defines 'farmers' from a community rights perspective as those who 'cultivate crops by cultivating the land', as well as those people who supervise cultivation directly or indirectly through other people. Further 'a farmer who is engaged in the conservation of genetic resources of land races and wild relatives of economic plants and their improvement through selection and preservation shall be entitled in the prescribed manner for recognition and reward from the Gene Fund.' Fourth, the PPVFR Act openly refers to farmers' varieties,[121] defined as a variety that 'has been traditionally cultivated and evolved by the farmers in their fields, or is a wild relative or land race of a variety about which the farmers possess the common knowledge'.[122] Finally, the PPVFR Act contains provisions for benefit sharing whereby local communities are acknowledged as contributors of land races and farmers' varieties in the breeding of 'new' plant varieties. In addition to the substantive provisions of the PPVFR Act that make it possible for farmers to be considered as plant breeders, there are also numerous administrative provisions that assist farmers to make use of the plant variety rights scheme. For example, Indian farmers are not required to pay a DUS testing or registration fee.[123]

Thailand, too, has *sui generis* plant variety protection that encompasses aspects of the UPOV Convention, the Plant Treaty and the CBD. The *Plant Variety Protection Act* (1999) (PVP Act) was introduced by Thailand's government in an attempt to promote plant breeding but also to recognise and protect the interests of local farming communities in developing and improving plant varieties.[124] One of the ways that this

[121] See also Chapter 4. [122] PPVFR Act, Section 2(1). [123] PPVFR Act, Section 44.

[124] *Plant Variety Protection Act, B.E. 2542* (1999) (PVP Act). Notably, in 2014, Bills were introduced by the Thailand Government that would amend Thailand's plant variety rights laws. For an overview and criticism of Thailand's plant variety protection scheme, see P. Lertdhamtewe, 'Protection of plant varieties in Thailand' (2014) 17(5–6) *The Journal of World Intellectual Property* 142; P. Lertdhamtewe, 'Plant variety protection in Thailand: The need for a new coherent framework' (2013) 8(1) *Journal of Intellectual Property Law & Practice* 35; P. Lertdhamtewe, 'Thailand's plant protection regime: A case study in implementing TRIPS' (2012) 7(3) *Journal of Intellectual Property Law & Practice* 186

was achieved was by allowing individual farmers as well as groups of farmers to hold rights over 'local domestic plant varieties'.[125] The introduction of local domestic plant varieties was 'intended to create "special and differential" (S&D) treatment favouring farmers and local communities, taking into account the large size of the farming population in the country'.[126] In order to be granted a right, local farmers need to satisfy the criteria of the variety needing to exist in a particular locality within Thailand that is not already registered and that is distinct, uniform and stable in accordance with the PVP Act.[127] Once granted a right, farmers or communities enjoy the same rights as those conferred to plant breeders under the PVP Act.

Although a number of countries, including India and Thailand, have granted rights to farmers and local communities there is limited research into the effects of these schemes. Most of the research to date has been more a commentary of particular schemes and provisions rather than their impacts on farmers and local communities. It is perhaps fair to say, however, that despite the best of intentions, there are doubts about whether farmers and local communities have benefitted from such *sui generis* schemes. In Thailand, for example, no farmers have made use of the PVP Act.[128] One of the reasons for this is that the plant varieties that farmers produce generally do not satisfy the requirements of distinctness, uniformity and stability.[129] Another set of issues relate to the practicality of farmer-specific plant variety protection. It has been said that Thailand's PVP Act 'constitutes insufficient and inadequate protection for the rights of farmers and local communities since it does not create any practical means for them to enjoy its benefits'.[130] For instance, a number of communities might be connected to a plant variety, so no one community can claim the right to register and benefit from plant variety rights. Yet another obstacle for farmers is cultural: traditional farmers have not generally sought intellectual property rights nor have they known or cared about intellectual property rights. Finally, it is possible that plant variety protection schemes such as India's and Thailand's that

[125] PVP Act, ss. 43–51.
[126] Lertdhamtewe, 'Plant variety protection in Thailand: The need for a new coherent framework', 39.
[127] Although Thailand's criteria is less strict than the UPOV Convention: see Lertdhamtewe, 'Plant variety protection in Thailand: The need for a new coherent framework'.
[128] Ibid.
[129] See Lertdhamtewe, 'Thailand's plant protection regime: A case study in implementing TRIPS'.
[130] P. Lertdhamtewe, 'Plant variety protection in Thailand: The need for a new coherent framework', 40.

try to achieve so much – by attempting to balance breeders' and farmers' rights – dilute of benefits of such schemes.

Not everyone thinks that traditional and collaborative farmers should be treated like plant breeders. Correa, for example, points out that it is illogical to protect farmers' rights through intellectual property, as it is the very system that the concept that farmers' rights was introduced to solve.[131] Plant varieties developed by farmers should remain free for everyone to use, and assuming that local farmers want to be included in intellectual property schemes underestimates:

> the likelihood that communities might not recognize individual rights over germplasm and might not wish to exercise community rights against neighboring communities. Assigning ownership for economic or financial returns runs against farmers' spirit of free exchange. More than the legal problems that would result from attempts to bring farmers' varieties under current intellectual property rights systems, these cultural motives will probably prevent the application of such IPR systems on farmers' varieties.[132]

4.4 Conclusion

This chapter has shown how UPOV 1961 elevated plant breeders to the position of 'creator' under intellectual property law. This was important because it endowed plant breeders – long thought of as discoverers and practitioners – with a legal right over new plant varieties. Despite plant breeding being left undefined in UPOV 1961 and 1978 clarification of the concept of the plant breeder was necessitated by two key developments. First, there was a fragmentation of plant breeding so that a 'traditional' plant breeder was no longer at the centre of plant variety development. In this way the identity of the plant breeder was reconfigured to include scientists, geneticists, biochemists, entomologists and others. Second, and perhaps more importantly, there was increasing political concerns over 'plant piracy' during the 1980s which meant that plant breeders, and the UPOV Convention more generally, had to engage with broader legal, regulatory and political frameworks. A notable consequence of this engagement with concerns over biological resources and other treaties such as the CBD and Plant Treaty was that the place of the 'discoverer' in intellectual property law was challenged.

[131] C. Correa (with S. Shashikant and F. Meienberg), *Plant Variety Protection in Developing Countries: A Tool for Designing Sui Generis Plant Variety Protection System, An Alternative to UPOV 1991* (APBREBES, 2015).

[132] Salazar, Louwaars and Visser, '"Protecting farmers" new varieties: New approaches to rights on collective innovation in plant genetic resources', 1523.

As a consequence, a definition of a plant breeder as a person who 'bred, or discovered and developed, a new variety' was introduced to UPOV 1991 to clarify the place of discoverers in the UPOV Convention.

More recently, questions have been raised over whether farmers are plant breeders for the purpose of the UPOV Convention. In addressing this question, I have argued that legally at least farmers who show that they have 'bred, or discovered and developed, a variety' are plant breeders under UPOV 1991. More practically, however, there are a number of challenges for farmers claiming to be plant breeders under UPOV-based plant variety rights schemes. These challenges stem from the collective and collaborative nature of traditional plant breeding, and the difficulty of satisfying the level of distinctness, uniformity and stability required for plant variety registration. More empirical or ethnographic research needs to be done to assess the effects of *sui generis* plant variety protection that specifically caters for famers' rights. It is perhaps fair to say, however, that despite the best of intentions, there are doubts about whether farmers and local communities have benefitted from *sui generis* schemes defining plant breeders so broadly.

5 The Proliferation, Politicisation and Legalisation of Plant Varieties

5.1 Introduction

The concept of the plant variety plays a central organising role in the UPOV Convention. Indeed, given the focus of protecting plant breeders' outputs, it was the obvious choice as the organising concept around which the UPOV Convention could be structured. In this way, the concept of plant variety was initially used to underpin and sustain much of UPOV 1961. It also acted as the thread that linked the various provisions of the Convention together including the purpose of the Convention, the rights protected and the scope of protection, the conditions required for protection, the duration of protection and the requirement of a denomination. Yet, despite the important role the plant variety plays in intellectual property law, it has been largely seen as unimportant and overlooked.[1] Even as UPOV 1961 was being negotiated, not a great deal of importance was attached to the subject of plant variety. This ambivalence appears to presume that the plant variety concept is unproblematic. Such thinking also tends to equate a 'natural' or 'real' plant variety with the concept of plant variety under the UPOV Convention.

Taking the concept of plant variety as its focus, this chapter argues that while the concept of the plant variety embodied in UPOV 1961 relied heavily on horticultural practice, and accommodated plant breeding practices, it has been rendered more political and legal under the UPOV Convention and *sui generis* plant variety protection laws alike. The concept of plant variety also became central in determining the relationship between UPOV-based national plant variety protection schemes and

[1] Some notable exceptions include M. Janis and S. Smith, 'Technological change and the design of plant variety protection regimes' (2007) 82 *Chicago Kent Law Review* 1557; M. Llewelyn and M. Adcock, *European Plant Intellectual Property* (Hart Publishing, 2006); J. Roselló, 'The UPOV Convention – The concept of variety and the technical criteria of distinctness, uniformity and stability' in UPOV, *Seminar on the Nature of and Rationale for the Protection of Plant Varieties Under the UPOV* Convention, UPOV Pub No 717(E) (UPOV, 1994), pp. 57–69.

other legal schemes such as patent law. Although TRIPS members can exclude plants from patent protection, they are required to apply some form of protection to 'plant varieties' either by patents, a *sui generis* system, or a combination of both.[2] Increasingly, too, we have seen an expansion in the kinds of plant varieties referred to in both the UPOV Convention and *sui generis* national plant variety protection laws. One example of a 'new' kind of plant variety is EDVs: introduced by UPOV 1991 to strengthen plant variety rights by addressing concerns over the breeder's exemption and trivial improvements in plant varieties.[3] Other examples of 'new' kinds of plant varieties are extant or farmers' varieties, which can be found in *sui generis* national laws of some developing countries such as India and Thailand.[4]

At the Diplomatic Conference of 1991, the Report of the Working Group on Article 1 stressed that it was necessary to 'avoid any concrete term which could represent, physical elements of the variety'.[5] Indeed, far from narrowing the concept of plant variety, we have seen a proliferation of the kinds of plant varieties referred to in both the UPOV Convention and *sui generis* (i.e. non-UPOV based) plant variety protection. Broadly speaking the introduction of different kinds of plant varieties has been with one of two intents: either to strengthen plant variety rights protection (e.g. EDVs) or to recognise the role played by farmers and local communities in the development of new plant varieties (e.g. farmers' varieties and local domestic varieties). The end result of this is that the plant variety concept has been rendered more political and legal, bringing with it the challenges of implementation, application and interpretation.

5.2 Plant Varieties as 'Outputs' and Protectable Varieties (UPOV 1961)

In crafting a new international treaty on plant varieties, it was necessary to establish an organising concept or common denominator that united the various provisions of the UPOV Convention. While it was generally accepted that the UPOV Convention would be structured around the notion of the plant variety, it was not immediately clear how the plant variety concept would be constructed or defined. The process of

[2] UPOV 1991, Article 27(3)(b).
[3] See Chapter 9 for more details on the concept of EDVs.
[4] Other countries to introduce concepts of farmer or local plant varieties include Malaysia, the Philippines and Indonesia.
[5] UPOV, *Records on the Diplomatic Conference for the Revision of the International Convention for the Protection of New Varieties of Plants 1991*, Publication No. 346(E) (UPOV, 1992), p. 138 [7].

including a definition of a plant variety in UPOV 1961 was complex and took approximately three years to complete.[6]

One area that was drawn upon to develop the notion of plant variety was taxonomy, particularly related to variety denominations.[7] In particular there are two codes that set out the principles, rules and recommendations related to the naming of plants:[8] the *International Code of Botanical Nomenclature* (*Botanical Code*) and the *International Code of Nomenclature of Cultivated Plants* (*Cultivated Plant Code*).[9] The *Botanical Code* dates back to 1753 and deals with botanical names given to plants, fungi and a number of other organisms. It also sets out the hierarchical ranks (or taxa), including the primary ranks of taxa (in descending order: kingdom, division, class, order, family, genus and species)[10] and the secondary ranks such as tribe, section, series, form and variety.[11] Despite the importance of the *Botanical Code*, it is generally the domain of wild plant taxonomists and focuses on naturally found plants.[12] By contrast, the *Cultivated Plant Code* was introduced in 1953 as a set of supplementary rules to cater for horticulture, forestry and agriculture. In so doing, the *Cultivated Plant Code* regulates the names of plants whose origin or selection is primarily due to intentional human activity. Under the *Cultivated Plant Code*, plant varieties or cultivars were defined as 'a plant that has been deliberately altered or

[6] UPOV, Publication No. 337(E), p. 33.

[7] See Chapter 6 for more details on variety denominations.

[8] The *International Code of Nomenclature of Botanical Nomenclature* (ICBN) also applies to fungi and a small number of other organisms.

[9] For a discussion of the relationship between the *International Code of Nomenclature of Botanical Nomenclature* (ICBN) and the *International Code of Nomenclature of Cultivated Plants* (ICNCP), see R. Spencer and R. Cross, 'The cultigen' (2007) 56(3) *Taxon* 938; J. McNeil, 'Nomenclature of cultivated plants: A historical botanical standpoint' (2004) 634 *Acta Horticulturae* 29.

[10] Botanists generally categorise plants with the use of the ranks of species (individual plants that are similar in appearance and can reproduce or breed among themselves and produce other individuals that resemble the parents) and genus (a group of related species). For discussions of taxonomic ranks, see O. Rieppel, 'Origins, taxa, names and meanings' (2008) 24 *Cladistics* 598; R. Spence, R. Cross and P. Lumley, *Plant Names: A Guide to Botanical Nomenclature* (3rd ed., CSIRO Publishing, 2007); O. Rieppel, 'The taxonomic hierarchy' (2006) 26 *The Systematist* 5; W.T. Stearn, 'Historical survey of the naming of cultivated plants' (1986) 182 *Acta Horticulturae* 19.

[11] *International Code of Botanical Nomenclature* (1958), Article 4.1.

[12] In botanical nomenclature, the term 'variety' has been used by taxonomists in a range of ways to include members of different populations that can interbreed easily (although not all traits will run true and the traits of a new variety will often be a combination of its parents so that the new variety will have an appearance distinct from other plant varieties) and plants that are often geographically separate from each other. See Hawkes, 'Infraspecific classification: The problems' in B. Styles (ed.) *Infraspecific Classification of Wild and Cultivated Plants* (Clarendon Press, 1986); McNeil, 'Nomenclature of cultivated plants: A historical botanical standpoint'.

selected by humans'.[13] Because the target audience of the *Cultivated Plant Code* was similar to those targeted by the proposed international treaty on plant varieties – viz. plant breeders – the *Cultivated Plant Code* had the most relevance for, and impact on, the development of UPOV 1961.

The *Cultivated Plant Code* was influential in establishing the framework around which plant varieties were given a denomination.[14] In relation to defining plant varieties, however, the impact of the *Cultivated Plant Code* was less noticeable. Because the overriding aim of those present at the Diplomatic Conference of 1957–1961 was to protect the work of the plant breeder, when attempting to define a plant variety, the focus was on ensuring that the definition captured the work of the plant breeder rather than in defining what a plant variety actually was. The culmination of this was a definition of plant variety that listed various outputs of plant breeding. Article 2(2) of UPOV 1961 defines a plant variety as 'any cultivar, clone, line, stock or hybrid which is capable of cultivation and which satisfies the provisions of subparagraphs (1)(c) and (d) of Article 6'.[15] While this definition of a plant variety reflected horticultural practice, it also accommodated plant breeding practices by listing the products from various modes of reproduction. That is, it lists various types of plant varieties. For example, Article 2(2) lists 'lines', which include pure line plant varieties that are homozygous, or near homozygous, and are largely produced in naturally self-pollinating species such as wheat, barley and soybeans.[16] Another example is that of a 'clone', which tends to be highly heterozygous and is vegetatively propagated by asexual reproduction including cuttings, bulbs and grafts.[17] In this way, the subject matter of UPOV 1961 – the plant variety – was defined as a tangible object in nature, produced by plant breeders. The concept of plant variety in UPOV 1961 correlates to a 'real' plant variety: it could be pointed to, touched and picked up.

Although the concept of the plant variety as subject matter focused on the end product of plant breeding, it was evident that it was not appropriate to protect all types of plant varieties. Indeed, the concept of the plant variety was given shape and limited by a number of conditions.

[13] *International Code of Nomenclature of Cultivated Plants 1958,* Principle 2.
[14] The topic of Chapter 6.
[15] The terms 'variety' and 'cultivar' are often used interchangeably, although the word cultivar is not interchangeable with the botanical rank of variety or with the legal term 'plant variety'.
[16] J. Brown and P. Caligari, *An Introduction to Plant Breeding* (Blackwell Publishing, 2008), p. 15.
[17] Ibid.

By limiting the concept of the plant variety in this way, it was hoped that the same plant variety could not be subject to a number of claims and that protected plant varieties expressed the characteristics described by the breeder. After a relative short period of debate, it was decided that to be eligible for protection a plant variety had to be homogenous, stable, distinct and new.[18] These requirements helped to qualify the concept of the plant variety in the context of subject matter. Indeed, as we see in the following paragraphs, for the purposes of UPOV 1961, a plant variety was synonymous with a protectable plant variety.

Most notably, perhaps, the requirement of homogeneity was included in the UPOV Convention so that growers could be confident they were purchasing a plant variety that exhibits the characteristics that it is said to.[19] However, because plants are by their very nature variable, there is generally some variation within a group of plant varieties. Because of this, the condition of homogeneity placed a limit on the amount of variation permitted in a plant variety protected under the UPOV Convention. So, the requirement of homogeneity applied to plant variety rights does not connote complete homogeneity, but rather it is a question of relative homogeneity.[20] The level of homogeneity required from a plant variety under UPOV 1961 was related to whether the plant variety was sexually reproduced or vegetatively propagated,[21] as Article 6(c) of UPOV 1961 required the new variety to be 'sufficiently homogeneous, having regard to the particular features of its sexual reproduction or vegetative propagation'.[22] Yet another factor that shapes the 'plant variety' is the requirement that the variety be stable. A stable plant variety is one that remains true to its description when it is reproduced or propagated. Article 6(d) of UPOV 1961 states that the 'variety must be stable in its essential characteristics, that is to say, it must remain true to its

[18] A plant variety must also be given a suitable name or other denomination; UPOV 1961, Article 13; UPOV 1978, Article 13; UPOV 1991, Article 20.

[19] Article 2(2) of UPOV 1961 explicitly tied the concept of the plant variety to the requirements of homogeneity (Article 6(1)(c)) and stability (Article 6(1)(d)).

[20] N. Byrne, *Commentary on the Substantive Law of the 1991 UPOV Convention for the Protection of Plant Varieties* (University of London, 1996), p. 39.

[21] While distinctness is assessed against other plant varieties of common knowledge, uniformity is assessed by comparing individual plants of the applicant variety. See UPOV, *General Introduction to the Examination of Distinctness, Uniformity and Stability and the Development of Harmonized Description of New Varieties of Plants*, TG/1/3 (UPOV, 2002); UPOV, *Examining Distinctness, TGP/9 (UPOV, 2008); UPOV, Examining Uniformity*, TGP/10 (UPOV, 2008).

[22] Sherman and Bently have stated that it is a 'truism that in intellectual property law the protected subject matter must be both reproducible and repeatable': B. Sherman and L. Bently, *The Making of Modern Intellectual Property Law* (Cambridge University Press, 1999), p. 51.

description after repeated reproduction or propagation or, where the breeder has defined a particular cycle of reproduction or multiplication, at the end of each cycle'.

In addition to homogeneity and stability, the concept of plant variety under UPOV 1961 was qualified by the condition of distinctness. The assessment of distinctness played a crucial role in shaping the plant variety concept under UPOV 1961. The requirement of distinctness has its origins in certification, trade mark and patent laws and was introduced to ensure that the same plant variety could not be subject to a number of claims. Trade mark or seed certification laws required plant breeders to show that their plant variety warranted protection by including notions of 'originality' or 'novelty' to distinguish new plants or seeds from old plants or seeds.[23] One example, and arguably the most significant for UPOV 1961, was the German *Seed Law 1953* which required new seeds to be 'individualised'. This ensured that only the best plant varieties were put on the market.[24] Because it was envisaged that making agronomic value a condition of certification would have the effect of improving seed quality, these early laws were targeted towards seed certification and qualitative control, and consequently they also placed importance on the merit or quality of the seeds granted protection. In doing so, *German Seed Law 1953* reserved protection for new plant varieties that had 'agronomic value'.[25] A second influence on the concept of distinctness in UPOV 1961 was the United States *Plant Patent Act 1930* which provided protection to 'any distinct and new variety of plant' that was asexually reproduced.[26] While the *Plant Patent Act 1930* did not define the term 'distinct', the United States Congress expressly stated that 'improvements' over existing plants were not required: this is in contrast to the European seed laws such as the German *Seed Law 1953* discussed above.[27] These earlier notions that a new plant variety had to be 'individualised' and 'distinct' were influential in the development of an international treaty on plant

[23] See A. Heitz, 'The history of plant variety protection' in *The First Twenty-Five years of the International Convention for the Protection of New Varieties of Plants* (UPOV, 1987), pp. 68–70; C. Fowler, *Unnatural Selection: Technology, Politics and Plant Evolution* (Gordon and Breach, 1994), p. 90.

[24] Committee on Transactions in Seeds, *Plant Breeders' Rights: Report of the Committee on Transactions in Seeds*, Cmnd. 1092 (London, July 1960), pp. 112–116.

[25] See Heitz, 'The history of plant variety protection', pp. 72, 75–76.

[26] *Plant Patent Act 1930* 35 U.S.C., § 161.

[27] C. Fowler, 'The Plant Patent Act of 1930: A sociological history of its creation' (2000) 82 *Journal of the Patent and Trademark Office Society* 621, 641. See also UPOV, 'Report of International Conference for the Protection of New Plant Products' [1961] *Industrial Property Quarterly* 104; M. Janis and J. Kesan, 'US plant variety protection: Sound and fury...?' (2002) 39 *Houston Law Review* 727, 739.

varieties. It was these aspects of the earlier laws (less the additional requirement of agronomic value) that 'correspond more or less' to the concept of distinctness included in Article 6(1)(a) of UPOV 1961.[28] According to Article 6(a) of UPOV 1961, a new variety is distinct if it is:

> clearly distinguishable by one or more important characteristics from any other variety whose existence is a matter of common knowledge at the time when protection is applied for...A new variety may be defined and distinguished by morphological or physiological characteristics. In all cases, such characteristics must be capable of precise description and recognition.

Despite the inclusion of the adjective 'important' in UPOV 1961, distinctness was not tied to the merit or quality of the plant variety. Because of the inherent difficulties of determining whether a plant variety was qualitatively better than other plant varieties, various proposals to link the distinctness criterion to 'usefulness' or 'superiority' were rejected at the Second Session of the Conference held in 1961.[29] An important rationale for rejecting a qualitative criterion was that it would have placed too heavy a burden on independent authorities such as a member states' Department of Agriculture or Intellectual Property Offices. This reluctance was summed up by the *Report of the Committee on Transactions in Seeds* presented to the British Parliament in 1960, which said that the 'authority's decision in any particular case would not only determine whether the breeder could claim royalties, but might in practice also settle the question for the breeder whether it might be worthwhile marketing the variety or abandoning work on it'.[30]

In addition to the qualifications on the concept of plant variety discussed above, a plant variety had to be new. While this is often referred to as novelty requirement, this can be misleading and confusing. Although there were some proposals to introduce a notion of novelty similar to that found in patent law, one of the overriding aims of those at the Diplomatic Conference of 1957–1961 was to reflect the commercial nature of plant breeding. This was based largely on the view that knowledge of the existence of the plant variety does not mean that the public has access to the plant variety in order to reproduce that plant variety. In protecting the commercial nature of plant breeding, UPOV 1961 regularly refers to the new plant variety. In relation to Article 5 of UPOV 1961 and the scope of protection, reference is made to the new plant variety. In

[28] Heitz, 'The history of plant variety protection', p. 76 (referring particularly to the *German Seed Law of 1953*).

[29] Ibid., pp. 86–87.

[30] Committee on Transactions in Seeds, *Plant Breeder's Rights: Report of the Committee on Transactions in Seeds*, p. 56.

addition, Article 6 of UPOV 1961 and the conditions of grant state that the new variety must be clearly distinguishable, sufficiently homogeneous and stable and must be given a denomination. Despite the repeated reference to the term new, UPOV 1961 does not define what it means to be 'new'. However, at the Diplomatic Conference of 1957–1961, it appears that there was overlap between the concepts of priority, distinctness and newness. The First Meeting of the Committee of Experts, held in 1958, discussed the issue of priority that existed in the Paris Convention but felt that this needed to be adapted (to suit plant breeding) to include certain timeframes in which 'disclosure or exploitation that occurred during [this] period could be held against him if it concerned his own variety'.[31] Other discussions concentrated on the novelty of the distinctive characteristics of the applicant plant variety and the importance of the differences between the new variety and the old varieties.[32] Despite the early focus on priority and distinctness, the principle of commercial newness was introduced by way of Article 6(1)(b) of UPOV 1961. To be eligible for protection, the plant variety could not have been 'offered for sale or marketed' in that member state or for more than four years in a territory other than the state in which the application was sought.

While the concept of plant variety as the subject matter of the UPOV Convention was widely accepted by UPOV Members, a number of problems soon arose. As we will see in the next part, plant breeding techniques underwent significant changes from the 1950s onwards and the identification of plant breeders fragmented, as did the potential outputs of plant breeding. Further difficulties arose with plant varieties being synonymous with protectable plant varieties, and questions were being raised about the interaction and potential overlap between the UPOV Convention and patent laws.

5.3 No Definition Is Better Than a Narrow One (UPOV 1978)

During the late 1960s and early 1970s, a number of problems arose with the definition of plant variety contained in UPOV 1961. One of the reasons why the plant variety concept in UPOV 1961 became problematic was that plant breeding was a dynamic endeavour. More specifically, the nature of plant breeding and the identity of the plant breeder changed in the mid-to-late twentieth century.[33] In the 1950s Crick and Watson

[31] Heitz, 'The history of plant variety protection', p. 87.

[32] UPOV, *Actes des Conférence Internationales pour la Protection des Obtentions Végétales 1957–1961, 1972*, UPOV Publication No. 316 (UPOV, 1972), pp. 33, 76.

[33] The concept of plant breeder is the topic of Chapter 4.

described the physical structure of DNA,[34] which started a transformation in the biological sciences that ultimately meant that plants could be viewed on a molecular level (as well as on a physical and chemical level). This meant that there was a degree of uncertainty over the direction that plant breeding might take. As plant breeding absorbed and utilised developments in science and technology, a definition of plant variety that focused on the type of work plant breeders carried out became less relevant. The concern was that plant breeding no longer merely produced cultivars, clones, lines, stocks or hybrids. As the notion of a plant variety became uncertain, potential issues were created for delegates of the Diplomatic Conference of 1978 who felt that such a definition was likely to be exclusionary.[35] It was suggested, for example, that the definition used in UPOV 1961 did not include multiline plant varieties which are mixtures or blends of a number of different plant varieties.[36]

As well as not wanting to exclude certain plant varieties from the UPOV Convention, a second concern for delegates at the Diplomatic Conference of 1978 was the difficulty of placing a scientific definition of plant variety into a legal scheme such as the UPOV Convention. While UPOV 1961 had focused on horticultural plant varieties (that resulted from the intervention of a plant breeder), there were questions over the (un)certainty of both botanical and cultivated taxonomic plant varieties. For example, below the category of species, there were, and continues to be, questions over the utility and validity of such taxa, with the variety taxon being referred to as a category of indecision.[37] At the Diplomatic Conference of 1978, questions were raised over whether such indecision was appropriate for an international convention such as UPOV. As Mr. Bustarret of France stated at the Diplomatic Conference, 'the word "variety" as it was used, without being defined in the Convention, had a meaning for everyone present. What was not absolutely certain was that the meaning was really the same for everyone'.[38]

[34] L. Kay, *Who Wrote the Book of Life? A History of the Genetic Code* (Stanford University Press, 2000).

[35] For the discussions of the plant variety definition, see UPOV, Publication No. 337(E), 134–135 [102]–[116], 141–142 [195]–[207].

[36] Byrne, *Commentary on the Substantive Law of the 1991 UPOV Convention for the Protection of Plant Varieties*. See, generally, Brown and Caligari, *An Introduction to Plant Breeding*, p. 16.

[37] For a discussion, see R. Spence and R. Cross, 'The international code of botanical nomenclature (ICBN), the international code of nomenclature for cultivated plants (IICNCP) and the cultigen' (2007) 56(3) *Taxon* 938; W. Camp and C. Gilly, 'The structure and origin of a species' (1943) 4 *Brittonia* 323.

[38] UPOV, Publication No. 337(E), p. 142 [199].

Yet another concern about the definition of plant variety as subject matter used in UPOV 1961 was that the definition stipulated that a plant variety 'satisfies the provisions of subparagraph 1(c) and (d) of Article 6'. This provided a direct relationship between the notion of a plant variety and the substantive requirements of homogeneity and stability.[39] Consequently, according to UPOV 1961, the subject matter and a protectable plant variety were effectively the same. One argument against linking the subject matter with the requirements of grant was that if a plant variety did not conform to the definition used in Article 2(2) of UPOV 1961 (because it was not homogenous or stable), it was not a plant variety.[40] By linking the notion of plant variety to the conditions of grant, it was felt that Article 2(2) of UPOV 1961 provided a narrow definition of a plant variety that was untenable because it was likely to be problematic with national patent laws that referred specifically to plant varieties and in some instance demarcated patent law and plant variety protection using the concept of plant variety.

Faced with these concerns, a number of proposals were presented at the Diplomatic Conference of 1978 in relation to the definition of plant variety. One suggestion was that the subject matter should be defined in terms of an 'assemblage of plants which is capable of cultivation and which satisfies the requirements of [protection]'.[41] It was also proposed that a plant variety be defined as an 'assemblage of cultivated plants'.[42] While these proposals were similar to the definition employed in the *Cultivated Plant Code*, concerns were raised over the use of the term 'cultivation' in the UPOV Convention. It was argued that, because not all plant varieties are cultivated, a definition of plant varieties that included 'cultivation' would exclude a number crops and plants from protection. This was particularly the case when 'a variety for which protection was granted was, for example, represented by its seed and by the seed sample deposited and there was no obligation…to actually cultivate a variety'.[43] Another problem of using the term 'cultivation' in the definition of plant variety was that, similar to the variety concept itself, the meaning of 'cultivate' was uncertain.[44]

Because of these concerns, a satisfactory definition of plant variety could not be found. The position of the delegates was that any definition of plant variety – whether it referred to cultivation, an assemblage of

39 UPOV 1961, Articles 1(c) and 1(d).
40 UPOV, Publication No. 337(E), p. 16.
41 Ibid., p. 102 (Netherlands, DC/14). See also UPOV, 'Summary of the main amendments to the Convention' (1979) *Industrial Property* 33.
42 UPOV, Publication No. 337(E), p. 102 (United Kingdom, DC/15).
43 Ibid., p. 137 [107] (Mr. D. Böringer, Federal Republic of Germany).
44 Ibid., p. 142 [198] (Mr. A.F. Kelly, United Kingdom).

plants, or was explicitly linked to the requirements of grant – would be too restrictive. Further, it was felt that providing a definition of plant variety could inhibit technical progress and potentially impair the adoption of the UPOV Convention by those countries not already members.[45] On a more practical level, despite that fact that the issue of defining plant variety had been on the 'agenda of the six sessions of the Committee of Experts in the Interpretation and Revision of the Convention, and at sessions of other bodies of UPOV, a satisfactory definition had not been found'.[46] Faced with these problems, delegates at the Diplomatic Conference of 1978 decided to remove the definition of plant variety from UPOV 1978.[47] It was hoped that removing the definition of plant variety would 'facilitate the adaption of the meaning of the word "variety" to scientific and technological progress, especially in the area of plant improvement, and thereby also the adoption of the system of protection of new plant varieties as a whole'.[48]

Despite the fact that the definition of plant variety was removed from UPOV 1978, there were no substantive changes made to the conditions of distinctness, homogeneity and stability. While UPOV 1978 removed the adjective 'new' from the text of the Convention, Article 5 of UPOV 1978 expressly sets out 'new' as a condition of grant. That said, there was no substantial change in the grace periods after which offering for sale or marketing the plant variety would destroy newness, although Article 6(1)(b) of UPOV 1978 qualified the grace period in which the applicant plant variety can be 'offered for sale or marketed' in another member state (six years for vines, forest trees, fruit trees and ornamental trees or four years for all other plants) before newness is destroyed.[49]

What is, perhaps, most remarkable about the decision in 1978 not to define plant variety is the total lack of interest in the removal of the definition. The removal of the definition of plant variety was not considered to be an essential amendment and, significantly, was all but ignored in the legal and scientific scholarship of the time.[50] This is

[45] Ibid., p. 142 [198]–[207]. [46] Ibid., p. 142 [205]. [47] Ibid., p. 142 [198].

[48] The absence of a definition of plant variety under UPOV 1978 meant that the term plant variety took on its natural and ordinary meaning: *Vienna Convention on the Law of Treaties*, Article 31(1): Ibid., p. 33 (Explanatory Notes).

[49] While these grace periods were (relatively) unchallenged at the Diplomatic Conference of 1978, Denmark argued for a standard four-year period for all plant varieties offered for sale or marketed in another country: See UPOV, Publication No. 337(E), p. 152 [365]. There was also discussion about the use of the term 'trees', p.152 [349]–[364].

[50] The focus of the discussions surrounding UPOV 1978 appears to be on the extension from four to six years of the period during which a variety may have been marketed abroad without its novelty being affected, in the case of vines, trees and their rootstock: Heitz, 'The history of plant variety protection', p. 92.

despite the fact that it was acknowledged at the Diplomatic Conference of 1978 that it was 'desirable from a legal point of view to have a definition of "variety"'.[51] The ambivalent attitude towards the plant variety concept began to change, however, in the 1980s, as the concept of plant variety was increasingly used in other legal schemes: most notably, in patent law.

5.4 Rendering Plant Varieties More Legal

As a consequence of the increasing reliance on the notion of plant variety, an ad hoc Working Group was formed to 'examine questions with respect to the definition of the term "variety" as laid down in Article 1 of the Basic Proposal [for UPOV 1991]'.[52] The Working Group considered a number of proposals including those put forward by Italy,[53] the United Kingdom,[54] Poland[55] and Sweden.[56] From these proposals, two main questions arose: was a definition of plant variety necessary? And if so, what form should it take?

At the Diplomatic Conference of 1991, some of the UPOV Members and Observers were opposed to the reintroduction of a definition of a plant variety. Arguing that a definition was either not necessary or would be in fact counter-productive, AIPPI observed that, on a practical level, the absence of a definition in UPOV 1978 had not led to any problems.[57] Despite arguments of this nature, the prevailing view was that a definition of plant variety was both 'desirable and necessary'.[58] One reason for including a definition of plant variety in UPOV 1991 was that without clearly establishing the subject matter of protection under the UPOV Convention, it would be difficult to prove what was protected, raising issues for both identifying and proving infringement. This argument was based on the idea that the plant variety concept had an important function for those 'working for the suppression of fraud' in distinguishing those plant varieties that were protected from those that were not.[59] In this sense, it was argued that it was important to be able to distinguish

[51] UPOV, Publication No. 337(E), p. 142 [204].
[52] UPOV, Publication No. 346(E), p. 137 (DC/91/106).
[53] UPOV, Publication No. 346(E), p. 104 (Italy, DC/91/22).
[54] Ibid., p. 105 (United Kingdom, DC/91/23). [55] Ibid., p. 106 (Poland, DC/91/26).
[56] Ibid., p. 107 (Sweden, DC/91/28). [57] Ibid., p. 195 [137.1].
[58] Rosselló, 'The UPOV Convention – The concept of variety and the technical criteria of distinctness, uniformity and stability', 58.
[59] M. Piatti and M. Jouffray, 'Plant variety names in national and international law: Part I' [1984] 10 *European Intellectual Property Review* 283; M. Piatti and M. Jouffray, 'Plant variety names in national and international Law: Part II' [1984] 11 *European Intellectual Property Review* 311.

between plant varieties so as to diminish confusion and discourage deceptive and misleading practices.[60]

However, by far the most significant concern for the Working Group was the need to establish a boundary between different forms of legal protection, particularly between UPOV-based systems and patent law. Prior to UPOV 1991, the task of managing legal categories occurred, in effect, automatically. While the industry-specific nature of the UPOV Convention meant that there was a demarcation between UPOV-based schemes and patent law, changes in science and technology (as well as in plant breeding and the marginalisation of the objections to the use of patent protection for plants) meant that this boundary was increasingly blurred. One of the consequences of this was that patents were being sought to protect plant varieties and other plant-related developments. For example, in the early 1980s, there was a sharp increase in the number of patents granted by the United States Patent and Trademark Office (USPTO) and the European Patent Office (EPO) for agricultural developments.[61]

While United States Courts had previously considered the meaning of variety under the *Plant Patent Act 1930*,[62] the interface between plant variety protection and patent law has been considered more extensively in Europe: where the term 'plant variety' is used as a way to distinguish and demarcate plant variety right and patent protection. Specifically, Article 53(b) of the *European Patent Convention 1973* ('EPC') provides an exception to patentability for 'plant or animal varieties or essentially biological processes for the production of plants or animals'.[63] The importance of defining and understanding the plant variety concept can be seen in a number of cases from the EPO. One of the earliest decisions to consider the scope of Article 53(b) of the EPC, and the nature of plant varieties under UPOV 1961, was the 1983 case of *Ciba-Geigy/Propagating*

[60] More recently, Janis and Smith pointed out that 'identifying plants as "varieties" owes its origins more to farmers and practical considerations of commerce than to botanists and the rigours of science': Janis and Smith, 'Technological change and the design of plant variety protection regimes', 1572.

[61] G. Graff, S. Cullen, K. Bradford, D. Zilberman and A. Bennett, 'The public-private structure of intellectual property ownership in agricultural biotechnology' (2003) 21(9) *Nature Biotechnology* 989; J. King and P. Heisey, 'Ag biotech patents: Who's doing what?' (2003) 1(5) *Amber Waves* 12.

[62] See, e.g., *Imazio Nursery* v. *Dania Greenhouse* 63 F. 3d 1560 (Fed Cir. 1995) where the parties disputed the meaning of the term 'variety' in § 161 of the *Plant Patent Act 1930*.

[63] Historically, the exclusion of plant varieties from the scope of patentable subject matter under the *European Patent Convention 1973* ('EPC') is explicable on the basis that the EPC was drafted in light of UPOV 1961, when dual protection was prohibited.

Material.[64] In this case, the disputed invention claimed any propagating material that had been chemically treated to make it resistant to other agricultural chemicals such as herbicides.[65] The examiner had rejected the disputed claims on the grounds that plants treated with the invention were 'plant varieties' and, thus, excluded under Article 53(b). However, the Board of Appeal reversed the examiner's decision, stating that the claimed invention 'does not lie within the sphere of plant breeding, which is concerned with the genetic modification of plants' and held that chemically treating seeds did not create new plant varieties.[66] In allowing the patent, the Technical Board of Appeal determined that the nature of plant varieties excluded under Article 53(b) is determined by that which is protectable under the UPOV Convention. At the time the term 'variety' was not defined in the EPC.[67] The rationale for giving the term 'plant variety' under the EPC the meaning of a protectable plant variety under UPOV 1961 – cultivated varieties, clones, lines, strains and hybrids which can be grown in such a way that they are clearly distinguishable from other varieties, sufficiently homogeneous and stable in their essential characteristics – was that plant innovations 'which cannot be given the protection afforded to varieties are still patentable if the general prerequisites [of novelty, inventive step, and industrial application] are met'.[68] Indeed, the Technical Board of Appeal reasoned that the history of Article 53(b) suggests that plant varieties were excluded from patent protection under the EPC mainly because an alternative form of protection was available under the UPOV Convention.

As a result of the concerns about the intersection of the UPOV Convention and patent law, the Working Group felt that it was necessary to develop a consistent definition of a plant variety for the purposes of both the UPOV Convention and patent law. The view was that there was a need to clarify the place of plant varieties within intellectual property law

[64] T0049/83 [1984] *EPOR* 112. For a discussion see Llewelyn and Adcock, *European Plant Intellectual Property*, p 289–319.

[65] The application contained two product claims which defined 'propagating material for cultivated plants' treated with an oxime derivative (claim 13) and characterised in that it consists of seeds (claim 14).

[66] T00049/83 (Propagating material), 26 July 1983, [4].

[67] Regulation No. 2100/94, the Biotechnology Directive and Implementing Rules were amended to include or refer to a similar definition of a 'plant variety' to that in UPOV 1991, Article 1.

[68] T49/83 [1984] *EPOR* 112. This interpretation was followed in *Lubrizol/Hybrid Plants* T320/87 [1990] *EPOR* 77, in which a claim to hybrid seeds and plants was held to be patentable on the basis that at least one of the parent plants was heterozygous with respect to a specific trait and therefore would never breed true, with the result that the subsequent generations of plants, considered as a whole population, were not stable and therefore could not be considered a 'plant variety'.

and that providing a definition of a plant variety was part of this process. Indeed, it was felt that the legitimacy of the UPOV Convention would be at issue, or even diminished, unless Member States were able to determine the subject matter of the UPOV Convention. As one commentator noted, 'how could we talk about protecting plant varieties if we couldn't even define them'.[69]

Once it was established that a definition of plant variety was required, the Working Group on Article 1 turned to look at the nature and substance of the definition. An important difference between attempts to define plant varieties in 1991, and the definition of plant varieties used in UPOV 1961, was that the Working Group agreed that the definition of plant variety in the UPOV Convention 'should make a clear distinction between a variety as an object which might be protected, which must be defined as a concept, and the scope of protection of a variety'.[70] In so doing, the Working Group agreed to avoid any concrete term which could represent physical elements of the plant variety and took into account a wide range of factors including (1) economics, where a plant variety is seen as 'a subdivision of the species which is distinguished for the purposes of the exploitation of the plant resources of the species'; (2) plant breeding, where a plant variety is 'an artificial population with a narrow genetic base, with rather well-defined agronomic characteristics, which is reproducible with more or less precision'; and (3) horticulture, where there are two important features of plant varieties – the level of a plant entity being recognised as a variety and the level of certain material being recognised as belonging to a specific variety.[71] The result of the Working Group's consideration was the adoption of Article 1(vi) of UPOV 1991 which defines a plant variety as:

a plant grouping within a single botanical taxon of the lowest known rank, which grouping, irrespective of whether the conditions for the grant of a breeder's right are fully met, can be:

defined by the expression of the characteristics resulting from a given genotype or combination of genotypes,

distinguished from any other plant grouping by the expression of at least one of the said characteristics and

considered as a unit with regard to its suitability for being propagated unchanged.

[69] J. Ardley, 'The 1991 UPOV convention, ten years on' in M. Llewelyn, M. Adcock and M. Goode (eds.), *Proceedings of the Conference on Plant Intellectual Property within Europe and the Wider Global Community* (Sheffield Academy Press, 2002).

[70] UPOV, Publication No. 346(E), p. 138.

[71] Roselló, 'The UPOV Convention – The concept of variety and the technical criteria of distinctness, uniformity and stability', 57–58.

While appearing to provide specific details of a plant variety, Article 1(vi) offers a conceptual definition that covers, in principle, a plant grouping that is not necessarily protectable under the UPOV Convention. This transformed the concept of plant variety under the UPOV Convention and departed from the position under UPOV 1961, which explicitly linked the concept of plant variety with the requirements of homogeneity and stability. Article 1(vi) is also different to UPOV 1961 because it does not attempt to directly reflect horticultural practice. Moreover, while UPOV 1991 uses the plant variety concept as the subject matter of the UPOV Convention, it attempts to formulate the plant variety concept in a way that is neither too broad nor narrow. This was borne out by a number of factors.

The draft definition of plant variety that was considered by the Working Group referred to a 'group of plants' because it was argued that 'group of plants' was more accurate and was commonly used in biometrical genetics, plant breeding and statistics.[72] However, in line with the discussion above, the Working Group concluded that it would be best to avoid any explicit reference to 'plants' and 'groups of plants' because this would make the notion of plant variety too broad.[73] For this reason, Article 1(vi) of UPOV 1991 includes 'a plant grouping within a single botanical taxon of the lowest known rank'. Maintaining reference to 'plant grouping' makes clear that neither a single plant nor trait is a variety[74] and implicitly excludes a single plant, trait, a chemical or other substance such as DNA or a plant breeding technology from the definition of plant variety under the UPOV Convention.[75] The reference to 'plant grouping', however, is mediated by the reference to 'taxon', which is a taxonomically neutral term (because a particular rank such as species or family is not employed) and is broad enough to indicate the rank of a group as well as the organisms contained within that group.[76] While the term taxon was used, the unqualified use of the term 'taxon' or 'botanical taxon' was seen to be inappropriate because it was too broad.[77] This was because it referred to any unit of the taxonomic classification and could refer to species, genus, family, order or even a plant kingdom in its entirety. The solution to these concerns was to use the phrase 'single

[72] UPOV, Publication No. 346(E), p. 16. [73] Ibid., p. 328 [990]–[993].
[74] UPOV, *Explanatory Notes on the Definition of Variety under the 1991 Act of the UPOV Convention*, UPOV/EXN/VAR/1 (UPOV, 21 October 2010), p. 4.
[75] Ibid.
[76] Spencer, Cross and Lumley, *Plant Names: A Guide to Botanical Nomenclature*. For a general discussion, see Ibid., pp. 193–194 [132.3].
[77] The initial proposal referred to a 'group of plants': UPOV, Publication No. 346(E), p. 16.

botanical taxon of lowest known rank', which it was thought would overcome most of the issues raised by the delegations at the Diplomatic Conference of 1991. In regard to the concern that the definition would be too inclusive, it was thought that this change would not only narrow the concept of plant variety (by placing a limit on the possible plant groups that can be included in the definition of plant variety), but importantly, for those present at the Diplomatic Conference of 1991, the conceptual definition of plant variety includes interspecific crosses such as triticale, a cross between *Triticum aestivum* (wheat) and *Secale cereale* (rye).[78] As the Chairman of the Working Group on Article 1 stated:

> The concept of "botanical taxon of the lowest know rank" constituted a first limitation on the concept of "plant grouping" whilst meeting the concern to encompass within the definition those varieties produced by inter-specific or intergeneric crossing. A simple reasoning had shown to the Working Group that the taxon in which a variety derived from such crossing was included could be very rapidly identified. In the case of triticale, for example, there was, initially, neither question of the species level nor the genus level (since triticale belonged to neither to the genus *Triticum* nor the genus *Secale*), but – speaking only of the major ranks – to the family of Gramineae or – to be more precise – the subtribe of Triticeae. A place could therefore always be found for such a variety and the aim of the Conference should of course be to ensure that it was covered by the system of protection for new plant varieties.[79]

Another significant change in the definition of plant variety used in UPOV 1991 is that unlike Article 2(2) of UPOV 1961, which required a plant variety to be 'sufficiently homogenous' and 'stable in its essential characteristics', Article 1(vi) of UPOV 1991 does not link the plant variety concept to the requirements of grant: making the definition of 'variety' broader than a protectable variety. Most importantly, Article 1 (vi) explicitly states a 'variety' is 'irrespective of whether the conditions for the grant of a breeder's rights are fully met'. In addition, the three indents of Article 1(vi) of UPOV 1991 correspond to a lower level of distinctiveness, uniformity and stability than is required to be granted protection under Articles 7 to 9 of UPOV 1991.[80] The decision to remove the link between the plant variety as subject matter and the requirements of grant was not unanimous, and the *International Federation of Industrial Property Attorneys* argued that they were opposed to a definition of the plant variety embracing entities that were not

[78] Ibid., p. 203 [185] (Mr. Gugerell, European Patent Office). [79] Ibid., p. 328 [991.3].

[80] See N. Byrne, *Commentary on the Substantive Law of the 1991 UPOV Convention for the Protection of Plant Varieties* (University of London, 1996); B. Greengrass, 'The 1991 Act of the UPOV convention' (1991) 12 *European Intellectual Property Review* 467.

protectable under the UPOV Convention.[81] According to the *International Federation of Industrial Property Attorneys*, there was a danger that such a definition could lead to the situation in which a plant was not protectable under a UPOV-based system of protection or patent law: meaning that there would be no protection at all. Despite these concerns, the Chairman of the Working Group on Article 1 stated that extending the definition of plant variety beyond protectable plant varieties extends the concept of plant variety and covers, in principle, a plant grouping that is not necessarily protectable under the UPOV Convention.[82] In addressing the relationship between the UPOV Convention and patent law, the Report of the Working Group on Article 1 stressed that it was necessary to 'avoid any concrete term which could represent, physical elements of the variety'.[83] Furthermore, the Chairman stated:

> Even with the restrictive indents that followed, it was still possible to discover plant groupings meeting the definition, but which were not protectable. The Working Group had considered it important to maintain that clause since it enabled anyone reading the definition to fully apprise the situation. The fact that a variety did not meet the criteria for protection, as defined subsequently in the Convention, did not mean that such variety did not exist.[84]

The interaction between the UPOV Convention and patent law has perhaps been considered most thoroughly in Europe, where the importance of the plant variety concept, not just as an organising concept for the UPOV Convention but also for the way in which the UPOV Convention interacts with other legal schemes, was affirmed. During the 1990s there were a series of disputes in Europe – including *Plant Genetic Systems* and *Novartis/Transgenic Plant* –[85] that considered the relationship between plant variety protection and the plant variety exception under Article 53(b) of the EPC. Specifically these cases considered the meaning of plant variety, namely whether the term 'plant variety' was the same for UPOV-based schemes as it was for patent law. In so doing they considered the exclusion of plant variety under Article 53(b) of the EPC with reference to Article 1(vi) of UPOV 1991.

In *Plant Genetic Systems* the Technical Board of Appeal considered whether the claimed patent protected a plant variety or a 'broad class of plants'.[86] In contrast to the earlier decision of *Ciba-Geigy/Propagating*

[81] UPOV, Publication No. 346(E), p. 196 [143]. [82] Ibid., p. 328 [991.3(iii)].
[83] Ibid, p. 138 [7]. [84] Ibid.
[85] *Plant Genetic Systems/Glutamine Synthetase Inhibitors (Opposition by Greenpeace)* T356/93 [1995] *EPOR* 357; *Novartis/Transgenic Plant* [2000] *EPOR* 303.
[86] *Plant Genetic Systems/Glutamine Synthetase Inhibitors (Opposition by Greenpeace)* T356/93 [1995] *EPOR* 357, 381.

Material (discussed in the preceding part of this chapter), the Technical Board of Appeal considered the plant variety concept from the perspective of UPOV 1991. In distinguishing *Ciba-Geigy/Propagating Material*, the Technical Board of Appeal held that the eligibility of protection under the UPOV Convention was not a precondition for consideration as a plant variety under the EPC.[87] The definition of plant variety used in the *Plant Genetic Systems* decision distinguished the scope of Article 53(b) EPC with respect to 'plant varieties' from the scope of plant variety protection afforded by the UPOV Convention. By removing the requirement that a claimed plant be eligible for plant variety protection under the UPOV Convention (the old UPOV 1961), the Technical Board of Appeal indicated that the EPO no longer regards the first half-sentence of Article 53(b) as merely carving out the legal space occupied by the UPOV Convention. In so doing some plant material may be excluded from plant protection and plant variety protection, for instance, a plant that is found to be an unpatentable variety under Article 53(b) but does not have sufficient distinctiveness, stability or homogeneity to qualify for UPOV-style plant variety protection.

In *Novartis/Transgenic Plant*, the EPO considered the patentability of anti-pathogenic transgenic plants and to methods of producing the transgenic plants.[88] In *Novartis*, the EPO refused to grant the patent because it deemed the claimed invention to be a plant variety and thus excluded due to Article 53(b) of the EPC. Novartis appealed to the EPO Technical Board of Appeal and argued that as they were patenting 'more than one variety', the claims fell outside the scope of the exclusion of Article 53(b) of the EPC.[89] The Technical Board of Appeal referred a number of questions to the Enlarged Board of Appeal, who found for Novartis on the basis that the method of invention did not necessarily result in a 'plant variety'. More specifically, after examining various definitions of 'plant variety' – including Article 1(vi) of UPOV 1991, Rule 23(b)(4) of the EPC Implementing Regulations, and earlier case law[90] – the Enlarged Board concluded that the expression of characteristics of a plant variety that results from a given genotype, or combination of genotypes, is a reference to the entire constitution of a plant or set of

[87] *Plant Genetic Systems* T356/93 [1995] *EPOR* 357, 375.

[88] *Novartis/Transgenic Plant* (G01/98) [2000] *EPOR* 303.

[89] Ibid, [30]. Novartis argued that this approach was supported by Article 4(2) of *Council Directive 98/44/EC on the legal protection of biotechnological inventions* [1998] OJ L213/13 ('*Biotechnology Patent Directive 1998*'), which states that plants 'shall be patentable if the technical feasibility of the invention is not confined to a particular plant or animal variety'.

[90] For e.g. T 0356/93 (Plant cells), 21 February 1995.

genetic information.[91] In contrast, a plant defined by single recombinant DNA sequences is not an individual plant grouping.[92] Therefore, the Enlarged Board concluded that the resulting products of the Novartis invention did not define a single variety, or a multiplicity of varieties,[93] concluding that '[i]n the absence of the identification of specific varieties in the product claims, the subject-matter of the claimed invention is neither limited nor even directed to a variety or varieties'.[94] In other words, product claims which do not identify or individually claim specific plant varieties but instead include subject matter that covers or embraces plant varieties are not plant varieties within the meaning of Article 53(b). More generally the Enlarged Board of Appeal held that Article 53(b) did not exclude anything other than plant varieties which can be protected under plant variety rights law, stating that this interpretation is consistent with the purpose of the exclusion in Article 53(b) of the EPC, which 'defines the borderline between patent protection and plant variety protection',[95] and that 'inventions ineligible for protection under the plant breeders' rights system were intended to be patentable under the EPC provided that they fulfilled the other requirements of patentability'.[96] So as long as the claimed invention is not directed to an individual plant variety, then it does not fall under the Article 53(b) exclusion.

So far we have seen how the concept of the plant variety embodied in UPOV 1961 (that relied heavily on horticultural practice) has been rendered more political and legal under subsequent versions of the UPOV Convention. In the next part, I want to highlight the expansion of the plant variety concept and the proliferation of new 'types' of plant varieties. Broadly speaking the introduction of different kinds of plant varieties has been with one of two intents: either to strengthen plant variety rights protection (e.g. EDVs) or to recognise the role played farmers and local communities in the development of new plant varieties (e.g. farmers' varieties and local domestic varieties).

5.5 More and More Varieties

While EDVs are the topic of Chapter 9, in the context of plant varieties, they are illustrative of the expansion and proliferation of plant varieties, as well as their increasing politicisation and legalisation. Most notably, the concept of EDVs was introduced in order to strengthen UPOV-based plant variety

[91] *Novartis/Transgenic Plant* (G01/98) [2000] *EPOR* 303, 313. [92] Ibid. [93] Ibid.
[94] Ibid.
[95] Ibid., 319. Although, it was suggested that the 'borderline' between patents and plant variety rights was in the fact in hands of those who draft the claims: see *Novartis/Transgenic Plant* [1999] *EPOR* 123, 133.
[96] Ibid.

protection. Briefly stated, EDVs were introduced due to concerns over the (relatively) low threshold of distinctness, the broad ranging breeder's exemption and the limited infringement provisions in the UPOV Convention. These concerns were exacerbated by the advent of molecular plant breeding techniques in the 1970s and 1980s as well as the absence of a comparable breeder's exemption under patent law. There was a perception, therefore, that plant breeders could and did abuse UPOV-based schemes because subsequent plant breeders had an unrestricted ability to protect the plant varieties that had been derived from existing varieties. While the breeder's exemption remained relatively unchanged by UPOV 1991, the concept of EDVs was introduced to mediate the breeder's exemption and to mitigate the effects of trivial third party plant breeding.

Up until this time, EDVs did not exist in agriculture, horticulture or plant breeding. The concept of EDVs was created to reduce the extent of trivial plant breeding in UPOV-based schemes. It has now become apparent that EDVs are a hybrid concept that are at least part scientific, legal and practical. Indeed, since the introduction of the concept of EDVs UPOV, scientists, lawyers and breeder organisations have all worked, at times together, to determine what constitutes an EDV. UPOV, for example, has contributed to understanding how to examine and identify EDVs by considering the use of molecular techniques such as DNA markers, establishing explanatory notes and guidelines,[97] and holding seminars on the topic.[98] Scientists have looked for useful measures of similarity and difference in an attempt to examine and identify EDVs in various plant species.[99] Breeder organisations have also contributed to the concept of EDVs. Most notably, the ISF has developed position papers,[100] guidelines[101] and lists of

[97] For example, Working Group on Biochemical and Molecular Techniques, and DNA-Profiling in Particular, Concepts of Dependence and Essential Derivation, *The Possible Use of DNA Markers*, BMT/11/24 (UPOV, 2008); Working Group on Biochemical and Molecular Techniques, and DNA-Profiling in Particular, Essentially Derived Varieties (EDV) in the Area of Asexually Reproduced Ornamental and Fruit Varieties, BMT/11/22 (UPOV, 2008).

[98] See UPOV, *Seminar on Essentially Derived Varieties*, Publication 358 (Geneva, 2013), www.upov.int/edocs/pubdocs/en/upov_pub_358.pdf.

[99] See, e.g., E. Noli, M. Teriaca and S. Conti, 'Identification of a threshold level to assess essential derivation in Durum Wheat' (2012) 29(3) *Molecular Breeding* 687; A. Kahler, et al., 'North American study on essential derivation in Maize: II. Selection and evaluation of a panel of simple sequence repeat loci' (2010) 50(2) *Crop Science* 486; E. Jones et al., 'Development of single nucleotide polymorphism (SNP) markers for use in commercial maize (Zea mays L.) germplasm' (2009) 24(2) *Molecular Breeding* 165.

[100] See, e.g., ISF, *ISF View on Intellectual Property* (ISF, 2012).

[101] See ISF, *Guidelines for Handling a Dispute on Essential Derivation in Ryegrass* (ISF,2009); ISF, *ISF Guidelines for the Handling of a Dispute on Essential Derivation of Maize Lines* (ISF, 2008); ISF, *Guidelines for the Handling of a Dispute on Essential Derivation in Oilseed*

arbitrators for disputes.[102] Importantly, the ISF's approach to EDVs combines science, law and cooperation. The legal nature of EDVs is further highlighted by the fact that it has been the subject of legal dispute and litigation that has attempted to reconcile scientific (quantitative) and legal (qualitative) questions around the concept, as well as distinctly legal issues such as the standard of proof required and whether the plaintiff or defendant has the burden of proof.[103] As we will see more thoroughly in Chapter 9, examining and identifying EDVs remains one of the important unresolved issues for UPOV, its Working Groups and Members.[104]

Other kinds of plant varieties have been introduced into national *sui generis* plant variety protection laws. Many of these 'new' kinds of plant varieties have been introduced, particularly in developing countries, in an attempt to balance the rights of plant breeders and farmers and to recognise the rights of farmer and local communities in the development of plant varieties. Some examples of these 'new' kinds of plant varieties are farmers' varieties, local and domestic varieties and indigenous varieties. Referring to Thailand's *Plant Variety Protection Act* (1999) (PVP Act), Lertdhamtewe views the introduction of local domestic varieties, general domestic varieties and wild plant varieties as a 'means to provide exclusive monopoly rights to farmers and local communities that take care of the existing plant variety found within Thailand's territory'.[105] In addition to providing protection to new plant varieties and EDVs, India's *Protection of Plant Varieties and Farmers' Rights Act 2001* (PPVFR Act) affords protection to extant varieties.[106] Under the PPVFR Act, extant varieties include 'a farmer's variety' or 'a variety about which there is common knowledge' or 'any other variety in the public domain'.[107] A 'farmers' variety' is defined in Section 2(l) of the PPVFR Act as a variety which 'has been traditionally cultivated and evolved by the farmers in their fields; or is a wild relative or land race of a variety about which the farmers possess the common knowledge'. Other kinds of plant varieties can be found in Thailand's PVP Act. As well as new plant varieties, the PVP Act provides protection to a range of extant varieties including local domestic varieties, general domestic varieties and wild plants. These kinds of varieties are defined in Section 3 of the PVP Act as:

Rape (ISF, 2007); ISF, *Issues to be Addressed by Technical Experts to Define Molecular Marker Sets for Establishing Thresholds for ISF EDV Arbitration* (ISF, 2010).

[102] ISF, *List of International Arbitrators for Essential Derivation* (ISF, 2010).

[103] *Danziger* v. *Astée* 105.003.932/01, Court of Appeal, The Hague (2009); *Danziger* v. *Azolay* 1228/03, District Court, Tel-Aviv-Jaffa (2009).

[104] UPOV, Publication 358.

[105] P. Lertdhamtewe, 'Protection of plant varieties in Thailand' (2014) 17(5–6) *The Journal of World Intellectual Property* 142, 142.

[106] PPVFR Act, s. 2(j). [107] PPVFR Act, 2(j).

- *local domestic plant variety*: a plant variety which exists only in a particular locality within the Kingdom and has never been registered as a new plant variety and which is registered as a local domestic plant variety under this Act;[108]
- *wild plant variety*: a plant variety which currently exists or used to exist in the natural habitat and has not been commonly cultivated; and
- *general domestic variety*: a plant variety originating or existing in the country and commonly exploited and shall include a plant variety which is not a new plant variety, a local domestic plant variety or a wild plant variety.

Civil society groups and academics have also called for the recognition and protection of different 'categories' of plant varieties. In 2015, for example, Correa and others set out a range of options for developing countries to protect plant variety rights.[109] According to Correa an essential element of *sui generis* plant variety protection is providing protection to different 'categories' of plant varieties, the purpose of which is to balance the rights of farmers and breeders and recognise farmers' rights, in light of the needs of particular countries. The three categories of plant varieties proposed by Correa are (1) new uniform plant varieties, (2) new farmer and other heterogeneous varieties and (3) traditional farmers' varieties. Each of the three categories of plant variety would have different requirements of protection and would confer different rights on the title holder (see Table 5.1).[110]

What are the consequences of having so many kinds of plant varieties? The additional plant variety types have been introduced for a variety of reasons and often include a political imperative and a desire to combine aspects of the UPOV Convention, the CBD and Plant Treaty. Yet despite the best of intentions, perhaps trying to satisfy the UPOV Convention, CBD and Plant Treaty in one piece of national legislation is trying to achieve too much and may lead to complications in interpretation and application. One of the challenges of having so many kinds of plant varieties is the amount and diversity of the varieties being conceptualised and introduced. Because of the political imperative of these plant varieties, they may or may not lend themselves to assisting local farmers and breeders. Thailand's local domestic varieties illustrate this problem well.[111] As we have seen, Sections 3 and 45 of Thailand's PVP Act require that a local domestic variety must be 'a plant variety existing only in a particular locality

[108] Also see *Plant Variety Protection Act B.E. 2542 1999*, s. 43.

[109] C. Correa (with S. Shashikant and F. Meienberg), *Plant Variety Protection in Developing Countries: A Tool for Designing a Sui Generis Plant Variety Protection Systems: An Alternative to UPOV 1991* (APBREBES, 2015).

[110] Ibid., p. 49.

[111] For a thorough discussion of Thailand's plant variety protection scheme, see Lertdhamtewe, 'Protection of plant varieties in Thailand'.

Table 5.1. *Three possible categories of varieties proposed for* sui generis *plant variety protection*[112]

Category of variety	Summary	Criteria for protection	Rights conferred
New uniform plant varieties	Similar to UPOV 1978, these are varieties that are intended for commercialisation and that are generally developed by breeders, companies and other institutions	New, distinct, uniform and stable[113]	Exclusive right to produce the variety for commercial purposes, the offering for sale and marketing
New farmer and other heterogeneous varieties	Includes varieties that may not satisfy the criteria: new, distinct, uniform and stable	New, distinct and identifiable[114]	A remuneration right
Traditional farmers' varieties	Includes varieties that are of common knowledge to farmers	Identifiable	A remuneration right

within Thailand' and that '[w]hen a plant variety exists in a particular locality and has been conserved or developed exclusively by a particular community, that community shall have the right to submit, to the local government organisation in whose jurisdiction such community falls, a request for initiating an application for registration of the local domestic plant variety in the name of such community'. There may, however, be a problem in the implementation and effectiveness of protecting 'local domestic varieties' and their ability to benefit local communities. One of the leading academics working on Thailand's plant variety protection, Lertdhamtewe, has argued that local domestic varieties, as defined in the PVP Act, do not exist in Thailand, stating that:

> since its inception, the lack of locally domestic plant variety registrations has proved that Thailand has no such plant varieties. Also, proposing that local domestic plant varieties should belong to local Thai communities overlooks the fact that a single community owner of a plant variety in Thailand cannot be identified.[115]

[112] Correa, *Plant Variety Protection in Developing Countries.*

[113] Similar to UPOV 1978. See Correa, *Plant Variety Protection in Developing Countries*, p. 52.

[114] The requirement of identifiability is that 'each generation of a plant variety be identifiable as the same distinct plant variety, without necessarily being uniform in all of its characteristics': see Correa, *Plant Variety Protection in Developing Countries*, p. 53.

[115] Lertdhamtewe, 'Protection of plant varieties in Thailand', 154.

5.6 Conclusion

The plant variety concept occupies a central position in the UPOV Convention. The ways in which plant varieties are constructed or defined have not only the potential to play a pivotal role in the interpretation and implementation of the UPOV Convention but also in managing the legal boundary between the UPOV Convention and other legal regimes. In exploring the question – what is a plant variety? – this chapter has demonstrated how the plant variety concept has changed over time. While the plant variety concept embodied in UPOV 1961 was clearly a legal idea, it relied heavily on and reflected horticultural and plant breeding practice. That said, because the overriding aim of those present at the Diplomatic Conference of 1957–1961 was to protect the work of the plant breeder, the focus was on ensuring that the definition of plant variety captured the work of the plant breeder rather than on defining what a plant variety was. In so doing, UPOV 1961 defined plant variety in relation to a list of the products of plant breeding such as cultivar, clone, line, stock and hybrid. Under UPOV 1961 a plant variety was also synonymous with a protectable variety.

As the UPOV Convention asserted itself in broader legal and regulatory frameworks however, the plant variety concept embodied in UPOV 1961 became problematic. First, as plant breeding is a dynamic endeavour, it became less clear about what was an output or end product of plant breeding. Second, during the 1980s it became apparent that there was a need to establish a clearer boundary between different forms of legal protection for plant varieties, particularly between UPOV-based systems and patent law. As a consequence, the plant variety concept embodied in UPOV 1991 occupies distinctly legal spaces, whose meaning operates within different, albeit interdependent, conceptual spaces.[116] The definition of a plant variety used in UPOV 1991 does not refer to a 'natural' or 'real' plant variety. Instead, the subject matter of UPOV 1991 is a juridical concept that 'retains sufficient flexibility to accommodate the various forms that the existing types of variety will take, at the same time allowing for and efficiently satisfying expectations and clearing up the assortment of situations that caused the inclusion of

[116] In *The Order of Things*, Michel Foucault argued that the classification of plants occupied different conceptual spaces that were formulated within particular frameworks or patterns. Indeed, Foucault suggests that all periods of history have possessed certain underlying conditions of truth and order including the classification of plants: M. Foucault, *The Order of Things: An Archaeology of the Human Sciences* (Routledge, 2002).

an express definition to be desirable'.[117] Importantly, it is broad enough to include a number of manifestations of the plant variety. Although the plant variety as subject matter remains limited and shaped by the conditions of grant, we have also witnessed an expansion and proliferation in the kinds of plant varieties referred to in both the UPOV Convention and *sui generis* plant variety protection schemes. While some of these varieties (most notably EDVs) strengthen plant variety rights protection, others have been introduced to recognise the role of farmers and local communities in the development of new plant varieties (e.g. farmers' varieties and local domestic varieties). The end result of this is that the plant variety concept(s) has been rendered more legal and political, bringing with it challenges in implementation, application and interpretation.

[117] Rosselló, 'The UPOV Convention – The concept of variety and the technical criteria of distinctness, uniformity and stability', 59.

6 Bringing Order and Stability to Variety Denomination

6.1 Introduction

For centuries it has been common practice for botanists and horticulturalists to name plants.[1] As far back as 160 B.C.E., valuable crops such as apples and figs were given names based on the origin of the propagating material.[2] Since then, there has been a long line of people who have proposed different methods of naming plants including Theophrastus, Dioscorides, Pliny, de Tournefort and most famously Linnaeus.[3] One of the key reasons for naming plants is to identify, distinguish and order them. Plant names act as generic identifiers and thus facilitate the identification of a plant and reduce confusion and deceptive practices around the trade of plants. More specifically ensuring that plants are precisely and consistently named can safeguard against problems around synonyms and homonyms: that is, one plant being given different names or different plants being given the same name, respectively. An example of the same plant having different names comes from France, where, during the 1920s, one variety of wheat was sold under four different plant names: *Vilmorin 23*, *St Michel*, *Hybride 23* and *Hybride d'automne*.[4] And the problem of the same name (Mountain Ash) being given to different trees, that are not necessarily related, is illustrated by Spencer, Cross and Lumley who explain:

> The Mountain Ash of the Australian state of Victoria, *Eucalyptus regnans*, is so called because its timber resembles that of the European Ash, *Fraxinus excelsior*.

[1] For a detailed history of botanical taxonomy, see A. Pavord, *The Naming of Names: The Search for Order in the World of Plants* (Bloomsbury, 2005).

[2] See W.T. Stearn, 'Historical survey of the naming of cultivated plants' (1986) 182 *Acta Horticulturae* 19.

[3] See Pavord, *The Naming of Names: The Search for Order in the World of Plants*; D. Gledhill, *The Names of Plants* (Cambridge University Press, 2002).

[4] B. Laclavière (quoted in M. Piatti and M. Jouffray, 'Plant variety names in national and international law: Part I' [1984] 10 *European Intellectual Property Review* 283). Also see M. Piatti and M. Jouffray, 'Plant variety names in national and international law: Part II' [1984] 11 *European Intellectual Property Review* 311.

In Tasmania, it is known as the Swamp Gum, a name in Victoria that is generally given to *Eucalyptus ovata*. In England, the Mountain Ash is a small upland tree with ash-like leaves and red berries, *Sorbus aucuparia*, which in Scotland is called Rowan. In America, Mountain Ash is *Sorbus americana*.[5]

To attain order and stability in the naming of plants, botanists and taxonomists have developed and refined a set of rules and procedures that govern the naming of plants. These rules and procedures include the *International Code of Botanical Nomenclature* ('*Botanical Code*') which was adopted in 1867 and the *International Code of Nomenclature for Cultivated Plants* ('*Cultivated Plant Code*') which was first published in 1953.[6] Further, cultivar plant names are generally registered under a voluntary system relying largely on International Cultivar Registration Authorities ('ICRAs'): which, as we saw in Chapter 3, operate under the *Cultivated Plant Code* and aim to reduce the duplication of plant names. Although the *Botanical Code*, *Cultivated Plant Code* and ICRAs have helped 'to transform local knowledge of plants, critical to the survival of indigenous people anywhere, into a comprehensive system of naming, of ordering and classifying, which now embraces every known plant in the world':[7] nonetheless aspects of plant names has remained unsettled and problematic.[8] The relevance and effectiveness of the *Codes* and ICRAs have been questioned as they rely largely on the goodwill of plant breeders, traders and marketers and have no legal effect or effective enforcement mechanisms.[9] These *Codes* and ICRAs are, therefore, open to disregard and misuse by plant breeders, traders and marketers. A lack of order and stability in plant names has been further exacerbated by the increasing number of new plant varieties being introduced by plant breeders.

Importantly, then, by including legally binding rules and practices on variety denomination, the UPOV Convention established a legal

[5] R. Spencer, R. Cross and P. Lumley, *Plant Names: A Guide to Botanical Nomenclature* (CSIRO Publishing, 2007), p. 11.

[6] The most recent versions of the *Botanical Code* and *Cultivated Plant Code* are 2011 and 2009, respectively. For a history see J. McNeill, 'Nomenclature of cultivated plants: A historical botanical standpoint' (2002) 634 *XXVI International Horticultural Congress: IV International Symposium on Taxonomy of Cultivated Plants* 29–36. P. Trehane, '50 Years of the International Code of Nomenclature for Cultivated Plants: Future prospects for the Code' (2002) 634 *XXVI International Horticultural Congress: IV International Symposium on Taxonomy of Cultivated Plants* 17–27.

[7] Gledhill, *The Names of Plants*, p. 4.

[8] As early as the nineteenth century botanists and taxonomists lamented the confusion of Latin plant names: see Stearn, 'Historical survey of the naming of cultivated plants'.

[9] See, e.g., E. Scott, 'Plant breeders rights trials for ornamentals: The international testing system and its interaction with the naming process for new cultivars' in S. Andrews, A. Leslie and C. Alexander (eds), *Taxonomy of Cultivated Plants: Third International Symposium* (Royal Botanical Gardens, 1999), pp. 89–94.

framework that facilitates the consistent and effective naming of plants: anyone that 'offers for sale or markets propagating material' of a protected variety must use the variety denomination, even after the plant variety right expires.[10] Since its introduction in 1961, the UPOV Convention has played an increasingly significant role in the naming of plants and, to some extent at least, has ameliorated some of the concerns and confusion over plant names stemming from the *Plant Codes*. In this chapter I argue that while the regulation of plant names is a mixture of taxonomy, seed certification schemes and trade mark law, it is the newest scheme (the UPOV Convention) that plays the most crucial role. In order to make my argument, this chapter begins by examining the system of variety denomination under the UPOV Convention. Here I focus on three notable features: (i) the generic and free use of denominations and their relationship with trademarks,[11] (ii) accuracy in denominations (e.g. the requirement that denominations are not misleading or confusing)[12] and (iii) uniformity, cooperation and legal effect. Taken together, the fact that plant variety rights protection has legal effect, UPOV's expanding membership and provisions on variety denomination have meant that UPOV and the UPOV Convention have been relatively effective in ordering and stabilising variety denominations. In some ways, then, UPOV's impact on variety denominations has surpassed that of the *Plant Codes* and the ICRAs. Specifically, UPOV and the UPOV Convention have helped to standardise, normalise and stabilise the use of variety denominations,[13] so that increasingly there is a single variety denomination for each variety worldwide.

6.2 Variety Denomination Under the UPOV Convention

A condition of protection under the UPOV Convention is that a variety is given a 'variety denomination'.[14] The main purpose of providing a variety denomination is the facilitation of plant identification and the concomitant reduction in confusion surrounding plant varieties in the marketplace.[15] The identification function of variety denominations is

[10] UPOV 1991, Article (20)(1)(b); UPOV 1978, Article 13(1); UPOV 1961, Article 13 (7), (8).

[11] UPOV 1991, Article 20(1).

[12] UPOV 1991, Article 20(2).

[13] The idea that plant intellectual property, including patent and plant variety protection, has played an important role in the naming of plants was articulated by Brad Sherman: see B. Sherman, 'Taxonomic property' (2008) 67(3) *The Cambridge Law Journal* 560.

[14] UPOV 1991, Articles 5(2) and 20; UPOV 1978 and UPOV 1961, Articles 6(1)(e) and 13.

[15] UPOV, *Explanatory Notes on Variety Denomination under the UPOV Convention*, UPOV/INF/12/4 (UPOV, 2012), p. 4.

clearly articulated in Article 20(2) of UPOV 1991 which states that '[t]he denomination must enable the variety to be identified' and 'must be different from every denomination which designates, in the territory of any Contracting Party, an existing variety of the same plant species or of a closely related species'. In order to ensure that variety denominations clearly, unambiguously and uniformly identify plant varieties, the UPOV Convention sets out numerous rules and principles. In summary, the criteria set out in Article 20 of UPOV 1991 are:

- the variety will be designated by a denomination that is generic;
- no rights in the denomination shall hamper its free use as the variety denomination, even after the expiry of the breeder's right;
- it must enable the variety to be identified;
- it may not consist solely of figures, unless this is an established practice;
- it must be different from all other denominations used for existing varieties in other members of UPOV for the same, or a closely related, species;
- a breeder must submit the same denomination to all members of UPOV and, unless this is considered to be unsuitable within a particular territory, this same denomination will be registered by all members of UPOV;
- it must not be liable to mislead or cause confusion concerning the nature of the variety or identity of the breeder;
- prior rights of third persons must not be affected and such rights can require a change of the variety denomination; and
- a trade mark, trade name or similar indication may be associated with the denomination for the purposes of marketing or selling, but the denomination must be easily recognisable.

Before examining some of the key features of variety denomination under the UPOV Convention, it is worth making some preliminary comments. Rules and principles associated with variety denominations are not new to UPOV and do not apply in isolation. Indeed, various rules, principles and guidelines have been introduced in order to achieve greater certainty in dealings with plant varieties, to prevent confusion and to streamline the process of distinguishing varieties. The presence of different rules, guidelines and so on around variety denominations means that UPOV's principles and practices operate in conjunction with other rules and principles including those found in the *Botanical Code*, *Cultivated Plant Code*, seed certification and national listing schemes and trade mark law. The principles for variety denomination set out in the UPOV Convention were to a large extent informed by the existing principles and practices in botany, horticulture and taxonomy. Most notably perhaps the underlying principles of all of the rules around variety denomination are to ensure that plant names are freely available for use and are not confusing.[16]

That the underlying principles of variety denomination under UPOV mirror those of the *Cultivated Plant Code* was acknowledged by the *Group of Legal Experts on the Relations between Protection of the Names of New Plant Varieties and Trademark Protection* (Group of Legal Experts), at the Diplomatic Conference of 1957–1961. The Group of Legal Experts set out the principles of variety denomination to include that plant names must be filed and registered in all Member States; used in marketing even after the period of protection has expired; not be misleading; once registered, a plant name may not be used, in any Member State, as the name of another variety of the same or closely related species; and trademarks cannot be used to hamper the use of the denomination.[17] While the principles on variety denomination in the UPOV Convention were based largely on the principles of the *Cultivated Plant Code*, the importance of plant names to the trade in plants meant that the needs of plant breeders, and the relationship between variety denomination and trade mark law, were also directly taken into account.[18] Compared to the *Cultivated Plant Code*, one of the improvements made by the UPOV Convention was that it provides a legal framework to compel plant breeders to comply with the rules and principles on variety denomination set out in the UPOV Convention. As we will see in Section 6.2.3, the ability of the International Commission for the Nomenclature of Cultivated Plants to impose the *Cultivated Plant Code's* rules on variety denominations is hindered because it does not have legal effect or enforcement mechanisms, relying instead on goodwill and consensus.[19] Finally, although the principles around variety denomination have remained consistent, and variety denomination is now largely uncontroversial, this was not always the case. For example, Dr. Büchting, the delegate for ASSINSEL, argued that UPOV's denomination provisions were a 'greater hindrance to the actual management of plant variety protection than any other provision in the Convention'.[20]

There were a number of issues related to variety denomination in the UPOV Convention. Most broadly there were complaints that the UPOV Convention was not the appropriate forum for rules about variety

[16] UPOV 1991, Article 20.

[17] A. Heitz, 'The history of plant variety protection' in *The First Twenty-Five years of the International Convention for the Protection of New Varieties of Plants* (UPOV, 1987), p. 87.

[18] U. Löscher, 'Variety denomination according to plant breeder's rights' (1986) 182 *Acta Horticulturae* 59.

[19] The International Commission for the Nomenclature of Cultivated Plants.

[20] Ibid., p. 133 [55].

denomination.[21] At the Diplomatic Conference of 1978, some breeder organisations argued that rules around variety denomination were superfluous to plant variety protection.[22]

Having set out some preliminary points on how the UPOV Convention deals with variety denomination, the next part examines some notable features of variety denomination as set out in the UPOV Convention.[23] These are (i) the generic and free use of denominations and their relationship with trademarks,[24] (ii) accuracy in denominations (e.g. the requirement that denominations are not misleading or confusing),[25] and (iii) uniformity, cooperation and legal effect.

6.2.1 *Generic, Free Use and Trademarks*

In order to facilitate variety denominations functioning as reliable identifiers, it is important that there are no restrictions on when and how the denomination can be used in connection with the variety. Ensuring that variety denominations remain generic and free to use enables the variety to be easily identified because breeders, traders and consumers can consistently use the same name when using, selling or buying plant varieties. This is achieved in the UPOV Convention by the requirement that the variety is designated by a denomination which is generic and that no other rights for that designation shall hamper its free use in connection with the variety.[26] The free and consistent use of variety denominations is further facilitated within the UPOV Convention by the requirement that a variety denomination is free to use even after the expiry of plant variety protection.[27]

One of the key ways in which the UPOV Convention facilitates the free use of variety denominations is through explicit contemplation of trademarks being associated with plant varieties. Delegates and observers at the various Diplomatic Conferences have given special consideration to the potential overlap between variety denominations and trade mark protection, acknowledging that on the one hand a trade mark may hinder the use of a variety denomination and on the other hand registration of

[21] See, UPOV, UPOV Publication No. 337(E). [22] Ibid.

[23] For an overview on the requirements of variety denominations under the UPOV Convention, see UPOV, *Explanatory Notes on Variety Denomination under the UPOV Convention.*

[24] UPOV 1991, Article 20(1). [25] UPOV 1991, Article 20(2).

[26] UPOV 1991, Article 20 (1)(b); UPOV 1978, Article 13(1); UPOV 1961, Articles 13 (7), (8).

[27] UPOV 1991, Article 20 (1)(b). Under the *Cultivated Plant Code,* this is reflected in the principle that the name must be universally available for use: *International Code of Nomenclature for Cultivated Plants* (2004), Principle 4.

variety denomination may affect trade mark registration. However, this was not the first time that the potential overlap between variety denominations and trade mark protection had been contemplated, with early attempts to protect plant varieties focusing on trade mark law. For example, in the United States, in 1906, it was proposed that breeders could register the name of their plant variety in order to provide 'the exclusive right to propagate for sale and vend such variety of horticultural product under the name so registered'.[28] And as we saw in Chapter 2, early seed certification laws and seed traders were often required to use the denomination entered into the Catalogue of Cultivated Species and apply for a trade mark licence for that use.[29] In France, French plant breeders could obtain protection by registering a trade mark for the name of their variety (which enabled plant breeders to license others who wish to use the trade mark when selling the variety), and any seed of that variety must be sold under the registered variety name and in bags labelled with that name, and it is an offence to sell seed under any other name.[30] And in West Germany, trade mark law could be used by West German plant breeders to protect generic designations:[31] the *Protection of Varieties and the Seeds of Cultivated Plant of 1953* stipulated that anyone who marketed seed of a protected variety had to use the variety denomination, and if the denomination was subject to a trade mark registration, this would not prohibit the use of the denomination.

While early attempts to protect plant varieties through trade mark law were largely unsuccessful, the relationship between variety denominations and trade mark was considered in the drafting of the UPOV Convention. At the Diplomatic Conference of 1957–1961, the *Group of Experts on the Relations Between Protection of the Names of New Plant Varieties and Trademark Protection* saw trade mark 'as a means of obtaining protection in countries that do not accede to the Convention'.[32] It was also acknowledged that if someone was permitted to obtain trade mark protection for a variety denomination (a generic name), they would obtain an unfair advantage over competitors, since it would make it difficult for competitors to adequately describe their plant variety. Acknowledging that a breeder may also have a trade mark for a plant name – and given the desire to balance the interests of plant breeders and consumers – it was decided that plant variety owners would be required to relinquish any trademarks over plant names. Specifically, Article 13(3) of

[28] H.R. 13570 § 28b cited in M. Janis and J. Kesan, 'U.S. plant variety protection: sound and fury...?' (2002–2003) 39 *Houston Law Review* 727, 731.

[29] See Heitz, 'The history of plant variety protection', p. 69 [30] Ibid.

[31] Ibid., p. 72–76. [32] Ibid., p. 87

UPOV 1961 required that a variety denomination must not be a trade mark, but if the denomination is also a trade mark then the plant variety owner cannot assert his trade mark from when the denomination is registered. In addition, an applicant who registered a trade mark as a plant name is obliged to renounce his trade mark. Not everyone was happy with this approach, however, and the requirement that a trade mark could not prohibit the free use of the variety denomination was 'severely criticised by the lawyers as being contrary to the fundamental principles of trade mark law'.[33] Unsurprisingly, the question over trademarks coincided with a desire for stronger plant variety protection. Put simply, plant breeders wanted both plant variety and trade mark protection.

A desire for stronger plant variety protection led to renewed discussion of the treatment of variety denomination and trademarks in the UPOV Convention at the Diplomatic Conference of 1978. Plant breeder organisations and the United States, who were both advocating stronger plant variety protection, led the criticisms and put forward two alternatives. First, it was suggested that provisions in the UPOV Convention on variety denomination were not required for plant variety protection. This, according to some, meant that any reference to variety denominations should be removed entirely from the UPOV Convention. Second, it was suggested that any reference to trademarks should deleted from the Convention. For example, Dr. Büchting, the delegate for ASSINSEL, argued that 'a clear separation was made between variety denominations and trade marks' and 'wished particularly to support the elimination from Article 13 of all references to trade marks'.[34] Despite the concerns of plant breeder organisations and the United States, the reference to trademarks remained in UPOV 1978. Although important amendments were made. UPOV 1978 no longer required the variety rights holder to renounce their trade mark; instead they were prevented from asserting their right to the trade mark over a plant name. UPOV 1978 also explicitly acknowledged that a variety denomination and trade mark might both be used on plant labels and the marketing of plant varieties. Specifically, Article 13(8) of UPOV 1978 required that variety denominations 'shall be permitted to associate a trade mark, trade name or other similar identification with a registered variety denomination. If such an indication is so associated, the denomination must nevertheless be easily recognizable'. These amendment acknowledged that trade marks were

[33] Ibid., p. 76 (referring to similar provisions in the *German Seed Law of 1953*).
[34] UPOV, Publication No. 337(E), p. 131 [55]. The proposal to remove any reference to variety denominations from the UPOV Convention was supported by other breeder organisations such as CIOPORA, FIS and AIPH.

crucial to the trade of plant varieties and were in fact not incompatible with the use and registration of variety denominations. In so doing UPOV 1978 also clarified that plant varieties could be marketed with both a generic variety denomination and a unique trade mark while maintaining that variety denominations are generic and free to use as descriptive names. Another reason for the change was to 'simplify the procedure before the plant variety rights offices of member States since such offices would no longer be required to compel the applicant to renounce his right in a trade mark and the applicant would no longer be required to attach a declaration of renunciation to his application'.[35]

At the Diplomatic Conference of 1991, the issue of variety denominations and trademarks was largely uncontentious. With discussions on the topic focusing on two main issues raised by the United States, the use of figures and the obligation to use the denomination.[36] However, as we will see in the next part, both of the United States' proposals were rejected by Members, and the provisions on variety denomination were adopted as appearing in the Basic Proposal. In terms of variety denominations and trademarks, Article 20(8) of UPOV 1991 requires that 'when a variety is offered for sale or marketed, it shall be permitted to associate a trade mark, trade name or other similar indication with a registered variety denomination. If such an indication is so associated, the denomination must nevertheless be easily recognizable'. Content with the clarity of Article 20 of UPOV 1991, in its *Explanatory Note on Variety Denomination*, UPOV merely states that '[t]his provision is self-explanatory'.[37]

Having summed up the key provisions that ensure that variety denominations are generic and free to use in connection with varieties, including those dealing with plant name trade marks, it is helpful to consider some of the practical considerations and challenges for plant breeders, traders and marketers. Thinking carefully about the uses and benefits of trade mark protection for plants allows plant breeders and markets to take advantage of the opportunities provided through trade mark protection.[38] The main function of a trade mark is to indicate the origin of goods with which the trade mark is associated. Importantly, a trade

[35] Ibid., p. 33 (Explanatory Notes). Although a UPOV Member could require the renouncement of any trade mark.

[36] UPOV, Publication No. 346(E), pp. 280–85.

[37] UPOV, UPOV/INF/12/4.

[38] For a discussion of how trademarks might be used to facilitate 'nutritional autonomy' see J. Sanderson, 'Can intellectual property help feed the world? Intellectual property, the PLUMPYFIELD network and a sociological imagination' in C. Lawson and J. Sanderson, *The Intellectual Property Food Project: From Rewarding Innovation and Creation to Feeding the World* (Ashgate, 2013), pp. 145–174.

name may be a variety denomination or a trade mark: but not both. In relation to ornamentals, for example, it was standard practice 'to take the first three letters of the breeder's name and to attach to them one of several syllables chosen to a large extent at random'.[39] Where a plant name is registered as a trade mark, the subsequent use of the name as a variety denomination may convert the trade mark into a generic name, and as a consequence the trade mark is open to removal or cancellation.[40] Generic terms are incapable of indicating source and are, therefore, not entitled to trade mark registration. In effect, then, when a plant breeder registers a plant variety and its denomination for plant variety protection, they are agreeing to not seek a trade mark for the same denomination.[41] This, however, does not mean that they have not, or will not, seek trade mark registration of another name.

While the name used for a variety denomination and trade mark cannot be the same, a variety denomination and trade name can be used on the same plant, as long as they are different and 'easily recognizable'.[42] A label, tag or other source of descriptive information concerning the plant must display the variety denomination (i.e. the generic name) and, if applicable, the trade mark. The distinction between the two can be maintained if variety names are enclosed in single quotation marks and carry a national plant variety symbol or the letters PVR/PBR. By contrast, trademarks are not used with single quotes and are followed by the appropriate symbol, either ™ or ®. For example, a trade mark and variety denomination can be presented as EVERGREENEDGER® 'Rotundifolia'.

Why should plant breeders utilise both plant variety and trade mark protection? A trade mark used on a plant variety can be a valuable marketing tool, particularly as it can be licensed. The licensing of intellectual property – whether it be a plant variety right or trade mark – is often critical to the commercial success of seed and nursery industries. Where a trade mark is used in addition to the variety denomination, the trade mark could be registered and subsequently licensed. There are a number of advantages to this strategy. Most notably perhaps, a trade mark may be used to develop a 'brand' around one or several plant

[39] A. Heitz, 'Plant variety protection and cultivar names under the UPOV Convention' in Andrews, Leslie and Alexander, *Taxonomy of Cultivated Plants: Third International Symposium*, pp. 59–65, 64.

[40] UPOV, UPOV/INF/12/4.

[41] Although some UPOV members (e.g. EU-CPVO) inform the breeders of their obligations and the risk of trademarks being cancelled: See UPOV, UPOV/INF/12/4 [1].

[42] UPOV 1991, Article 20(8).

varieties or indicate that a plant has been grown by a particular grower or company. Because a trade mark is a designation of origin, the holder of the trade mark is allowed to market different varieties (i.e. of the same or of a different species) under one and the same trade mark. In addition, trade mark registration is potentially of unlimited duration and thus does not expire when the plant variety protection rights expire. Further, the owner of a trade mark can licence others to use the mark subject to various conditions including in relation to the quality of the plant.

In order to further illustrate the potential interactions between variety denominations and trademarks, two examples are provided in the following paragraphs: first, *Rebel grass seed (Re Pennington Seed, Inc),*[43] which reaffirmed the idea that variety denominations are in fact generic designations that should not be registered as trademarks; second, Pink Lady apples, which highlights why and how trademarks (different to variety denomination) might be used, and some possible challenges of a trade mark becoming generic and, therefore, incapable of trade mark registration.

Variety Denominations Are Generic: 'Rebel' Grass Seed (Re Pennington Seed, Inc) In 2001, KRB Seed Company ('KRB') attempted to register 'Rebel' as a trade mark for grass seed with the United States Patent and Trademark Office ('USPTO'). However, the examiner refused to register the 'Rebel' mark on the ground that it was a variety denomination for a type of grass seed and thus should be treated as a generic designation. In 2005, KRB unsuccessfully appealed to the United States Patent and Trademark Office Trademark Trial and Appeal Board (the 'Board').[44] In dismissing the appeal, the Board applied case law that had established that variety denominations are generic designations and are thus not capable of trade mark registration. Notably the Board referred to *Dixie Rose Nursery* v. *Coe* in which the District of Columbia Circuit determined that the term 'Texas Centennial' could not be registered as a trade mark because it was the variety denomination for an individual rose variety.[45] The Board also relied on other sources to confirm that variety denominations are generic including the *Trademark Manual of Examining Procedure* which advised trade mark examiners to reject the registration of variety denominations because they do not function as an indication of source; the UPOV Convention which requires each new plant variety to be given a variety denomination that will be the generic designation for the

[43] 466 F.3d 1053 (Fed. Cir. 2006).
[44] *In re KRB Seed Co.*, 76 U.S.P.Q. 2d 1156 (T.T.A.B. 2005).
[45] 131 F.2d 446 (D.C. Cir. 1942), cert. denied, 318 U.S. 782 (1943).

plant and Section 52 of the *Plant Variety Protection Act 1970* (PVPA) which requires applicants seeking protection for new plant varieties to designate names for the varieties.[46] Taking all of these sources together, the Board concluded that 'Rebel' became a generic designation for a type of grass seed when KRB designated it as the variety denomination of its grass seed in its plant variety application.

On appeal to the United States Court of Appeals for the Federal Circuit, Pennington Seed, Inc ('Pennington') – who had acquired KRB's trade mark application – argued that 'a blanket refusal to register a varietal name fails to consider this court's test for genericness, viz., the primary significance of the mark to the purchasing public'.[47] However, the Federal Circuit agreed with the examiner's finding that the name 'Rebel' was a variety denomination and therefore generic. According to the Federal Circuit, the policy of treating variety denominations as generic was 'an established principle' based on sixty years of case law, public policy (new plant varieties must be given names, and those variety denominations may be the only names that purchasers use to designate particular plant varieties) and was reflected in the PVPA and UPOV Convention, stating that '[w]hile the requirements of the UPOV do not control this case, they underlie and are consistent with the conclusion that a varietal name is generic and hence support the PTO's refusal to register the term "Rebel" as a trade mark'.[48] Further, the Federal Circuit found that there was no evidence to suggest that the claimed mark functioned as an indication of source, stating that:

> The evidence consists of excerpts of articles from a variety of sources, including one from the Germplasm Resources Information Network web server, wherein "Rebel" is listed as a cultivar name for tall fescue grass seed, an excerpt from the database maintained by UPOV listing "Rebel" as the denomination of a tall fescue variety, and an excerpt from a listing on plant varieties kept by the Seed Regulatory and Testing Branch of the United States Department of Agriculture. The evidence thus demonstrates that the term "Rebel" is the name of a variety of grass seed and as such is the variety's generic designation.[49]

Using Variety Denomination Alongside Trade Mark: 'Cripps Pink' Pink Lady™/® What is the difference between a Cripps Pink and Pink Lady apple? In 1973, John Cripps bred a new variety of apple by

[46] 7 U.S.C. § 2422. In 1981, KRB had obtained protection for a new variety of grass seed from the United States Plant Variety Protection Office and in so doing had designated the term Rebel as the variety denomination for its grass seed.

[47] *Rebel grass seed (Re Pennington Seed, Inc.)* 466 F.3d 1053 (Fed. Cir. 2006), 4.

[48] Ibid., 10. [49] Ibid., 7–8.

crossing the Australian apple 'Lady Williams' with a 'Golden Delicious'.[50] The new variety of apple – which combined the storing properties of 'Lady Williams' and the sweetness and lack of storage damage of 'Golden Delicious' – was given the variety denomination 'Cripps Pink'. Despite the fact that consumers have increasingly purchased and eaten 'Cripps Pink' apples, it is unlikely that they are aware of this. Instead, consumers are almost certainly aware of and seek out 'Pink Lady' apples.[51] 'Pink Lady' is a trade mark that is used on various apple varieties including Cripps Pink.[52] The circumstances surrounding 'Pink Lady' apples are instructive on a number of points around trademarks and variety denominations under the UPOV Convention.

Apple and Pear Australia Limited ('APAL') is the peak representative body in Australia for commercial apple and pear growers. It is the registered owner of 'Pink Lady' trade marks in more than 80 countries and uses the 'Pink Lady' trade mark primarily as an indication of quality. To be able to use the 'Pink Lady' trademarks, apples must meet certain quality specifications including for sugar content, firmness, blemishes and colour. Inspections are regularly performed to ensure both the quality and traceability of the apple from the orchard to consumers. Significantly, 'Pink Lady' apples are not a particular variety of apple but rather it is a trade mark, and, therefore, numerous plant varieties (variety denominations) of apple are sold under that trade mark including Cripps Pink, Barnsby, Maslin, Rosy Glow, Ruby Pink and Lady-In-Red. While there are a number of advantages of using a trade mark such as 'Pink Lady', there are also potential complications that plant breeders should be aware of. If there is a conflation or confusion of the putative trademark and variety denominations, the mark is at risk of becoming generic and therefore unregistrable as a trade mark. In fact, so effective has APAL been at promoting 'Pink Lady' apples that it has been refused trade mark registration in a number of countries. In Australia, for example, repeated attempts to register the mark as a plain word mark ('Pink Lady') were unsuccessful. The Australian Trade Marks Office viewed 'Pink Lady' as descriptive; primarily due to the apple being widely known by the industry and the public as 'Pink Lady' rather than the variety denomination such as 'Cripps Pink'.

[50] J. Cripps, L. Richards, and A. Mairata, '"Pink Lady" apple' (1993) 28(10) *HortScience* 1057.

[51] In 2012, 'Pink Lady' apples replaced 'Granny Smith' from the number three spot on the list of apples sold, by value, in the United Kingdom: S. Godwin, 'Pink Lady apples shine', *The Weekly Times*, 4 April 2012.

[52] John Cripps apparently took the Pink Lady® name from his favourite novel, Nicholas Monsarrat's *The Cruel Sea* (1951), in which the main character drinks a cocktail called 'Pink Lady': See *The Pink Lady® Story*, www.pinkladyapples.co.uk/the-pink-lady-story.

All is not lost for breeders or growers wanting to trade mark a generic or well-known name however. Despite not being able to register the words 'Pink Lady' as a trade mark in Australia, APAL has registered two trademarks over a composite version of the trade mark including a composite mark.[53] The trade mark was, however, endorsed so that the trade mark registration is limited to the colours *pink* and *white* as shown in the representation of the trade mark attached to the application form. In addition to this, APAL registered the word mark ('Pink Lady') for fruits other than apples.[54] While apples are not covered by this registration,[55] it is possible that a trade mark registration for 'closely related' fruits might be enough to deter those wanting to use the name 'Pink Lady' on apples without a licence, because competitors may not know that apples are not contained in the genus listed. Moreover, this approach could provide de facto trade mark protection because in many countries trade mark infringement occurs when third parties use a registered trade mark in relation to goods and services that are 'closely related' to the goods and services of the registered trade mark.[56]

6.2.2 Characteristics of the Variety

Another feature of UPOV's rules on variety denomination is that a variety denomination must enable the variety to be accurately identified.[57] Most broadly this requirement reflects a desire to balance the interests of plant breeders with those of consumers and is similar to restrictions employed by consumer and trade mark laws. Specifically, the UPOV Convention establishes that a denomination should not convey the impression that the variety has particular characteristics which in fact it does not have (e.g. 'a variety denomination for "dwarf" for a variety which is of normal height, when a dwarfness trait exits within the species, but is not possessed by the variety');[58] give the impression that it is the only variety to possess a character when it is

[53] Australian Trade Mark Number 1580181 (accepted 14 December 2015; registered owner: Apple and Pear Australia Limited) in class 31 '[a]pples being of the variety Cripps Pink and mutations of the Cripps Pink variety'. Also Australian Trade Mark Number 1409743 for class 31, apples and apple trees (accepted 23 August 2011; registered owner: Apple and Pear Australia Limited).

[54] Australian trade mark registration number 1280838 (accepted 12 May 2011; registered owner: Apple and Pear Australia Limited) in class 31 for '[f]Fruits, plant material and trees; all being of the genera: Citrus, Prunus, Pyrus or Vitis'.

[55] Apples, on the other hand, are in the Malus genus.

[56] See *Trade Mark Act 1995* (Cth), s. 120.

[57] See UPOV 1991, Article 20(2); UPOV 1978, Article 13(2); UPOV 1961, Article 13(2).

[58] UPOV, UPOV/INF/12/4, p.4.

not (e.g. '"Sweet" for a fruit variety');[59] is derived from another variety when it is not (e.g. '"Southern cross 1"; "Southern cross 2"...giving the impression that these varieties are a series of closely related varieties with similar characteristics, when, in fact, this is not the case');[60] or is bred by a particular breeder when it is not.[61] Furthermore, variety denominations will not be accepted if they contain superlatives such as best, bigger or sweeter. Nor will simple descriptive phrases or denominations which are too similar to another existing denomination by granted. For example, a difference of a single letter or number between two denominations may be rejected because it is likely to cause confusion and error around variety identification.[62] The denomination must also be different from any existing variety of the same genus or species or a closely related species.

Perhaps the main issue on the use of denominations in identifying plant varieties has been the use of figures as variety denominations. UPOV 1961 was the strictest in this regard as Article 13(2) prohibited a denomination being made up solely of figures (e.g. 23890). However, restricting the use of figures was seen to be contrary to the practice of breeders in many countries, and at the Diplomatic Conference of 1978, the removal of this restriction was proposed, with some arguing that allowing designations to consist solely of numbers would be less confusing, as the use of numbers and figures for a variety denomination ameliorates problems association with translation.[63] After some discussion it was agreed to reduce the restriction on figures, although not entirely. Article 13(2) of UPOV 1978 discourages the use of figures for variety denominations 'except when this is established practice in the designation of varieties'. UPOV 1978, therefore, provided a 'limited opening for variety denominations consisting of figures'.[64]

Not to be deterred, the United States again tried to have the restriction on denominations not consisting solely of figures deleted at the Diplomatic Conference of 1991,[65] arguing that, in the United States, there had been *no* confusion and issues with figures being used to distinguish plant varieties and that:

> this practice was established in the United States of America; an American breeder who filed an application for protection in another country and was to comply with the spirit of Article 20—which was that the variety denomination

[59] Ibid. [60] UPOV, UPOV/INF/12/4, p.5. [61] Ibid. [62] Ibid.
[63] UPOV, Publication No. 337(E), p. 185 [997.4] (Mr. Gfeller, Chairman of the Working Group on Article 13).
[64] Ibid., p. 185 [997.4] (Mr. Gfeller, Chairman of the Working Group on Article 13).
[65] Ibid., p. 103 (DC/91/17).

should be the same in all countries—would thus immediately run into a problem in the countries which did not accept variety denomination.[66]

However, the United State proposal for removing the restriction on figures did not find support from the other delegations and was rejected. As an alternative, the United States wanted plant breeders to sell material of a protected variety without a denomination: giving the example of allowing a breeder who had overproduced to sell that material at a price lower that of the same variety identified by its denomination.[67] This too was rejected: in part because it was felt that any relaxation on the requirement of a denomination would imply that anybody could sell a protected variety without using its denomination and would, therefore, open the door to infringement.[68]

6.2.3 Uniformity, Cooperation and Legal Status

The need for a uniform set of internationally acceptable rules and principles on variety denominations has long been evident. In order to achieve uniformity and stability in the naming of plants, a set of principles, rules and recommendations around variety denominations was established – the *Botanical Code* since 1867 and the *Plant Cultivated Code* since 1953 – as well as a of system registration of cultivar names relying on various ICRAs. With only some exceptions, a variety name consists of a Latin part governed by the *Botanical Code* and a cultivar name governed by the *Cultivated Plant Code.* Notably the key principle of the *Cultivated Plant Code* is that plant names are freely available for use and not to be confused with marketing names (such as trademarks) or common names. With cultivated plants the process of deciding whether a name is valid is performed by a nominated ICRA.[69] The ICRAs prepare and maintain registers of cultivar names, register new and acceptable cultivar names and provide information to plant breeders, marketers and plant variety rights offices.[70] The agencies appointed as ICRAs represent a wide range of specialist societies and institutions and are located in many countries

[66] Ibid., p. 281 [694.1] (Mr. Hoinkes).
[67] Ibid., UPOV Publication No. 346(E), p. 282 [702.2] (Mr. Hoinkes).
[68] Ibid., pp. 282.
[69] *Cultivated Plant Code* (2009), Principle 8, Article 11.5. The ICRAs are appointed by the International Society for Horticultural Science on the recommendation of the International Commission for Nomenclature and Registration of the International Society. See A. Leslie, 'International plant registration' in B.T. Styles (ed.), *Infraspecific Classification of Wild and Cultivated Plants* (Oxford, 1986), p. 359.
[70] C. Brickell, 'The International Code of Nomenclature for Cultivated Plants: Its role in stabilizing the nomenclature of cultivated plants' in Styles (ed.), *Infraspecific Classification of Wild and Cultivated Plants*, p. 352.

around the world.[71] For example, the Royal Horticulture Society is the ICRA for nine groups of cultivated plants including clematis, conifers and dahlia. One of the strategies that these societies have used to stabilise plant names is to produce lists of standardised names that define both how and when a particular name is to be used.[72]

Yet despite the best attempts of the various ICRAs, there are problems with the accurate and consistent naming of plants. Importantly, registration of cultivar plant names with the relevant ICRA is optional and voluntary, so that the system relies on the goodwill and acceptance of plant breeders and marketers. The problems of a voluntary system of plant names were acknowledged by W.T. Stearn, the Secretary of the International Committee on Horticultural Nomenclature and Registration, in 1952, where he urged plant breeders and others 'to give names which are in accordance with the Code' and:

> Seedsmen and nursery-men should try to bring their catalogues into line with it. Registering authorities should refuse to register names not in accordance with the Code; no awards should be given to plants not named in accordance with the Code. Specialist societies should use it as a basis for their own codes of nomenclature. Writers on cultivated plants should endeavour to employ only names correct according to the Code; by frequently mentioning the Code as a standard of procedure they will help to make it known and appreciated. Instructors should bring it to the notice of their students and explain its provisions to them. Government agricultural and horticultural departments should take note of its provisions when drafting legislation. By action of this kind its utility and any weaknesses which need emendation will become evident.[73]

One of the consequences of optional registration is a level of disregard, unawareness and, ultimately, non-compliance with the *Cultivated Plant Code* and ICRAs.[74] And because not all plant breeders and marketers followed the *Cultivated Plant Code* and registered names with ICRAs, there was a lack of consistency and stability in the use of plant names. Plants are often given common or vernacular names that arise from common use by people in contact with the plants. While these common

[71] These include the United Kingdom, France, Germany, North America, China, India, Singapore, Australia, New Zealand, South Africa and Puerto Rico.

[72] For background see J. Kempton, 'What's in a plant name?' (1942) 33 *Journal of Heredity* 133.

[73] W.T. Stearn, The address given by the Secretary of the International Committee on Horticultural Nomenclature and Registration at the opening meeting on 7 September 1952: available at The Bromeliad Society International, www.bsi.org/brom_info/cultivar/ICNCP.html.

[74] See, E. Scott, 'Plant breeders rights trials for ornamentals: The international testing system and its interaction with the naming process for new cultivars'.

names are often stable and useful in their place of origin, they are often unhelpful outside of this area. In other places the same plant might have a different common name, or the same name might apply to a different plant. A good illustration of this is the Mountain Ash example provided in the introduction to this chapter: explaining how the same name (Mountain Ash) has been used on different plants: in Victoria, Australia (*Eucalyptus regnans*), England (*Sorbus aucuparia*), Scotland and the United States (*Sorbus americana*).[75] Taking a rather dim view of plant breeders and marketers, it has been suggested that 'users of cultivated plant names are generally not the public spirited, democratic, rational people that scientists and even some plant taxonomists are. Many of them in fact do not care what the name is as long as they can make a profit from it'.[76] Given the limitations of the *Cultivated Plant Code* and the ICRAs, UPOV's provisions on variety denominations fill a crucial gap. The potential of the UPOV Convention to play a leading role in the ordering and stabilisation of variety denominations was not lost on those involved in ensuring stabile variety denominations. At the Diplomatic Conference of 1978, Mr. Laclavière, in supporting the formation of a working group on variety denominations, stated that he 'would like a representative from the International Commission for the Nomenclature of Cultivated Plants to be a member of the group' because he claimed that 'the sight of the very purpose of the variety denomination was occasionally somewhat lost'.[77] And at the Third International Symposium of *Taxonomy of Cultivated Plants*, it was made clear that UPOV, the International Commission for the Nomenclature of Cultivated Plants and ICRAs must work 'to find solutions for the problems we are facing'.[78]

There are a number of reasons why UPOV and the UPOV Convention has helped to order and stabilise the naming of plants. Perhaps most significantly UPOV and the UPOV Convention, unlike the *Cultivated Plant Code* and ICRAs, have legal effect. As we saw in Chapter 3, the status of UPOV is set out in Article 24 of UPOV 1991 including that UPOV has legal personality and enjoys the legal capacity 'necessary for the fulfilment of the obligations of UPOV and for the exercise of its functions' in each member state.[79] This means that UPOV can make

[75] Spencer, Cross and Lumley, *Plant Names: A Guide to Botanical Nomenclature*, p. 11.
[76] I. Dawson, 'Cultivar registration in Australia' in Andrews, Leslie and Alexander, *Taxonomy of Cultivated Plants: Third International Symposium*, pp. 107–111, 110.
[77] UPOV, Publication No. 337(E), p. 132 [69].
[78] K. van Ettekoven, 'The future of the taxonomy of cultivated plants' (2002) 634 *XXVI International Horticultural Congress: IV International Symposium on Taxonomy of Cultivated Plants* 211–216.
[79] UPOV 1991, Article 24(2).

and amend rights and obligations around plant variety protection and has the capacity to exercise its functions in each of its Member States. Perhaps more importantly for the adaptation of uniform and consist variety denomination rules is the fact that many UPOV Members make failure to comply with certain requirements, including those on variety denomination, an offence. Pursuant to Article 22(1)(b)(iii) of UPOV 1991 in certain circumstances, a variety right can be cancelled because 'the breeder does not propose, where the denomination of the variety of the variety is cancelled after the grant of the right, another suitable denomination'. In a number of UPOV Members, for example, incorrect use of a variety denomination is punishable by a fine or even by imprisonment. One example of this is Article 94 of the *EC Regulation on Plant Variety Protection* which makes it possible to enforce correct use of the established variety denomination in a civil procedure and to claim damages from the person who does not use the denomination correctly.[80] In addition, the *EC Regulation on Plant Variety Protection* makes it possible, in certain circumstances, for the Community Plant Variety Office to cancel a Community plant variety right if the holder of that right does not propose a suitable denomination.[81]

The fact that UPOV's work on variety denominations has been taken up by so many countries means that in certain situations, UPOV's practices, which are themselves drawn from taxonomy, have been embraced by scientists, plant breeders and lawyers and accepted by taxonomists and horticultural bodies. For instance, the existence of plant variety rights legislation was used to justify the need to update and revise taxonomic practices. Referring to the crucial role played by UPOV in stabilising plant names, the Director of the Kew Royal Botanic Gardens said in the preface to the Third International Symposium of Cultivated Plants, in 1999, it 'is good to see this healthy growth in an area that seemed to be on the wane until recently, yet which is so vital as the legal complications of names become more significant and horticulture becomes an ever more popular pursuit' and '[t]he increasing worldwide trade in cultivated plants together with stronger legal protection of new cultivars demands that names be precise, accurate and stable'.[82] The impact of the UPOV Convention on variety denominations is strengthened further by cooperation between UPOV

[80] *Council Regulation (EC) No 2100/94 of 27 July 1994 on Community plant variety rights.*
[81] Ibid. The same provision can be found in UPOV 1991, Article 22(1)(b)(iii).
[82] G. Prance, 'Preface' in *Taxonomy of Cultivated Plants: Third International Symposium of Cultivated Plants* (Kew, 1999), v. See also J. Hawkes, 'Infraspecific classification: the problems' in Styles (ed.), *Infraspecific Classification of Wild and Cultivated Plants*, p. 6.

Members and between UPOV, national plant variety offices and the ICRAs. Generally a national plant variety office will accept denominations submitted in other UPOV Members, so that there is a single denomination for each variety worldwide. Further, UPOV's requirements, rules and practices for variety denominations are recognised under the *Cultivated Plant Code* as providing de facto registration of a plant name by horticultural authorities; once a plant variety is registered for plant variety rights protection under national UPOV-based schemes, the designated name is recognised in the relevant ICRA as the name of the plant, and often details are shared and the examination of variety denomination under UPOV is carried out in conjunction with the lists of the ICRA.[83] In some situations, the relationship between UPOV and plant taxonomy is also 'reinforced by the fact that the work of Registration Authorities and that of national plant variety rights offices have been fused into what is effectively a single process'.[84] For example, in Australia, the Australian Cultivar Registration Authority (ACRA; the ICRA for Australian natives)[85] advises the Australian Plant Breeder's Rights Office, and all applicants, on plant names for Australian native plants such as *Acacia* and *Banksia*. Furthermore, ACRA registers all Australian varieties accepted by the Plant Breeder's Rights Office.[86]

UPOV has also helped to order and stabilise variety denominations in more practical ways. By developing and maintaining the UPOV Code System (UPOV Code) and various databases (e.g. GENIE and PLUTO), UPOV has established a repository of the UPOV Code and information about variety denominations including botanical and common names.[87] This provides easily accessible information about

[83] U. Loscher, 'Variety denomination according to Plant Breeders' Rights' (1986) 182 *Acta Horticulturae* 61.

[84] Sherman, 'Taxonomic property', 582. In many countries, the checking of variety denomination under UPOV is carried out in conjunction with the lists of the International Registration Authority; Loscher, 'Variety denomination according to Plant Breeders' Rights', 61.

[85] *Plant Breeder's Rights Act 1994* (Cth), s 44(2). See also ACRA, www.anbg.gov.au/acra/.

[86] There are similar arrangements in numerous countries including Poland, South Africa and the United Kingdom: See J. Borys, 'DUS testing of cultivars in Poland' in *Taxonomy of Cultivated Plants: Third International Symposium of Cultivated Plants* (Kew, 1999), p. 199; J. Sadie, 'Cultivar Registration for Statutory and Non-Statutory Purposes in South Africa', in *Taxonomy of Cultivated Plants: Third International Symposium of Cultivated Plants*, p. 101; C. Thomson, 'Classification of brussels sprout cultivars in the UK' in *Taxonomy of Cultivated Plants: Third International Symposium of Cultivated Plants*, p. 439.

[87] See, UPOV, *Guide to the UPOV Code System* (22 February 2013), www.upov.int/genie/en/pdf/upov_code_system.pdf.

variety denominations to the authorities of UPOV members and in so doing has helped to resolve some of the concerns around synonyms and homonyms by providing by enhancing the usefulness of UPOV's databases. Specifically, the UPOV Code requires that botanical taxon has an alphabetical element of five letters indicating genus and where necessary three characters to indicate species and a further three characters to indicate subspecies.[88] This ensures that, for example, species are given a consistent code with other members and that synonyms for the same plant taxa are given the same code. While the UPOV Code is not a necessarily new concept – it was informed largely by the International Seed Testing Association's list of stabilised plant names and the information on taxa contained therein is based on the Germplasm Information Network databases – the fact that it is specifically adapted to, and adopted by, UPOV Members is easily accessible and is supported by a legal framework is crucial to its success.

Given UPOV's expanding membership, it appears that UPOV will continue to play a crucial role in ordering and stabilising variety denominations. One example of this is a proposed central approval system for variety denominations.[89] In 2013, the International Seed Federation ('ISF') presented to the members of UPOV a wish list aimed at further harmonising the application, examination and granting of plant variety protection.[90] In doing so the ISF posited a range of recommendations which aimed at bringing all of the different bodies, organisations and groups dealing with plant varieties together and that the creation of 'a Patent Cooperation Treaty-like system for PBR will enable a one-stop-shop approach and facilitate filing in more countries, which in turn will strengthen the UPOV system'.[91] One of the specific recommendations was the establishment of a UPOV quality assurance programme and a central approval system for variety denominations. While the UPOV Council considered the ISF's recommendation on variety denominations and invited the ISF to 'elaborate its ideas', there have not been any further developments.[92] That said, in proposing a centralised approval

[88] Ibid.

[89] Third World Network, *UPOV: To consider industry wish list on plant breeders' right*, 1 November 2013, www.twn.my/title2/biotk/2013/biotk131101.htm.

[90] UPOV, *Matters Raised by the International Seed Federation (ISF)*, UPOV CC/86/11 (UPOV, 8 August 2013), ISF's letter is reproduced in Appendix 1. ISF represents the interest of the mainstream of the seed industry, including multinational seed companies such as Monsanto, Syngenta, Bayer, DuPont Pioneer and DowAgroSciences, which continue to control about 75 per cent of all private sector plant breeding research and 60 per cent of the commercial seed market.

[91] See Third World Network, *UPOV: To consider industry wish list on plant breeders' right.*

[92] UPOV, CC/86/11, Annex 1, p. 7.

system for variety denominations, the ISF drew attention to UPOV's role in stabilising and ordering variety denominations and highlighted continuing issues around variety denomination by arguing that:

> Name-giving rules differ within and between different countries. It is a known fact that name conflicts arise frequently due to the fact that several countries have inconsistent internal naming rules. For example a name that is accepted in a country or region XYZ for national listing, can at a later date be refused for national or region-wide PBR. At the minimal there should be consistency in the naming rules for national listing and for PBR in the same country. In general it is felt that more harmonization is needed, in other words, naming rules should be standardized across the globe, and where possible there should be a central approval system.[93]

Taken together, UPOV's legal status, expanding membership and provisions on variety denomination have meant that the UPOV Convention has been effective in ordering and stabilising variety denominations. The consequences of the Convention's provisions on variety denominations are significant, and in some ways UPOV's impact on variety denominations has surpassed that of the *Plant Codes* and the ICRAs. Specifically, UPOV and the UPOV Convention have helped to standardise, normalise and stabilise the use of variety denominations.

6.3 Conclusion

Plants require names. More precisely, plants require names that are consistently applied and used and that facilitate their identification and distinction. While the conventions around cultivar denominations have a long history, in this chapter I have argued that UPOV and the UPOV Convention play a crucial role in ordering and stabilising variety denominations. Notably, a condition of protection under the UPOV Convention is that a variety is given a 'variety denomination'.[94] The specific criteria for establishing a variety denomination as set out in Article 20 of UPOV 1991 facilitate plant identification and the concomitant reduction in confusion surrounding plant varieties in the marketplace. While the rules and principles associated with variety denominations are not new to the UPOV Convention, UPOV's rules, guideline, practices and databases on variety denomination are now commonly used by plant breeders and marketers to describe and identify plant varieties.

Compared to the *Cultivated Plant Code* and ICRAs, one of the biggest advantages of the UPOV Convention is that it provides a legal framework

[93] Ibid., p. 5.
[94] UPOV 1991, Articles 5(2) and 20; UPOV 1978 & UPOV 1961, Articles 6(1)(e) and 13.

that compels plant breeders and marketers to comply with the rules and principles on variety denomination. Because the UPOV Convention has legal effect, it ensures that a variety denomination accepted in one UPOV Member is also used in other UPOV Members. Specifically, a variety must be submitted to all UPOV Members under the same denomination, unless the denomination is unsuitable for that Member.[95] And if accepted, the same denomination must be used in all UPOV Members. Increasingly, as a consequence of the UPOV Convention, there is a single variety denomination for each variety worldwide.

[95] UPOV 1991, Article 20(5); UPOV 1978, Article 13(5); UPOV 1961, Article 13(5).

7 Science Isn't Enough

Genotypes, Phenotypes and the Utilitarian Nature of Plant Variety Rights Schemes

> 'science is a necessary condition for understanding the world, but not a sufficient one'.[1]

7.1 Introduction

The need to be able to describe, distinguish and demarcate what is being protected is central to intellectual property law. It can be seen in the notions of novelty and inventive step in patent law, originality in copyright law and the need to distinguish marks in trade mark law. During the nineteenth and early twentieth centuries, satisfactorily describing and demarcating plants for the purpose of intellectual property law was an issue for those wanting to include plants in intellectual property law. Indeed, an early obstacle to protecting plants using patent law was that it was difficult, if not impossible, to identify and demarcate the property that was being protected.[2] Writing about the difficulty of defining plants, Robert Cook, in 1932, explained that while:

> [a] lever is always a lever, a cam is always a cam, and even a complex chemical compound stays the same in molecular structure. But not so with plants. Change the conditions and the plant changes...It follows that a verbal description, and even high-quality color plates are not sufficient when a new plant variety has to be defined with the necessary exactness. [3]

Plants have increasingly been identified and described using molecular or genotypic information such as deoxyribonucleic acid (DNA) and

[1] I. Stengers, 'Whitehead's account of the sixth day' (2005) 13(1) *Configurations* 35, 37–38.

[2] See, e.g., M. Janis and J. Kesan, 'US plant variety protection: Sound and fury...?' (2002) 39 *Houston Law Review* 727, 734; R. Cook, 'The first plant patent decision' (1937) 19 *Journal of the Patent Office Society* 187, 190.

[3] R. Cook, 'Patents for New Plants' (1932) 27 *American Mercury* 66, 66. Also cited in A. Heitz, 'The history of plant variety protection' in *The First Twenty-Five years of the International Convention for the Protection of New Varieties of Plants* (UPOV, 1987), p. 65.

ribonucleic acid (RNA). Techniques such as DNA fingerprinting and molecular marker identification have allowed plant scientists to map thousands of genes, and the presence or absence of a certain gene or genes-rather than the visual identification of expressed traits-has been used to describe and distinguish plants.[4] Molecular techniques and the ability of plant scientists to identify, describe and distinguish plants based on molecular or genotypic information have brought into question the use of physically observable traits to describe and classify plants that has been favoured throughout history. And, while these developments have raised questions for taxonomists and botanists, they have also raised concerns about the way in which plants are described and distinguished in intellectual property law: particularly in the UPOV Convention and UPOV-based plant variety protection schemes – specifically, molecular techniques and the ability of plant scientists to identify, describe and distinguish based on molecular or genotypic information that has brought into question the use of visually observable traits in the assessment of distinctness.[5] These questions have led some commentators to criticise plant variety rights schemes' reliance on phenotypic traits because plant variety rights schemes are ill-suited to modern plant innovation.[6]

Can molecular or genotypic information be used to distinguish plants? To answer this question, this chapter examines the way in which plants are described and distinguished for the purposes of the UPOV Convention.[7] The chapter begins by establishing the central (and historical) role of characteristics in the UPOV Convention and by noting the use of growing trials in the assessment of distinctness as well as documents – including UPOV's *General Introduction to the Examination of Distinctness,*

[4] See, e.g., G. Rouhan and M. Gaudel, 'Plant taxonomy: A historical perspective, current challenges, and perspectives in molecular plant taxonomy' in P. Besse (ed.), *Molecular Plant Taxonomy: Methods and Protocols* (Springer, 2014), pp. 1–37; J. Borevitz and J. Ecker, 'Plant genomics: The third wave' (2004) 5 *Ann Review Genomics Human Genetics* 443.

[5] These techniques are also related to the assessments of uniformity and stability.

[6] M. Janis and S. Smith, 'Technological change and the design of plant variety protection regimes' (2007) 82 *Chicago Kent Law Review* 1557, 1566–1570, who cite a number of sources to suggest that phenotypic assessments are subjective, costly, dependent on environmental interactions and that relying on phenotypic characteristics is problematic because, as reference collections become larger and market pressures on breeders to produce phenotypically similar plants continues, distinctness assessments are more subtle.

[7] While the deposit of plant material is also required for the UPOV Convention, this Chapter focuses on the description of plant varieties: UPOV 1991, Article 12. In relation to growing trials also see UPOV, *Experience and Cooperation in DUS Testing*, Document TGP/5 (UPOV, 2008).

Uniformity and Stability (*General Introduction*) and specific *Test Guidelines* for plant varieties and species – that set out detailed principles for the conduct of the examination of new plant varieties, which provide guidance about the development and observation of characteristics. In the final part of the chapter, I explore some of the reasons why molecular or genotypic information (as a means of describing and distinguishing plants) are not suited to plant variety protection schemes. While various technical and operational concerns are discussed, this part primarily focuses on the unexplored underlying antecedents to why molecular techniques are not used in the assessment of distinctness and why physically observable characteristics – particularly morphological and physiological – have remained central to the description and demarcation of plants. Molecular or genotypic information is highly reductionist, and, while it is one thing to reformulate plant varieties in terms of molecular structures, it does not necessarily follow that the identification and description of plant varieties have been reduced to the same level.[8] In the context of plant variety rights schemes, plants are not simply equal to their smallest part, nor can they be described in this way.[9] Indeed, the utilitarian nature of plant varieties and plant breeding, more generally, means that end-users – such as farmers, gardeners and consumers – play an important role in the way in which plant varieties are described and distinguished. When thinking about the role that molecular or genotypic information might play in the assessment of distinctness, we must not overlook the aim of the UPOV Convention, which is to 'provide and promote an effective system of plant variety protection, with the aim of encouraging the development of new varieties of plants, for the benefit of society'.[10]

7.2 Distinguishing Plant Varieties Using Morphological or Physiological Characteristics

Under the UPOV Convention, a plant variety needs to be described so that putative new plant varieties can be compared to plant varieties of 'common knowledge' and evaluated against the requirement of

[8] For various perspectives on, and criticisms of, technological determinism, see M. Smith and L. Marx (eds.), *Does Technology Drive History? The Dilemma of Technological Determinism* (MIT Press, 1994).

[9] B. Sherman, 'Taxonomic property' (2008) 67(3) *Cambridge Law Journal* 560, 578, stating that 'many of the techniques used to identify plant inventions have been subsumed within the legal process'.

[10] UPOV, *Mission Statement*, www.upov.int/about/en/mission.html.

distinctness.[11] While there have been changes in the language used in relation to the conditions of grant, the distinctness requirement has remained relatively unchanged since the introduction of UPOV 1961. In relation to how plant varieties are described and distinguished, a number of points from the discussions at the Diplomatic Conference of 1957–1961 and the text of Article 6(1)(a) of UPOV 1961 need to be drawn out. During the Diplomatic Conference of 1957–1961, there was a range of discussions on the kinds of features or traits that could be used to determine the suitability for protection under the UPOV Convention. One possibility was to require the assessment of qualitative criteria of the applicant plant variety in determining whether it was distinct. However, because of the inherent difficulties of determining whether a plant variety was qualitatively 'better' than other plant varieties, various proposals to link the distinctness criterion to 'usefulness' or 'superiority' were rejected at the Second Session of the Conference held in 1961.[12] An important rationale for rejecting a qualitative criterion was that it would have placed too heavy a burden on independent authorities such as a member states Department of Agriculture or Intellectual Property Offices. By rejecting the inclusion of qualitative criteria, the Diplomatic Conference of 1957–1961 borrowed from existing laws in the United States and Germany. First, the United States *Plant Patent Act 1930* set out the requirements for the protection of asexually reproduced plants in Section 161 and stipulated that new varieties of plants had to be 'distinct'.[13] While the United States *Plant Patent Act* did not define the distinct requirement, the accompanying Senate Committee Report explained the requirement in the following way: 'in order for a new variety to be distinct it must have characteristics clearly distinguishable from those of existing varieties'.[14] Second, to be granted protection under the German *Seed Law 1953*, breeders needed to show that their new variety of plant was both useful and individualised.[15] Significantly, under the German *Seed Law*, the useful requirement – which was taken to equate to

[11] The requirement of examination was set out in Article 7 of UPOV 1961, which required applicants to 'furnish all the necessary information, documents, propagating material or seeds'.

[12] UPOV, *Actes des Conférence Internationales pour la Protection des Obtentions Végétales 1957–1961*, 1972, UPOV Publication No. 316 (UPOV, 1972), pp. 86–87.

[13] *Patent Act 1930* 35 U.S.C. § 161.

[14] Senate Report No. 71–315 (1930). For a discussion see J. Rossman, 'The preparation and prosecution of plant patent applications' 17 *Journal Patent Office Society* 632 (1935). Also see C. Fowler, 'The Plant Patent Act of 1930: A sociological history of its creation' (2000) 82 *Journal of the Patent and Trademark Office Society* 621, 641; Janis and Kesan, 'US plant variety protection: Sound and fury. . .?', 739.

[15] The other requirements of grant included stability and uniformity.

'agronomic value' such as yield, quality or specific resistances – was separate to the individualised requirement which was based on unique features or traits. Having decided that it was not appropriate to test new plant varieties for usefulness or value, delegates at the Diplomatic Conference of 1957–1961 agreed to set out the distinctness requirement in the following way:

> clearly distinguishable by one or more important characteristics from any other variety whose existence is a matter of common knowledge at the time when protection is applied for ... A new variety may be defined and distinguished by morphological or physiological characteristics. In all cases, such characteristics must be capable of precise description and recognition.[16]

As such, it was the description of morphological or physiological characteristics that was central to the distinctness requirement under UPOV 1961. So, while a plant variety must be 'clearly distinguishable by one or more important characteristics', the inclusion of the adjective 'important' did not import a qualitative criterion. In this context, the word 'important' refers to important differences for the purpose of distinguishing the plant variety, not important in the sense of yields or qualities.[17] Article 6(1)(a) of UPOV 1961 expressly states that it is the 'characteristics' of the plant variety that are essential in assessing whether a plant variety is distinct. Furthermore, the notion of characteristics is further mediated by the requirement that a plant variety is defined and distinguished by morphological or physiological characteristics and that these characteristics 'must be capable of precise description and recognition'. In this way, therefore, Heitz has stated that the United States and German laws (minus the additional requirement of agronomic value) 'correspond more or less' to the concept of distinctness in Article 6(1)(a) of UPOV 1961.[18]

While the focus on the plant variety's characteristics did not change substantially with the amendments made by UPOV 1978, there was debate on the types of characteristics used in the assessment of distinctness. Some delegates were of the opinion that relying on 'morphological or physiological' traits for distinctness was too restrictive.[19] Mr. Duyvendak of the Netherlands, for instance, asked whether delegates could:

[16] UPOV 1961, Article 6(1)(a).

[17] N. Byrne, 'The agritechnical criteria in plant breeders' rights law' *Industrial Property* (1983) 293.

[18] Heitz, 'The history of plant variety protection', p. 76 (referring particularly to the German *Seed Law 1953*).

[19] UPOV, Publication No. 337(E), pp. 150–151 [326]–[347].

support the deletion of the words "morphological or physiological" which, although they might be correctly understood by the Conference, might lead to misunderstanding by other people who might wrongly interpret the omission from the Convention of the additional kinds of characteristics.[20]

However, these concerns over the use of 'morphological or physiological characteristics' did not amount to any change in the text of UPOV 1978 or the practice of UPOV Members. It was not until UPOV 1991 that the requirement of distinctness was amended. The text of UPOV 1991, Article 7 states:

The variety will be deemed to be distinct if it is clearly distinguishable from any other variety whose existence is a matter of common knowledge at the time of filing of the application. In particular, the filing of an application for the granting of a breeder's right or for the entering of another variety in an official register of varieties, in any country, shall be deemed to render that other variety a matter of common knowledge from the date of the application, provided that the application leads to the granting of a breeder's right or to the entering of the said other variety in the official register of varieties, as the case may be.

The first of the changes made by UPOV 1991 was the removal of the adjective 'important' from the assessment of distinctness.[21] On the surface, then, Article 7 of UPOV 1991 is less onerous than the earlier versions of the UPOV Convention because to be distinct a new plant variety has to be 'clearly distinguishable from any other variety' rather than be distinguished by one or more 'important characteristics'.[22] However, as discussed earlier, the term 'important' was not based on merit or quality in either UPOV 1961 or UPOV 1978, rather it was intended to reflect important differences for the purposes of distinguishing the variety. Despite assurances that the adjective 'important' did not import a qualitative test for distinctness, at the Diplomatic Conference of 1991, it was thought that the term 'important' was ambiguous and, as a result, that there was confusion about whether there was a requirement of merit or quality in the assessment of distinctness.[23] The intent of removing 'important' from the assessment of distinctness was to clarify the fact that

[20] Ibid., pp. 151 [339].

[21] There were other proposals; for example Poland proposed the inclusion of 'significantly distinguishable', arguing that these words were used in genetics, biometrics and applied statistics: UPOV, Publication No. 346(E), pp. 246 [463].

[22] The word 'clearly' was retained because it was opined that it had been used in the UPOV Convention since 1961 and had been given a clear meaning in the practical operation of the plant variety protection system. UPOV suggests that 'clearly' refers to consistent (over two seasons) and clear (depending on the type of expression of the characteristics) differences: UPOV, *Examining Distinctness*, Document TGP/9 (UPOV, 2008).

[23] B. Greengrass, 'The 1991 Act of the UPOV convention' (1991) 12 *European Intellectual Property Review* 467.

the assessment of distinctness was 'general' and did not purport to 'relate the degree of difference to some significance level which was not actually defined'.[24] A second change made by UPOV 1991 was to the way in which a plant's characteristics had to be described. While UPOV 1961 and UPOV 1978 required that the characteristics which permit the plant variety to be defined and distinguished were to 'be capable of precise description and recognition', this was removed in 1991. One of the consequences of this change was to broaden the ways in which plant varieties could be described. More specifically, it has been suggested that the removal of the requirement of 'precise description and recognition' effectively opened the door for molecular techniques to be used in the description of plant varieties and therefore in the assessment of distinctness.[25]

Before we look more closely at what characteristics can be used in the assessment of distinctness, it is important to point out that characteristics are also vital for the assessment of uniformity and stability. According to Article 6(1)(c) of UPOV 1961, the plant variety had to be 'sufficiently homogenous, having regard to the particular features of its sexual reproduction or vegetative propagation'. While Article 6(1)(c) did not refer directly to characteristics, it can be inferred that plant varieties have to be sufficiently homogenous in relation to the characteristics used to distinguish the plant variety. Finally, according to Article 6(1)(d), the plant variety had to be stable 'in its essential characteristics'. There was no change in either of these requirements until UPOV 1991. Though these changes merely reiterated the centrality of the plant's characteristics to the assessment of the conditions of grant rather than changed the substance of the assessment.[26] More specifically according to Article 8 of UPOV 1991, a 'variety shall be deemed to be uniform if, subject to the variation that may be expected from the particular features of its propagation, it is sufficiently uniform in its relevant characteristics'. While this did not make a substantial change to the UPOV Convention, it more clearly sets out the importance of the plant variety's characteristics in determining uniformity.[27] Similarly, the requirement of stability was

[24] UPOV, Publication No. 346(E), p. 264 [464] (Mr. Bould, United Kingdom).

[25] M. Camlin, 'Plant cultivar identification and registration – The role of molecular techniques' (2003) 625 *Acta Horticulturae* 37, 39.

[26] The Secretary-General emphasised that there was no change to the substance of the condition of homogeneity: UPOV, Publication No. 337(E), p. 253 [511].

[27] For vegetatively or self-pollinated varieties, it is possible to assess uniformity by the number of obviously atypical plants, known as 'off-types'. By comparison, for cross-pollinated varieties it is necessary to consider the overall range of variation to determine whether the applicant plant variety is similar to comparable varieties. UPOV, *General Introduction to the Examination of Distinctness, Uniformity and Stability and the Development*

amended by Article 9 of UPOV 1991 to reiterate the importance of characteristics. Finally, the language changed slightly in UPOV 1991 (from 'essential' to 'relevant' characteristics) with the introduction of Article 9 of UPOV 1991, which states that 'the variety shall be deemed to be stable if its relevant characteristics remain unchanged after repeated propagation or, in the case of a particular cycle of propagation, at the end of each cycle'. The effect of this amendment was to focus the assessment of stability on the same (relevant) characteristics that are used in determining distinctness.

Having clearly established that it is a plant variety's characteristics that are central to describing, identifying and distinguishing plant varieties for the purpose of the UPOV Convention, it is necessary to examine how UPOV has defined characteristics. Despite the reliance on characteristics in the UPOV Convention, the term has not been defined in any of the versions of the UPOV Convention. Nonetheless, it is generally accepted that it is a plant's phenotype that determines a plant variety's characteristics for the purposes of the UPOV Convention.[28] This is largely because the UPOV Convention has focused on the 'expression' of morphological or physiological characteristics, and, as a result, visually observable characteristics (especially those least affected by environmental factors) are used in the assessment of distinctness, uniformity and stability. The prominence of visually observable characteristics to the assessment of distinctness, uniformity and stability is reinforced by the methods used to assess distinctness and the features or traits compared during this assessment.

How is the examination of distinctness carried out? Significantly the basis for most of the technical examinations involves a growing test to determine the morphological or physiological characteristics of the putative plant variety, in comparison with reference varieties to establish its distinctness. In addition to setting out the conditions for granting the breeder's right, the UPOV Convention requires that the members of UPOV conduct an examination for compliance with these conditions. Article 12 of UPOV 1991 states that:

> [i]n the course of the examination, the authority may grow the variety or carry out other necessary tests, cause the growing of the variety or the carrying out of other

of Harmonized Descriptions of New Varieties of Plants, TG/1/3 (UPOV, 2002) 19. See also R. Löpez de Haro y Wood, 'The UPOV approach to the examination of application for protection – Past, present and future', *Seminar on the Nature of and Rationale for the Protection of Plant Varieties under the UPOV Convention*, UPOV Publication No 727(E) 1 (UPOV, 1994), 71.

[28] See, e.g., M. Llewelyn and M. Adcock, *European Plant Intellectual Property* (Hart Publishing, 2006), p. 40 (stating that 'phenotypic qualities go to the heart of UPOV').

necessary tests, or take into account the results of growing tests or other trials which have already been carried out.

Article 12 specifies that the Members of UPOV have different options for carrying out the examination. For example, an authority of UPOV might choose to conduct the examination itself, arrange for a third party to conduct the examination or arrange for the plant breeder to conduct some or all of the examination. Further, the ability to use 'results' that 'have already been carried out' allows members of UPOV to use the results from other Member countries which have previously conducted an examination for the variety concerned. As UPOV points out, this allows one Member of UPOV to purchase the report on the examination which has been conducted by another member of UPOV.[29]

What is compared during the examination of distinctness, uniformity and stability? In seeking to harmonise the testing of distinctness, uniformity and stability, UPOV has established two main guidelines which can be applied by its Members. The first is the *General Introduction to the Examination of Distinctness, Uniformity and Stability* (*General Introduction*) which provides guidance about the development and observation of characteristics by setting out detailed principles for the conduct of the examination of new plant varieties. The *General Introduction* outlines general aspects of examination that apply across all plant species or plant varieties. The second source of guidance on what constitutes a plant's characteristics is the specific *Test Guidelines* for particular plant varieties or species (*Test Guidelines*).[30] The specialised *Test Guidelines* extrapolate on the principles in the *General Introduction* into detailed practical guidance for the harmonised examination of distinctness, uniformity and stability. Both *Guidelines* must be read together.[31] At the time of publication, there were nearly 300 specific *Test Guidelines including for maize, wheat, rice, barley, oats, chrysanthemum, persimmon and Echinacea.* The *Test Guidelines* consist of highly technical data relevant to each species against which any new plant variety can be compared and may include written technical information, drawings and photographs. Generally speaking the *Test Guidelines* set out the relevant characteristics and their states, numerical values for states for electronic processing, and example varieties. Rather than attempt to define the term characteristics, these *Guidelines* set out the types of characteristics

[29] See UPOV, *The UPOV System of Plant Variety Protection*, www.upov.int/about/en/upov_system.html#P231_24902.

[30] UPOV, *Development of Test Guidelines*, Document TGP/7 (UPOV, 2004). Currently there are more than 300 *Test Guidelines*, www.upov.int/en/publications/tg_rom/tg_index.html.

[31] Ibid.

that are to be used when describing and distinguishing plant varieties for the purposes of UPOV. The *Guidelines* expand the notion of 'morphological or physiological' characteristics and outline, where possible, specific characteristics for particular plant varieties or species that should be used when describing plant varieties for the purpose of examination. The *Test Guidelines* set out how each characteristic or 'state' should be measured and includes measurements and visual assessments. The *Test Guidelines* for wheat (*Triticum aestivum* L.), for example, list 26 characteristics including growth habit, time of ear emergence, grain colour and seasonal type.[32] The *Test Guideline* for roses (*Rosa* L.) includes 54 characteristics including plant height, the number of prickles on the stem, the number of flowering laterals, the number of petals and the colour.[33]

Taken together, the *General Introduction* and the *Test Guidelines* identify, develop and elaborate the kinds of characteristics to be used in the assessment of distinctness, uniformity and stability. In so doing, the *General Introduction* and the *Test Guidelines* help to establish a list of characteristics which are useful to describe and distinguish between plant varieties. Both the *General Introduction* and *Test Guidelines* focus on characteristics that are physically observable, such as morphological characteristics (e.g. the colour of flowers, fruit; the shape or length of the leaves, pods or petals) or physiological characteristics (e.g. early maturity, resistance to environment, disease or chemical sprays). Furthermore, for the purpose of description and assessment, each characteristic in the *General Introduction* and *Test Guidelines* is divided into a number of states which are influenced by the type of expression for that particular characteristic, with each characteristic generally being grouped according to one of three types. These are:

- *qualitative characteristics*, usually evaluated by visual inspection as variation is categorical rather than continuous. This means that a plant's characteristics are described according to nonoverlapping categories. Take, for example, Lettuce (*Lactuca sativa* L.), where qualitative characteristics include seed colour (white, yellow or black), seedling colouration (absent or present) and the kind of anthocyanin distribution present on the leaf (diffused only, in spots only, or diffused and in spots).[34]

[32] A list of UPOV's *Test Guidelines* is available at www.upov.int/test_guidelines/en/list.jsp.
[33] UPOV, *Rose: Guidelines for the Conduct of Tests for Distinctness, Uniformity and Stability*, TG/11/8 Rev (UPOV, 2010).
[34] UPOV, *Lettuce (Lactuca sativa L.): Guidelines for the Conduct of Tests for Distinctness, Uniformity and Stability*, Document TG/13/10 (UPOV, 2006).

- *pseudo-qualitative characteristics*, also evaluated visually but are at least partially continuous. Unlike qualitative characteristics, pseudo-qualitative characteristics vary in more than one dimension and cannot be defined by simple description of extremes and may be described as a value or measure within a range of data. Again, using Lettuce (*Lactuca sativa* L.) as our example, pseudo-qualitative characteristics include leaf blade division (entire, lobed or divided), plant head formation (no head, open head, closed head) and leaf shape (narrow elliptic, medium elliptic, broad elliptic, circular, transverse broad elliptic, transverse narrow elliptic, obovate, broad obtrullate, triangular).[35]
- *quantitative characteristics*, those traits whose expressions are continuous and cover a full range of variation from one extreme to the other. Therefore, quantitative characteristics are divided into a number of states for the purpose of description. Examples of a quantitative characteristics for Lettuce (*Lactuca sativa* L.) include plant diameter (very small, small, medium, large, very large), leaf thickness (thin, medium, thick) and intensity of colour of outer leaves (very light, light, medium, dark, very dark).[36]

While the *General Introduction* and the *Test Guidelines* explicitly list visually observable traits, both documents are largely silent about the use of genotypes. This does not mean, however, that genotypic information is prohibited from being used to describe plant varieties and the UPOV Convention does not expressly 'require or forbid the use of molecular makers'.[37] Indeed, it has been suggested that the genomic structure of a plant variety may qualify as a form of 'expression' if both internal and external expression are taken into account.[38] Importantly, too, it is accepted that the types of characteristics used in the assessment of distinctness, uniformity and stability are not static and, in certain circumstances, 'special or additional characteristics' are used as supporting evidence when arguing that a new plant variety is distinct.[39] While in the past chemical and biochemical analysis have been incorporated into the assessment of resistance to disease and pathogens in order to determine distinctness, the question of whether molecular or genotypic information can be used to describe and distinguish plant varieties is still being

[35] Ibid. [36] Ibid.

[37] UPOV, *Progress Report of the Work of the Technical Committee, the Technical Working Parties and the Working Group on Biochemical and Molecular Techniques and DNA-Profiling in Particular*, Document C/34/10 (UPOV, 2000), p. 5.

[38] Camlin, 'Plant cultivar identification and registration – The role of molecular techniques', 39; Llewelyn and Adcock, *European Plant Intellectual Property*, p. 208.

[39] See, e.g., UPOV, *Guidance on Certain Physiological Characteristics*, Document TFP/12 (UPOV, 2009).

considered by UPOV, industry bodies and by various plant breeding organisations.

7.3 Can Molecular Techniques Be Used to Describe Plant Varieties?

Progress in science and technology has meant that plants can be described using a range of features, traits or relationships. These developments have led to the suggestion that the UPOV Convention and UPOV-based plant variety protection schemes, which generally rely on physically observable characteristics to describe plant varieties, are obsolete and outdated. This is expressed either explicitly or implicitly in the argument that the UPOV Convention is rooted to a particular technological paradigm and is not keeping pace with innovations in plant science. The view that science is outpacing plant variety rights schemes has arisen in relation to the description and demarcation of plant varieties and is characterised by a number of specific criticisms. For example, it has been suggested that the UPOV Convention is 'obsolete' because the legal concept of a plant variety has become ill-suited to deal with plant innovation. Specifically, Janis and Smith argue the requirements of grant which generally relate to 'characteristics' and 'features' under the UPOV Convention are no longer relevant as plant breeding has moved towards a genotypic approach which utilises genetic modification and molecular breeding techniques that are based on genetic 'data' and 'codes'.[40] Another criticism made by those in favour of using molecular or genotypic information to describe and distinguish plant varieties is that the use of phenotypic characteristics is subjective, costly and dependent on environmental factors.[41] Additionally, proponents of molecular or genotypic information suggest that a consequence of relying on phenotypic characteristics is that reference collections (i.e. those plants that make up the common knowledge) are becoming too large.[42]

[40] Focusing on the rise of molecular marker technologies, Janis and Smith propose the (re) conceptualisation of plant varieties as genetic datasets: Janis and Smith, 'Technological change and the design of plant variety protection regimes'. For further elaboration of the claims of Janis and Smith, see L. Helfer, 'The demise and rebirth of plant variety protection: A comment on technological change and the design of plant variety protection regimes' (2007) 82 *Chicago Kent Law Review* 1619.

[41] For a summary of issues, see Janis and Smith, 'Technological change and the design of plant variety protection regimes', 1583–1590. See also Camlin, 'Plant cultivar identification and registration – The role of molecular techniques'.

[42] Camlin, 'Plant cultivar identification and registration – The role of molecular techniques', 38.

The question of using molecular techniques to assess distinctness has been considered by UPOV at all three of its Diplomatic Conferences. At the Diplomatic Conference of 1957–1961, there was a proposal that the 'mere selection of a genotype from among those included in a pre-existing variety' would be sufficient to satisfy the need for distinctness.[43] The Swiss Delegation argued that 'the invention itself (i.e. the discovery) of a genotype should be protected'[44] and that '[i]t is possible that from the simple choice of a genotype some variety comes to life which would fulfil the conditions defined in Articles 6(1)(b) and 6(1)(c) and that deserves to be protected'.[45] Despite these proposals, at the Second Session of the Diplomatic Conference of 1957–1961, the specific qualification on selecting genotypes was removed. The emphasis was instead placed on morphological or physiological characteristics.

Questions over the use of genotypes in the assessment of distinctness were raised again during the 1970s. However, despite concerns that the reliance on morphological or physiological characteristics was restrictive and that it excluded other kinds of information being used in the assessment of distinctness, being raised during the Diplomatic Conference of 1978, there was no change. Most recently, and as was noted earlier, UPOV 1991 removed the need for the 'precise description and recognition' of applicant plant varieties which has been used to argue for expanding the kinds of characteristics that can be used in the assessment of distinctness. In circumstances where there have been problems in determining distinctness using morphological characteristics, the use of biochemical characteristics, examined, for example, by electrophoresis, has become accepted by UPOV.[46] The removal of the need for 'precise description and recognition' has likewise been used to support the possibility of using molecular information in the assessment of distinctness.[47]

In 1993, faced with persistent questions over the possibility of altering the way in which plant varieties are described and distinguished, UPOV set up a Working Party on Biochemical and Molecular Techniques to examine the use of molecular techniques in assessing distinctness, uniformity and stability. In 2002, the Working Party clarified the position of UPOV in respect of possible options for the use of molecular

[43] UPOV, Publication No. 316, p. 85. [44] Ibid. [45] Ibid.

[46] See UPOV, *Reports on Developments in UPOV Concerning Biochemical and Molecular Techniques*, BMT/13/12, prepared by the Working Group on Biochemical and Molecular Techniques and DNA-Profiling in Particular (UPOV, 2011), www.upov.int/edocs/mdocs/upov/en/bmt_13/bmt_13_2.pdf.

[47] Camlin, 'Plant cultivar identification and registration – The role of molecular techniques', 39.

techniques[48] and, in so doing, identified a range of ways in which molecular techniques could be used in the assessment of distinctness, uniformity and stability.[49] One option considered by the Working Party was whether molecular characteristics could be used as predictors of traditional characteristics. In this scenario, molecular characteristics could be used as gene specific markers that are tied to traditional characteristics such as flower colour or yield. A more radical option canvassed by the Working Party called for the development of a new system of protection that would fundamentally alter the way in which the UPOV Convention operates. After considering these options presented by the Working Party on Biochemical and Molecular Techniques, UPOV did not support the use of molecular techniques in the assessment of distinctness, uniformity and stability.

Since the early 2000s, UPOV has released a range of information documents, guidelines and options for the possible use of biochemical and molecular markers in the examination of distinctness, uniformity and stability.[50] Significantly, according to UPOV the use of molecular markers should be limited and is still largely dependent on the variety's physically observable characteristics. UPOV has only approved the use of molecular markers in the examination of DUS if there is a 'reliable link between the marker and the characteristic' and, when used in conjunction with a phenotypic differences, 'if the molecular distances are sufficiently related to phenotypic differences and the method does not create an increased risk of not selecting a variety in the variety collection which should be compared to candidate varieties in the DUS growing trial'.[51] The justifications for such limits are:

> in some cases, varieties may have a different DNA profile but be phenotypically identical, whilst, in other cases, varieties which have a large phenotypic difference may have the same DNA profile for a particular set of molecular markers (e.g. some mutations).

In relation to the use of molecular markers that are not related to phenotypic differences, the concern is that it might be possible to use a

[48] UPOV, *Recommendations of the BMT Review Group and Opinion of the Technical Committee and the Administrative and Legal Committee concerning Molecular Techniques*, TC/38/Add.-CAJ/45/5/Add (UPOV, 2002).

[49] Ibid.

[50] See, e.g., UPOV, *Guidance on the Use of Biochemical and Molecular Markers in the Examination of Distinctness, Uniformity and Stability (DUS)*, TGP/15 (UPOV, 24 October 2013);UPOV, *Guidelines for DNA-Profiling: Molecular Marker Selection and Database Construction ('BMT Guidelines')*, UPOV/INF/17/1 (UPOV, 24 May 2011); UPOV, *Possible use of Molecular Markers in the Examination of Distinctness, Uniformity and Stability (DUS)*, UPOV/INF/18/1 (UPOV, 20 October 2011).

[51] UPOV, UPOV/INF/18/1, p. 4.

limitless number of markers to find differences between varieties at the genetic level that are not reflected in phenotypic characteristics.[52]

That molecular or genotypic information is not yet generally appropriate for assessments of distinctness, uniformity and stability can be argued on practical, scientific and theoretical grounds. One problem of relying on molecular information to describe plant varieties is the (un)reliability and (in)validity of using molecular techniques for the description of plant varieties.[53] This problem was recognised by the Working Party on Biochemical and Molecular Techniques who acknowledged that, if the existing UPOV scheme is to accommodate links between molecular markers and traditional characteristics, then establishing valid and reliable relationships between molecular markers and the expression of the characteristic is paramount. The problem is that the relationship between genotype and phenotype is best described as ambiguous and the dividing line between genotype and phenotype is hard to draw.[54] While there are studies that support a correlation between phenotype and molecular markers, there are also studies that suggest that there is no correlation at all.[55] The issues of reliability and validity are further complicated because, even if a relationship exists between genotype and phenotype, it may not be a linear one.[56] Establishing a relationship between genotype and phenotype is also problematic since similar phenotypes can be observed for different genetic combinations.[57]

Another concern about including molecular information in the assessment of distinctness, uniformity and stability relates to the nature of the right granted under the UPOV Convention. According to UPOV's Working Party on Biochemical and Molecular Techniques, one of the consequences of basing distinctness assessments on molecular or

[52] UPOV, *Does UPOV Allow Molecular Techniques (DNA-Profiling) in the Examination of Distinctness, Uniformity and Stability (DUS)?* FAQs, C(Extr)/31, April 2014, www.upov.int/about/en/faq.html#Q27.

[53] ISF, *ISF View on Intellectual Property* (adopted June 2003, the paragraph on 'The Case of DNA Markers' was adopted on 27 May 2009), www.worldseed.org/cms/medias/file/PositionPapers/OnIntellectualProperty/View_on_Intellectual_Property_2009.pdf; European Seed Association, *Position: Access to Plant Genetic Resources and Intellectual Property in the European Union* (2004 and revised in 2009), www.euroseeds.org/position-papers/2009/ESA_04.0056.4.pdf.

[54] See, e.g., J. Borevitz, and J. Ecker, 'Plant genomics: The third wave' (2004) 5 *Ann Review Genomics Human Genetics* 443; J. Burstin and A. Charcossett, 'Relationship between phenotypic and marker distances: Theoretical and experimental investigations' (1997) 79 *Heredity* 477.

[55] Ibid.

[56] Burstin and Charcosett, 'Relationship between phenotypic and marker distances: Theoretical and experimental investigations'.

[57] Ibid.

genotypic information could be that the distance between protectable plant varieties could be reduced. This has the potential to effectively change the nature of protection provided by the UPOV Convention and, thus, 'jeopardise the value' of any plant variety right granted.[58] The ISF, for example, has argued that a possible consequence of using molecular information in the assessment of distinctness is that the rights under UPOV-based schemes would be marginalised because it may reduce the minimum distance for distinctness to a difference of only one base pair or by leading to impractical standards for uniformity and stability.[59] Other problems with using molecular information to assess distinctness, uniformity and stability are associated with the resources required. So, despite the fact that the cost of the techniques, tools and laboratories needed to perform these kinds of molecular assessments has reduced, these resources are prohibitive for many UPOV Members, particularly those in developing countries.[60] As a consequence, the use of molecular techniques in the assessment of distinctness, uniformity and stability is likely to be financially restrictive and prohibitive. In addition to the cost involved in the assessment of molecular or genotypic information, these tests require specific knowledge and skills that may not be universally available.

While practical and operational concerns have been central in the arguments against using molecular or genotypic information in the assessment of distinctness, uniformity and stability, there are some serious theoretical grounds for excluding molecular or genotypic information from the assessment of distinctness, uniformity and stability. Perhaps, most importantly, it is the contingent nature of plant breeding and of plant varieties that undermines the use of molecular or genotypic information. In practice neither genotypic nor phenotypic descriptions are complete: instead, they are restricted to a subset of the characteristics of the plant variety that regarded as relevant for a particular explanatory or experimental purpose. So while molecular techniques such as plant transformation and molecular markers provide a useful source of information, these techniques generally only assist to recognise and identify particular features or traits that are of importance. Such desirable traits may include greater yield, resistance to insects or pests, tolerance to heat

[58] UPOV Council, *Progress Report on the Work of the Technical Committee, the Technical Working Parties and the Working Group on Biochemical and Molecular Techniques and DNA-Profiling in Particular*, C/34/10 (UPOV, 2000), p. 6. See also European Seed Association, *Position: Access to Plant Genetic Resources and Intellectual Property in the European Union.*

[59] For a discussion see, UPOV, UPOV/INF/18/1.

[60] See, e.g., P. Poczai et al., 'Advances in plant gene-targeted and functional markers: A review' (2013) 9(6) *Plant Methods* 1.

and drought, better agronomic quality, higher nutritional value, growth rate and fruit properties. In this sense, while genotypic plant breeding provides insights and advances: nonetheless, the plant breeding techniques of selection and multiplication remain central to the development of new plant varieties. Often it is only after multiple measurements of various aspects of a plant's phenotype (such as yield or plant height) are performed that further investigations of the genotype are made. Indeed, there is a low probability that a poor phenotype will result in a good genotype, and as a consequence phenotypic evaluation is often the first, and last, stage in selecting parents as the 'new parent has proven to have merit on its phenotypic performance'.[61]

Within the framework of the UPOV Convention, molecular information presents a decontextualised view of plant breeding. Molecular techniques are highly reductionist, and, while it is one thing to reformulate plant varieties in terms of molecular structures, it does not necessarily follow that the identification and description of plant varieties for the purposes of the UPOV Convention must be reduced to the same level. This is what philosopher Graham Harman refers to as undermining: a process of reducing or dissolving objects (such as plant varieties) down into some component element (such as genes or DNA).[62] For advocates of molecular techniques, the real action of plants unfolds at a deeper layer than the plants themselves. On some level they try to have us believe that characteristics and phenotype are naïve. Furthermore, as Lewontin says:

> When scientists break up the world into bits and pieces, they think they're breaking things up into some natural entities, but they're not. They're breaking them up into bits and pieces that they see in the world by their own training, by their own ideology; by the whole way in which they've been taught to see the world. But that doesn't mean those bits and pieces are the real bits and pieces. So, an antireductionist is someone who believes that in order to understand processes of nature, you must not import into them some artificial notion of "real" bits and pieces.[63]

Most fundamentally, then, molecular techniques ignore the richness, complexity and context of the existing modes of description that involves both growing trials and visually observable characteristics. I am not

[61] J. Brown and P. Caligari, *An Introduction to Plant Breeding* (Blackwell Publishing, 2008), p. 150.

[62] See G. Harman, *The Quadruple Object* (Zero Books, 2011), pp. 7–19.

[63] R Lewontin, 'Not all in the genes' in L. Wolpert and A. Richard (eds.), *The Passionate Mind: The Inner World of Scientists* (Oxford University Press, 1997), p. 103. See also F. Browning, *Science, Seeds and Cyborgs: Biotechnology and the Appropriation of Life* (Verso, 2003).

denying that the discovery of the structure of DNA by James Watson and Francis Crick transformed a number of scientific disciplines, nor that it led to some remarkable developments including the elucidation of the genetic code, the first cloned sheep and the construction of genetic maps.[64] Or that from these (and other) discoveries, a set of techniques have also been applied to the development of new plant varieties including DNA profiles, plant transformations and molecular markers.[65] Nonetheless, despite the significance of these technological developments on their own, they have arguably only resulted in a modest impact on plant breeding. From the earliest twentieth century, when Mendel's genetics was rediscovered, plant scientists and plant breeders had differing views on the value of genetic information for farming. On the one hand, for example, Sir Rowland Biffen, the first director of the Plant Breeding Institute, believed genetic principles were essential to developing improved plant varieties, a view supported by the success of his influential Yeoman wheat variety. On the other hand, John Percival of the Department of Agriculture at University College of Reading insisted that characteristics of interest to farmers, such as yield and strength, were influenced by such a complex array of physiologic and environmental factors that they could not be reduced to Mendelian principles.[66] Throughout this debate, there was a tension between the aims of the academic scientist and the needs of the farmer. A similar tension is evident in the debates between scientists, lawyers, academics and plant breeders over the use of molecular information in the assessment of distinctness, uniformity and stability.

In addition to being reductionist, molecular information does not reflect the utilitarian nature of plant variety development in which the end-users (such as farmers, gardeners and nurserymen) play an important function. The purpose of the UPOV Convention is to 'provide and promote an effective system of plant variety protection, with the aim of encouraging the development of new varieties of plants, for the benefit of society'.[67] So, in the context of the UPOV Convention, the description of new plant varieties is mediated by the context of plant breeding: both by UPOV's objectives and the fact that plant breeding is a commercial enterprise in which the end product (i.e. the plant variety) is the

[64] See P. Shewry, H. Jones and N. Halford, 'Advances in biochemical engineering/biotechnology' (2008) 111 *Food Biotechnology* 149.

[65] For a general discussion of molecular techniques in plant breeding, see Brown and Caligari, *An Introduction to Plant Breeding*.

[66] See P. Palladino, *Plants, Patients and the Historian: (Re)membering in the Age of Genetic Engineering* (Manchester University Press, 2002).

[67] UPOV, *UPOV Mission Statement.*

commodity. The end-users of new plant varieties are not particularly interested in the genetic information of plant varieties such as the DNA sequence of their new wheat, apple, rose or barley variety. Instead, end-users of plant breeding are concerned with the yields of their wheat variety, the crispness of their apples, the colour of their roses and the drought resistance of their barley variety. Because the essence of plant breeding is the discovery or creation of desirable traits that can be inherited in a stable fashion, we cannot, or should not, separate out the genetic variation from the desirable traits or characteristics. Relevantly, the Vice Secretary-General of UPOV, Mr. Peter Button, stressed the importance of end-users such as farmers and growers to the UPOV system, by stating:

> it should not be forgotten that the work of plant breeders at the molecular level is aimed at improvements in the phenotype, whether that is at the level of plant morphology, development or biochemical or physiological properties. What is relevant is the resulting characteristics of the variety. Regardless of the processes involved, a farmer or grower will ultimately require the work of the plant breeder to be encapsulated in a new plant variety—which is the subject matter of protection in the UPOV system. That has not changed since 1961, when UPOV was established.[68]

A final point to make is that despite fundamental differences between the UPOV Convention and patent law, when claiming a patent over plant variety or cultivar, morphological and physiological characteristics are used to define and describe the plant variety.[69] This means that the same characteristics are often used to embody plant variety inventions in patent law. In patent law the description and claims for a novel plant variety generally explain how the novel plant variety differs from existing plant varieties and may include details about what has been created, why the invention is important and the problems that it has solved. While a patent might claim a specific DNA sequence or promoter gene, reference is often made to the resulting morphological and physiological characteristics. For example, the WIPO Application for Lettuce Line P2428-5006090 contains claims to a lettuce plant, pollen, ovules, plant cells, tissue cultures, seeds, regenerated lettuce plants, hybrid lettuce seed and methods for producing such lettuce seed.[70] Significantly, the plant is

[68] J. Sanderson, 'Why UPOV is relevant, transparent and looking to the future: a conversation with Peter Button' (2013) 8(8) *Journal of Intellectual Property Law and Practice* 615.

[69] G. Myers, 'From discovery to invention: The writing and re-writing of two patents' (1995) 25 *Social Studies of Science* 57.

[70] Lettuce Line P2428–5006090, International Application Number PCT/US2008/051903 (Publication date 31 July 2008).

both defined and described with reference to its physically observable characteristics. The invention is a 'new' lettuce that is characterised by: superior head shape, weight, leaf index, plant height, head diameter, ratio of plant height to head diameter, levels of Vitamin A and C, disease resistance, suitability for mechanical harvesting, and post-harvest fitness when compared to competing lines.[71] The acceptance of morphological or physiological characteristics as reasonable descriptors of plant varieties for patent law is also evident in disputes over utility patents in the United States. For example, in *Ex parte C*[72] the United States Board of Patent Appeals heard a dispute involving a variety of soybean. The invention concerned a novel variety of soybean plant, seeds produced, and a method of producing seeds by self-pollinating the soybean plant. The claimed soybean differed from the prior art in pod colour, pubescence colour and phytophthora root rot resistance. While there were a number of issues put before the Board of Patent Appeals including issues of obviousness, enablement and prior sale, the examiner rejected the invention under the *Patent Act* for lack of description.[73] The examiner noted that the specification was 'somewhat cryptic and provided little more detail than would be provided in a specification submitted to the Department of Agriculture under the Plant Variety Protection Act'.[74] In reversing the examiner's rejection, the Board of Appeals accepted morphological and physiological traits as sufficient descriptors of a plant variety. As the Board of Appeals said, the

> specification sets forth a reasonable description of the characteristics of the seed and plant including, flower color, plant type, maturity group, bacterial resistance, nematode resistance, etc. and that]... there is sufficient information of record to establish that the language objected to is accepted by the art as descriptive of the characteristics of a soybean variety.[75]

7.4 Conclusion

Given the importance of physically observable traits to botany and taxonomy, it is not surprising that the UPOV Convention is based on the evaluation of phenotypic characteristics such as leaf shape, plant colour and height. There is no denying the increasing use of molecular techniques in plant science and plant breeding more generally; nonetheless,

[71] Ibid., [48].

[72] *Ex parte C 27*, USPQ 2d 1492 (Board of Patent Appeals and Interferences, 1992). The Board of Appeal affirmed the examiner's rejection on a number of grounds including a lack of nonobviousness, differences were of practical significance and public use.

[73] *Plant Patent Act 1930* 35 U.S.C. § 112.

[74] *Ex parte C 27*, USPQ 2d 1492, 1493 (Board of Patent Appeals and Interferences, 1992).

[75] Ibid., 1495.

molecular or genotypic information is not suitable for the assessment of distinctness, uniformity and stability. This chapter has shown that despite advances in science and technology, physically observable characteristics have played and continue to play a key role in the identification, description and comparison of plants. Even though there have been debates in the UPOV forum about the use of molecular techniques to describe and distinguish plant varieties, UPOV 1991 reinforced the pivotal role of physically observable characteristics in the assessment of distinctness, uniformity and stability.

Further, the suggestion that the use of physically observable characteristics to describe plant varieties is obsolete is driven by a mode of speculative determinism that assumes that plant breeders have a specification of the biological entity or process they are trying to find.[76] Rather than be beholden to advances in science and technology, the way in which plant varieties are described and distinguished for the purpose of plant variety rights protection is contingent on a wide range of elements and factors. Perhaps and most importantly, the utilitarian nature of plant variety development means that the end-users (such as farmers, gardeners and nurserymen) play an important role determining how plant varieties are described and distinguished. The end-users of new plant varieties are not particularly interested in the DNA sequence of their new wheat, rose or lettuce variety. It is the expression of visually observable characteristics—such as bigger or brighter flowers, increased yields, or resistances to drought and salt—that end-users such as farmers and other growers are interested in.[77] Moreover, the aim of plant breeding is to develop improved plant varieties which are 'adapted to specific environmental conditions and suitable for economic production in a commercial cropping system'.[78] The way in which plant varieties are currently described and distinguished in the UPOV Convention embodies a level of richness, complexity and context that technological determinism ignores. Importantly, too, phenotype and genotype are not contradictions but are part of the same process, and molecular or genotypic information is brought into being by the characteristics expressed by the plant variety. The aim of plant variety rights schemes is to promote plant variety development for the benefit of society, not to merely develop genetically distinguishable plants.

[76] S. Franklin, *Dolly Mixtures: The Remaking of Genealogy* (Duke University Press, 2007).

[77] As Llewelyn and Adcock put it, 'it is still the plant as a whole which needs to be understood, and appreciated if a particular end result is to be achieved': Llewelyn and Adcock, *European Plant Intellectual Property*, p. 39.

[78] Brown and Caligari, *An Introduction to Plant Breeding*, p. 1.

8 Expanding Protected Material

Embedding Legal Language and Practices in the UPOV Convention

> [The Expert Committee of 1957–1961] 'often wondered whether the protection granted in favour of a variety could go, at least in some cases, right down to the product obtained from the seed or plants'.[1]

8.1 Introduction

The scope of protection granted has been referred to as the 'cornerstone' of the UPOV Convention.[2] In general terms, the scope of protection granted by the UPOV Convention is characterised by an attempt to balance the two imperatives of protecting the work of plant breeders and not unreasonably restricting the use of plant varieties. Unlike the situation in patent law, where the scope of protection is primarily limited by the valid claims of the granted patent, the UPOV Convention specifically defines the scope of protection that is appropriate for plant breeding. In so doing, the UPOV Convention has generally placed limits on the plant material that is protected and the uses to which protection applies. The effect of this is that protection is not extended to the plant in its entirety or to all uses of the plant variety.

This chapter explores the plant material that has been, and continues to be, protected under the UPOV Convention. Reflecting the involvement of the nursery industry to the development of the UPOV Convention, UPOV 1961 restricted the right of the plant variety owner to 'reproductive and vegetative propagating material' such as seeds, cuttings and bulbs. At the time, an important part of the reasoning for such a restriction was that the harvested material of agricultural plant varieties (such as wheat and maize) was an integral part of the food system and,

[1] UPOV, *Actes des Conférence Internationales pour la Protection des Obtentions Végétales 1957–1961*, 1972, Publication No. 316 (UPOV, 1972), p. 48.

[2] UPOV, UPOV Publication No. 337(E), p. 120 (Report of the Working Group on Article 5, DC/82).

thus, should not be restricted. However, the question of whether to extend protection beyond the propagating material – to, for example, 'consumption material' – was one of the most contested issues. While UPOV 1961 restricted protection to the 'reproductive or vegetative propagating material' of the protected plant variety,[3] many plant variety owners and professional organisations believed that limiting protection in this way was legally and practically inadequate. Yet it was not until the Diplomatic Conference of 1991 that there was a convergence of the professional interest in extending protection and a desire to strengthen the plant variety rights scheme. This convergence meant that the scope of protection was extended to harvested material and to products directly derived from the harvested material. Significantly, under UPOV 1991 plant variety owners were not given absolute discretion to exercise their rights at any stage of production[4] but were instead given a cascading right that stipulates that protection is only extended to the harvested material and products derived from the harvested material when the plant variety owner has not had a 'reasonable opportunity' to exercise their right in relation to the propagating material or the harvested material respectively.[5]

In addition to extending the level of protection granted to plant breeders, the extension of protection under UPOV 1991 has a number of other consequences. First, the UPOV Convention has become more legal. The concept of 'reasonable opportunity' has introduced more legal language and with it a degree of indeterminacy to the UPOV Convention that plant breeders and plant breeding organisations have had difficulty reconciling. This has facilitated and encouraged the use of contractual or licensing arrangements in relation to plants protected by plant variety protection schemes. This extension in protection has also increased the possibility of disagreement and legal dispute and increased the reliance on legal structures and lawyers. Thus bringing an increased interaction with the judiciary, legislature and the executive, as well as an increased reliance on legal process, legal language and legal argument. Second, the extension of plant variety protection to harvested material and products derived from harvested material has given the UPOV Convention a new

[3] Although, Article 5(1) of UPOV 1961 provides Member States with the possibility of extending protection to 'ornamental plants or parts thereof normally marketed for purposes other than propagation when they are used commercially as propagating material in the production of ornamental plants or cut flowers'.

[4] UPOV 1991, Articles 14(2), (3).

[5] For general guidance on harvested material see UPOV, *Explanatory Notes on Acts in Respect of Harvested Material Under the 1991 Act of the UPOV Convention*, UPOV/UXN/HRV/1 (UPOV, 24 October 2013).

lease on life. As we will see, the case of *Monsanto Technology LLC* v. *Cefetra BU and others*[6] – in which the European Court of Justice considered the nature and scope of protection granted to genetic information particularly in relation to Article 9 of the European *Biotechnology Directive* – highlights that, while UPOV-based plant variety protection has been criticised as outdated and obsolete, in some respects at least, it is better able to deal with the protection of plant-related developments than patent law. Though further consideration needs to be given to practical, operational and administrative issues-such as eligibility for protection, variety identification and establishing the origin of the plant material- in the push to promote agricultural research and to encourage the development of new plants, it is vital to acknowledge the role that plant variety protection can play, even in cases in which plant varieties contain specific genetic information.

8.2 Protecting Propagating Material

In determining the scope of protection for the UPOV Convention, delegates at the Diplomatic Conference of 1961–1957 were confronted with a range of issues. Of particular concern was whether protection should be limited to propagating material (such as seeds, bulbs and cuttings) of the protected plant varieties or whether it should extend to products obtained from the propagating material (such as flowers, fruit and perfume). Central to the arguments of those pushing for protecting products derived from propagating material was the idea that the destination of a plant crop was not always known at the time of production. This concern was encapsulated by professional groups such as the International Community of Breeders of Asexually Reproduced Ornamental and Fruit Plants (CIOPORA), who argued that limiting protection to propagating material was 'illusive' for vegetable production because plant breeding was a heterogeneous activity and the destination of the crop was not always known.[7] Further concerns were also raised about the use of 'consumption material' as propagating material: because a 'plant may often be reproduced from such material just as readily as from seed or propagating material'.[8] For these reasons it was argued that the plant

[6] *Monsanto Technology LLC* v. *Cefetra BU and others* Case (C-428/08) [2011] FSR 6. *Council Directive 98/44 on the legal protection of biotechnological inventions* [1998] OJ L213/13 ('Biotechnology Directive'), which entered into force on 30 July 1988.

[7] UPOV, UPOV Publication No. 316, p. 92.

[8] Committee on Transactions in Seeds, *Plant Breeder's Rights: Report of the Committee on Transactions in Seeds* (London, 1960), p. 40. Although the Committee on Transaction in Seeds did not recommend extending the breeder's right beyond propagating material as

material that is protected by the UPOV Convention should be 'broadened to the commercialised finished product, and generally, to any part of the plant which can, whether on a major or minor basis, be used for propagation'.[9] Specifically, at the Diplomatic Conference of 1957–1961, Spain proposed that protection should be extended to products in the following way:

> Any member State of the Union can, within the framework of its national legislation or by way of agreements with one or several other states, establish for the authors and some kinds of species a much larger right than the one defined in the present article, a right which can extend all the way to the commercialized product.[10]

Despite arguments in support of increasing the scope of protection, there was a reluctance to extend plant variety protection further than the propagating material of protected plant varieties. The number of countries invited to the Diplomatic Conference of 1957–1961 was relatively small and the agricultural and famer lobby groups were well represented, with a strong united voice. As such, States were not willing to extend the scope of protection beyond propagating material because, in part at least, 'the harvested material is frequently an element of the food supply and participating countries were not willing to be required to extend the right of the breeder to the end product of the variety on a mandatory basis'.[11] So while it appears that there was in-principle support for protecting products, any such extension had to be done 'without excessive difficulties'.[12] Extending protection to products was considered a question for individual countries by way of bilateral and multilateral arrangements and was also considered by UPOV Members looking to implement plant variety protection. For example, although the United Kingdom's Committee on Transactions in Seeds recognised the concern over 'consumption material', they did not recommend extending the breeder's right beyond propagating material as this would 'involve an unnecessary degree of interference with production and trade'.[13]

The outcome of these discussions was that a balance was struck between the needs of plant breeders, researchers and growers. Instead of protecting the plant variety as a whole, it was decided that the best way

this would 'involve an unnecessary degree of interference with production and trade', p. 40.

[9] Ibid. [10] UPOV, UPOV Publication No. 316, p. 77.

[11] M. Thiele-Witting and P. Claus, 'Plant variety protection – A fascinating subject' (2003) 25 *World Patent Information* 243, 245.

[12] UPOV, UPOV Publication No. 316, p. 85 (Switzerland).

[13] Committee on Transactions in Seeds, *Plant Breeder's Rights: Report of the Committee on Transactions in Seeds*, p. 40.

to achieve control over the propagation of the reproductive organs was to provide protection over the propagating material. The scope of protection was established by Article 5(1) of UPOV 1961 which stated that:

> The effect of the right granted to the breeder of a new plant variety or his successor in title is that his prior authorization shall be required for the production, for purposes of commercial marketing, of the reproductive or vegetative propagating material, as such, of the new variety, and for the offering for sale or marketing of such material. Vegetative propagating material shall be deemed to include whole plants. The breeder's right shall extend to ornamental plants or parts thereof normally marketed for purposes other than propagation when they are used commercially as propagating material in the production of ornamental plants or cut flowers.

As we have seen, however, limiting protection in this way was far from unanimous and by no means was it a popular outcome. Limiting protection to the propagating material was seen to be a compromise – better than nothing –[14] and meant that the nature of the plant material protected by the UPOV Convention was one of the most contentious issues for UPOV: with further questions about the appropriate scope of protection were raised at the Diplomatic Conference of 1978.

8.3 But Where Does Propagating Material End Up?

It was again the professional organisations such as ASSINSEL, FIS and CIOPORA that led the discussions on the scope of protection at the Diplomatic Conference of 1978. Arguing that Article 5(1) of UPOV 1961 was easily misunderstood, the professional organisations wanted it amended or, at the very least, to be clarified. One argument was that because of the heterogeneous nature of plant breeding, restricting protection to propagating material was not appropriate for all plant breeding activities. CIOPORA argued that restricting protection to propagating material was 'illusory and that the problem arising is not only one of "extending" this right but also of ensuring that the minimum right may be normally exercised'.[15] More specifically, CIOPORA argued that for numerous ornamental species (such as chrysanthemum, carnation and roses), the aim was to produce cut flowers, not propagating material. Furthermore, the trade in cut flowers often occurred across international borders, and, because there were so few UPOV Members, this trade

[14] See, e.g., R. Royon, 'Scope of protection for the breeder' in M. Llewelyn, M. Adcock and M.J. Goode (eds.), *Proceedings of the Conference on Plant Intellectual Property within Europe and the Wider Global Community* (Sheffield Academy Press, 2002).

[15] UPOV, UPOV Publication No. 337(E), p. 90.

often involved countries that were not members of the UPOV Convention.[16]

Plant breeding organisations also argued that the scope of protection provided by UPOV 1961 was deficient because the destination of a crop was not always known at the time of production.[17] While a plant variety may not have been originally produced for the purposes of commercial marketing, as UPOV 1961 required, the plant could end up being commercially marketed in a number of ways, not all of which were intentional or premeditated. As one commentator pointed out, the issue of knowing the destination of the crop was exacerbated because the product of the plant variety contained the 'means of production in a single fused form'.[18]At the Diplomatic Conference of 1978, FIS discussed the example of peas or beans that were being produced for canning.[19] FIS argued that when the production of peas or beans exceeded the canneries' handling capacity, it was not unusual for the canneries to reserve the surplus production for use as seed in the following year. The implication of this was that the destination of the peas or beans was not always known in advance: primarily because it was only after the cannery had 'found that they could not use for canning all the peas or beans produced then they changed the destination of the samples into that of use as seed in the following year'.[20] In a strict sense, then, FIS argued that the canneries were not producing peas or beans 'for the purposes of commercial marketing of the reproductive or vegetative propagating material' as set out by UPOV 1961.

Despite the pleas of the breeder organisations at the Diplomatic Conference of 1978, the Secretary-General of UPOV stated that at least part of the agitation on Article 5 of UPOV 1961 was 'based on misunderstandings'.[21] In addressing this issue, a Working Group on Article 5 was established to 'examine questions with respect to the scope of protection as laid down in Article 5 of the Convention'.[22] However, an obstacle to the success of the argument for extending protection to other plant material was that, while consistent in message, the arguments were specifically related to the interests of particular plant breeders and

[16] CIOPORA proposed the extension of the plant variety right to 'vegetatively reproduced ornamental plants' and to 'plants or parts thereof which are normally marketed for purposes other than propagation': Ibid., 91.

[17] See, e.g., UPOV, UPOV Publication No. 337(E), pp. 112–113, (the International Association of the Plant Breeders for the Protection of Plant Varieties (ASSINSEL)); p. 117 (the International Organization for Horticultural Produces (AIPH)).

[18] D. Rangnekar, *GATT, Intellectual Property Rights, and the Seed Industry: Some Unresolved Problems*, Economics Discussion Paper 96/5 (Kingston University, 1996).

[19] UPOV, UPOV Publication No. 337(E), p. 144 [253]. [20] Ibid.

[21] Ibid., p. 179 [892]. [22] Ibid., p. 120 (*Report of the Working Group on Article 5*, DC/82).

industries, notably the ornamental and cut flower industries. Further, any decision on extending the right was mediated by the overall aim of the Diplomatic Conference of 1978: that is, to make UPOV 1978 more attractive to potential members. As a consequence, it was generally felt that extending the scope of protection would make UPOV less attractive to potential members and that 'an "extension" of the minimum protection provided for in Article 5(1) could compromise ratification of the revised text or accession to it'.[23] A key factor here was the fear that any extension of the plant variety owners' right would impact on the protected uses, most notably the farmers' privilege or farm-saved seed exception which had been read into Article 5 of UPOV 1961.[24] This meant that UPOV Members were not politically supportive of changing either the protected plant material or the protected uses set out by the UPOV Convention.[25]

Presenting its report, the Working Group reiterated that Article 5 was the 'cornerstone' of the UPOV Convention[26] and argued that, in regard to UPOV 1978, there should not be substantive amendments to the wording of Article 5. The Working Group suggested that Article 5 should be rearranged to 'make it more clear that all three activities requiring prior authorisation by the breeder: production for purposes of commercial marketing, offering for sale and/or marketing related equally to the reproductive and vegetative propagating material as such'.[27] The suggestions of the Working Group were accepted by UPOV Members and adopted in Article 5(1) of UPOV 1978. Article 5(1) of UPOV 1978 provides that:

> The effect of the right granted to the breeder is that his prior authorisation shall be required for
> - the production for purposes of commercial marketing
> - the offering for sale
> - the marketing of the reproductive or vegetative propagating material, as such, of the variety.
>
> Vegetative propagating material shall be deemed to include whole plants. The right of the breeder shall extend to ornamental plants or parts thereof normally marketed for purposes other than propagation when they are used commercially as propagating material in the production of ornamental plants or cut flowers.

[23] Ibid., p. 90 (International Association of Plant Breeders for the Protection of Plant Varieties (CIOPORA)).
[24] Saving seed is the topic of Chapter 10.
[25] UPOV, UPOV Publication No. 337(E), pp. 144–150.
[26] Ibid., p. 120 (*Report of the Working Group on Article 5*, DC/82).
[27] Ibid.

While there was some clarification of the protected uses under Article 5 (1), there was no change to the plant material protected by the UPOV Convention.[28] It was not until there was a convergence of the professional interest in extending the plant material protected and the UPOV Members' support for strengthening the UPOV Convention that the reluctance to extend protection disappeared.[29]

8.4 Cascading Rights: Making the UPOV Convention 'More Legal'

While professional breeder organisations continued to argue for extending the scope of protection,[30] the Diplomatic Conference of 1991 was notable for the support of UPOV Members. The proposed text outlined a number of alternatives for extending protection to the harvested material of the protected plant variety and to products made directly from harvested material of the protected variety.[31] This prompted heated discussion at the Diplomatic Conference of 1991 as the scope of protection was one of the main provisions designed to strengthen the position of the plant variety owner. There were two main points for discussion. First, the Spanish delegation argued that any extension to harvested material should be optional. It was argued that this would 'permit the special circumstances of each county – social and political – to be taken into account';[32] nonetheless this proposal received widespread criticism and was quickly rejected. A second point of discussion concerned whether plant variety owners should be granted absolute discretion to exercise their right at any stage of production, namely, either over the protected propagating material, the harvested material or products derived from the harvested material. To this end, the

[28] UPOV 1978 retained the option to extend protection to 'plants or parts thereof' when they are used commercially as propagating material in the production of ornamental plants or cut flowers. Although, few countries appeared to implement the extended version of the right: See, S. Hassan, 'Ornamental plant variety rights: A recent Italian judgement' (1987) 18(2) *International Review of Industrial Property* 219.

[29] UPOV, UPOV Publication No. 337(E), p. 165.

[30] See UPOV, UPOV Publication No. 337(E). Also see, J. Straus, 'Protection of inventions in plants' (1989) 20(5) *International Review of Industrial Property* 619; J. Straus, 'The relationship between plant variety protection and patent protection for biotechnological inventions from an international viewpoint' (1987) 18(6) *International Review of Industrial Property* 723; B. Greengrass, 'UPOV and the protection of plant breeders – Past developments, future perspectives' (1989) 20(5) *International Review of Industrial Property* 622–636; B. Roth, 'Current problems in the protection of inventions in the field of plant biotechnology: A position paper' (1987) 18(1) *International Review of Industrial Property* 41.

[31] UPOV, Publication No. 346(E), pp. 28–31.

[32] Ibid., p. 311.

Japanese delegation suggested a so-called cascade principle so that the relationship between plant variety owners and the users of plant varieties would not be unpredictable and inconsistent.[33]

The cascading principle became central to the extension of the scope of protection under UPOV 1991, which was outlined in Article 14. Significantly, though, the plant variety owners' right remains primarily over the propagating material of the protected plant variety, as set out by Article 14(1)(a) of UPOV 1991:

[*Acts in respect of the propagating material*] *(a)* Subject to Articles 15 and 16, the following acts in respect of the propagating material of the protected variety shall require the authorization of the breeder:

production or reproduction (multiplication),
conditioning for the purpose of propagation,
offering for sale,
selling or other marketing,
exporting,
importing,
stocking for any of the purposes mentioned in (i) to (vi), above.

Although Article 14(1)(a) grants protection to propagating material, it amends the scope of protection in a number of ways. One way in which Article 14(1)(a) differs from Article 5(1) of UPOV 1961 and UPOV 1978 is that it does not limit protection to 'commercial marketing'. The aim of Article 14(1)(a) is to cover 'all the acts we could imagine which could apply to plant varieties'.[34] In doing so it enumerates the exclusive activities over which owners are able to exercise control in relation to the propagating material of protected plant varieties, and, therefore, protection covers most uses of the plant variety.[35] As well as outlining the uses to which protection is granted, UPOV 1991 extends protection in two ways. The first is to harvested material obtained through the unauthorised use of protected propagating material. This is achieved by Article 14(2), which states:

Subject to Articles 15 and 16, the acts referred to in items (i) to (vii) of paragraph (1)*(a)* in respect of harvested material, including entire plants and parts of plants, obtained through the unauthorized use of propagating material of the protected variety shall require the authorization of the breeder, unless the

[33] Ibid., 314–315. The 'cascading' principle was supported by Spain, Canada and Australia. The United States, however, were opposed to the 'cascading' principle.

[34] J. Ardley, 'The 1991 UPOV Convention, Ten Years On' in Llewelyn, Adcock and Goode (ed.), *Proceedings of the Conference on Plant Intellectual Property within Europe and the Wider Global Community*.

[35] Although the right is subject to Article 15 (Exceptions to the Breeder's Right) and Article 16 (Exhaustion of the Breeder's Right) of UPOV 1991.

breeder has had reasonable opportunity to exercise his right in relation to the said propagating material.

Article 14(2) potentially overcomes the problem of knowing the destination of crops at the time of production. As we have seen, in the case of ornamental plant varieties that are bred to produce cut flowers, limiting the right to propagating material was problematic as a plant variety can be taken to a country where no protection is available and then be propagated to produce the end product, in this case cut flowers. The cut flowers can then be exported to countries where the plant variety owner has legal rights. While plant variety rights owners did not have recourse under UPOV 1961 or UPOV 1978 to protect the cut flowers, under UPOV 1991 the plant variety owner can exercise their rights over the cut flowers (as harvested material).

A further (optional) extension to the scope of protection under UPOV 1991 is to products derived directly from the harvested material. The extension of the scope of protection to products derived directly from harvested material generated a great deal of opposition. One concern was that exercising rights in relation to such products was 'impossible or at least extremely difficult' because the identification of the protected plant variety would be difficult.[36] Despite such concerns, it was generally agreed that any problems associated with proof were not sufficient reason to refuse extending the right granted. As a result, UPOV Members were given the option of extending protection to products made directly from harvested material in Article 14(3) of UPOV 1991 which states:

Each Contracting Party may provide that, subject to Articles 15 and 16, the acts referred to in items (i) to (vii) of paragraph (1)(a) in respect of products made directly from harvested material of the protected variety falling within the provisions of paragraph (2) through the unauthorized use of the said harvested material shall require the authorization of the breeder, unless the breeder has had reasonable opportunity to exercise his right in relation to the said harvested material.

By extending protection to products in certain circumstances, Article 14 (3) gives plant variety owners the ability to exercise their rights in relation to products derived directly from harvested material. Although this is only going to be relevant in a small number of cases, it is still important. This is the case, for example, when propagating material is exported without authorisation to a country where it is reproduced and used to produce products which are then imported into the country of origin. In addition, Llewelyn and Adcock suggest that Article 14(3) provides

[36] UPOV, UPOV Publication No. 346(E), p. 406 [1551].

'protection for such derivatives of plants as essential oils (used in the perfume and aromatherapy industries) and medicines (for example, herbal remedies and vaccines)'.[37] Although plant variety protection was extended by Articles 14(2) and (3) of UPOV 1991, the intention was not to provide a right over the entire plant variety or the whole plant. In this way, UPOV 1991 does not extend unconditional protection to harvested material and products derived from harvested material. It was felt that granting unconditional protection would lead to uncertainty and be problematic for the users of protected plant varieties. To overcome this problem, the 'cascading principle' was introduced by Articles 14(2) and (3). This 'cascading principle' qualifies when the scope of protection will be extended beyond the propagating material of the protected plant variety. During the Diplomatic Conference of 1991, the Basic Proposal put forward was that the right would only be extended if there was 'no legal possibility' to exercise their rights at an earlier stage.[38] It was, however, felt that this language would be too narrow, and instead UPOV 1991 adopts the language of 'reasonable opportunity'. More specifically, the plant variety right will only extend to the harvested material or products derived directly from the harvested material if the plant variety owner has not had a 'reasonable opportunity' to exercise their right in relation to the propagating material or the harvested material.[39] Importantly, the concept of 'reasonable opportunity' has imbued the UPOV Convention with legal language.

During the Diplomatic Conference of 1991, there were two notable proposals made on the notion of 'reasonable opportunity'. The Japanese delegation argued that there should be a question of 'due care' when determining the cascading rights found in Articles 14(2) and (3).[40] While the Working Group accepted 'in principle' the requirement of due care, they stated that this was implicitly included in the concept of reasonableness and did not need to be explicitly stated.[41] Another suggestion was to extend the right to harvested material and to products derived directly from the harvested material if it had been 'impossible' to exercise the right in respect of the propagating material.[42] But this would have seriously limited when the plant variety owner could exercise their rights

[37] M. Llewelyn and M. Adcock, *European Plant Intellectual Property* (Hart Publishing, 2006), p. 224.

[38] UPOV, UPOV Publication No. 346(E), p. 30.

[39] There is also another condition the harvested material or product has to be obtained through unauthorised use of propagating material or the harvested material respectively. See UPOV, UPOV/UXN/HRV/1.

[40] UPOV, UPOV Publication No. 346(E), pp. 119–120. [41] Ibid., pp. 144–148.

[42] Ibid., pp. 128 (Spain DC/91/82), 119 (Japan, DC/91/61).

over anything more than the propagating material and was rejected by those present at the Diplomatic Conference of 1991. In adopting Article 14 into UPOV 1991, delegates noted that 'consequences arising from the extended scope of protection in the 1991 Act can be envisaged'.[43] Indeed, since the introduction of Article 14 of UPOV 1991, and its subsequent implementation into national and regional laws, there have been a number of disputes about the meaning of 'reasonable opportunity' and whether protection extends to the harvested material or to products derived directly from the harvested material.

One case where the question of reasonable opportunity arose was in a German Supreme Court decision involving two *Calluna vulgaris* plant varieties: one was protected by a Community plant variety right, the other was protected by a German plant variety right.[44] The holder of plant variety rights brought an infringement action against a German retailer who allegedly infringed the owners' rights by selling both of the *Calluna vulgaris* varieties. In deciding whether plant variety infringement had occurred, the German Supreme Court had to determine whether the alleged infringement fell within the scope of protection of Article 13(3) of the *European Commission Regulation on Community Plant Variety Rights* and Article 10(2) of the German *Plant Variety Protection Law*.[45] These provisions are based on Article 14 of UPOV 1991 and extend protection to the harvested material if the harvested material was obtained through the unauthorised use of variety constituents of the protected variety, and the rights holder did not have a reasonable opportunity to exercise their right in relation to the propagating material of the protected *Calluna vulgaris* plant varieties.

The German Supreme Court concluded that the plant variety owner did not have a 'reasonable opportunity' to act on the propagating material because the plant varieties were not protected in France. This meant that the plants could be reproduced in France where the plant variety rights owner did not have any rights and was only when the plants were sent back to and traded in Germany (as harvested material) that the plant variety owner could exercise their rights. On this basis, the Court concluded that the plant variety owner did not have a 'reasonable opportunity' to exercise their right over the propagating material. In fact, they did not have any opportunity to exercise their rights in relation to the

[43] B. Greengrass, 'The 1991 Act of the UPOV convention' (1991) 12 *European Intellectual Property Review* 467, 470.
[44] No. X ZR 93/04, 14 February 2006.
[45] *Council Regulation (EC) No. 2100/94 of 27 July 1994 on Community Plant Variety Rights*; *German Plant Variety Protection Law 1997*.

propagating material. The decision provides an example of a situation where the plant variety owner has not had an opportunity, let alone a reasonable opportunity, to exercise their rights in relation to the protected propagating material.[46] In such cases, protection is likely to extend to the harvested material.

A second case to consider what constitutes a 'reasonable opportunity' is the Australian case of *Cultivaust* v. *Grain Pool Pty Ltd.*[47] The decision of *Cultivaust* v. *Grain Pool Pty Ltd* considered a range of issues under the *Plant Breeder's Rights Act 1994* (Cth), including whether the plant variety owner – Cultivaust, as licensee of the State of Tasmania – had a 'reasonable opportunity' to exercise its rights over the propagating material. Justice Mansfield interpreted and applied Section 17 of the *Plant Breeder's Rights Act 1994 (Cth)* subject to the 'cascading' principle. Section 17 of the *Plant Breeder's Rights Act 1994*, which deals with plant breeders' rights in relation to harvested material, provides:

> (1) If:
>
> (a) propagating material of a plant variety covered by PBR is produced or reproduced without authorisation of the grantee; and
> (b) the grantee does not have a reasonable opportunity to exercise the grantee's right in relation to the propagating material; and
> (c) material is harvested from the propagating material;
>
> Section 11 operates as if the harvested material were propagating material.
>
> (2) Subsection (1) applies to so much of the material harvested by a farmer from propagating material conditioned and reproduced in the circumstances set out in subsection 17(1) as is not itself required by the farmer, for the farmer's own use, for reproductive purposes.

Justice Mansfield examined the conditions under which Section 14 would operate in relation to farm-saved seed, stating that saved seed would be categorised as 'harvested material' if two factual conditions were satisfied.[48] First, Section 14(1)(a) of the *Plant Breeder's Rights Act 1994* requires that the second or subsequent generation crop not be authorised by the grantee of the *Plant Breeder's Rights Act 1994*. That is, plant breeders holding a plant breeder's right can give permission to farmers to harvest and sell crops grown from farm-saved seed.

[46] The German Supreme Court also concluded that retailers specialising in trading plants have an obligation to check the existence of intellectual property rights and to ensure no infringement of such rights.

[47] (2004) 62 IPR 11. For a discussion see J. Sanderson, 'Back to the future: Possible mechanisms for the management of plant varieties in Australia' (2007) 30(3) *University of New South Wales Law Journal* 686.

[48] *Cultivaust Pty Ltd* v. *Grain Pool Pty Ltd* (2004) 62 IPR 11, 51.

Second, Section 14(1)(b) of the *Plant Breeder's Rights Act 1994* requires that the grantee does not have a reasonable opportunity to exercise a plant breeder's rights in relation to the propagating material.[49] The Federal Court of Australia found that the rights' owner had known that farmers were saving the seed and harvesting second and subsequent generation crops without authority, and so had a reasonable opportunity to exercise their rights. Unfortunately for the rights holder, they had not been able to reach an agreement with the respondent. The court did not go into any further detail on how the plant variety owners would have exercised their rights over the saved seed, other than to say that they had had a 'reasonable opportunity' to do so. In finding that Cultivaust did have a reasonable opportunity to exercise its right in relation to the propagating material, Justice Mansfield identified possible *indicia* of what constitutes a 'reasonable opportunity'. According to Justice Mansfield, the *indicia* of what constitutes a 'reasonable opportunity' were knowing that such crops were being grown and harvested, understanding that the crops were themselves subject to plant variety rights protection and, if relying on another entity (e.g., to obtain some form of payment or royalty), knowledge that there had been no agreement in relation to the payments or royalties on the harvested material.[50] These *indicia* suggest that it might only be in limited circumstances that a plant variety owner does not have a 'reasonable opportunity' to exercise their rights on the propagating or harvested material. In fact, Justice Mansfield stated that '[a]t the time of the initial sale of certified Franklin barley, Tasmania and Cultivaust could have imposed conditions upon the disposition of second and subsequent generations of crop'.[51]

[49] On the facts presented, Justice Mansfield held that Cultivaust and the Tasmanian government did have a reasonable opportunity to exercise their plant breeder's rights in relation to the propagating material leading to each harvest, and therefore s. 14 did not apply. According to Justice Mansfield, the *indicia* of what constitutes a 'reasonable opportunity' were knowing that such crops were being grown and harvested, understanding that the crops were themselves subject to the *Plant Breeder's Rights Act 1994* by reason of Section 14 and if relying on another body (e.g., to obtain end point royalties) knowledge that there had been no agreement: Ibid., 50–51.

[50] *Cultivaust Pty Ltd* v. *Grain Pool Pty Ltd* (2004) 62 IPR 11, 50–51. In 2010, the Advisory Council on Intellectual Property (ACIP) recommended that Australia's *Plant Breeder's Rights Act 1994* (Cth) be amended to clarify that harvested material that is also propagating material is to be considered as propagating material for the purposes of s. 11, even if it is not being used for that purpose: ACIP, *A Review of Enforcement of Plant Breeder's Rights*, 18 January 2010.

[51] Cultivaust appealed the decision to the Full Federal Court. The Full Federal Court expressed doubt about Justice Mansfield's construction of Section 14 due to confusion over the distinction between the primary rights under Section 11 of the *Plant Breeder's*

As these cases show, the notion of 'reasonable opportunity' has imbued the UPOV Convention with legal language and opened the door to disagreements and disputes.[52] They have also provided a means for plant variety rights owners, in some circumstance, to seek a return on harvested material or products derived directly from harvested material. While the German decision suggests that when there is no opportunity to exercise rights over the propagating material it will be a relatively straightforward determination, this is likely to be the outcome in only a small number of cases. It is more likely that plant variety owners have some degree of opportunity to exercise their rights over the propagating material of their protected plant varieties. The question of whether a rights holder had a 'reasonable opportunity' will ultimately depend on the particular facts of the case.

While the requirement of a 'reasonable opportunity' has introduced a level of indeterminacy that plant breeders and plant breeding organisations have had difficulty reconciling, this is not new to law.[53] Expecting certainty on what constitutes 'reasonable opportunity' is unrealistic and ignores the increasingly juridical nature of the UPOV Convention. On the issue of reasonableness, the Australian Plant Breeder's Rights Advisory Committee made the point that there will always be some element of uncertainty:

> The concept of reasonableness is well known to the law. It concerns an objective assessment as to what is or what is not reasonable in all the circumstances. There has never been legislation codifying the meaning of "reasonable" with good reason. For example the circumstances in any two cases are rarely the same.[54]

A second observation that can be made is that the concept of 'reasonable opportunity' facilitates and normalises the use of contracts in UPOV-based plant variety protection schemes. While the freedom to contract was implicit in UPOV 1961 and 1978,[55] Article 14(1)(b) normalises and

Rights Act 1994 and the secondary rights that arise by way of infringement under Section 53(1): *Cultivaust Pty Ltd* v. *Grain Pool Pty Ltd* (2005) 67 IPR 162, 174.

[52] The terms 'propagating material' and 'harvested material' are not defined. Indeed, the United Kingdom delegation to the Diplomatic Conference of 1991 stated that 'in the context of a variety of wheat, the harvested material is usually the seed (rather than the straw), whereas harvested material would include the seed and the straw (which might be valuable for other uses). In addition, the notion of what constituted a 'product' was also unclear: UPOV, UPOV Publication No. 346(E), p. 407 [1555]; Llewelyn and Adcock, *European Plant Intellectual Property*, p. 222.

[53] The issue of indeterminacy in law has been the subject of discussion and debate for some time, with scholars such as Hart, Dworkin, legal realists and critical legal studies: see K. Kress, 'Legal indeterminacy' (1989) 77 *California Law Review* 283; B. Bix, *Law, Language and Legal Determinacy* (Oxford University Press, 1996).

[54] Advisory Council on Intellectual Property, *A Review of Enforcement of Plant Breeder's Rights: Final Report* (2010), p. 46.

[55] UPOV, UPOV Publication No. 346(E), pp. 144–148.

entrenches the contracting or licensing of plant varieties by stating that '[t]he breeder may make his authorization subject to conditions and limitations'.[56] Therefore, while the use of contracts is not necessarily new in the protection of plant varieties, UPOV 1991 places the use of contracts squarely within the contemplation of the UPOV Convention. The result of this is that the notion of a 'reasonable opportunity' – and the indeterminacy contained therein – places an obligation on plant variety owners to enter into contracting or licensing agreements with the users of their plant variety. In this way, there is a strong argument that Article 14 of UPOV 1991 places an obligation on the plant variety owner 'to exercise his existing right at the earliest possible opportunity in the trade chain'[57] and to carefully consider the 'conditions and limitations' upon which they are willing to distribute or trade their plant varieties.[58]

8.5 Patents, Non-Functional DNA and Harvested Material

As well as imbuing UPOV with legal language and normalising the use of contracts and licensing agreements, extending protection beyond the propagating material has had consequences on the relationship between UPOV-based schemes and patent law. The potential of plant variety protection to provide protection where patent law cannot is illustrated by the decision of the European Court of Justice in the case of *Monsanto Technology LLC* v. *Cefetra BU and others*[59] In this case the Court considered the nature and scope of protection granted to genetic information particularly in relation to Article 9 of the European *Biotechnology Directive.*[60]

In trying to manage and control the use of its Roundup Ready® glyphosate tolerant agricultural products in Argentina, and by extension in other countries that do not grant patent protection to the same or similar technology, Monsanto alleged that a number of soy meal

[56] Some delegations wanted this provision to be optional. For example, UPOV, UPOV Publication No. 346(E), pp. 311–314.

[57] ISF, *Implementation of Article 14(2) and 14(3) of UPOV 1991 in Relation to the Phrase 'reasonable opportunity'* (Christchurch, May 2007).

[58] While the basic rules of contract law apply, there are also technical aspects particular to plant breeding and the distribution of plant varieties such as clarifying the scope of the right, restricting uses such as farm-saved seed and establishing payments (royalties) while possibly taking into account reproductive qualities of plant varieties. See J. Kesan, 'Licensing restrictions and appropriating market business benefits from plant innovation' (2005) 16 *Fordham Intellectual Property, Media and Entertainment Law Journal* 1081; M. Janis, 'Supplemental forms of intellectual property protection for plants' (2004) 6 *Minnesota Journal Law Science and Technology* 305.

[59] *Monsanto Technology LLC* v. *Cefetra BU and others* Case (C-428/08) [2011] FSR 6.

[60] *Council Directive 98/44 on the legal protection of biotechnological inventions* [1998] OJ L213/13 ('Biotechnology Directive'), which entered into force on 30 July 1988.

importers had infringed their European Patent 0 546 090 ('Glyphosate Tolerant 5-Enolpyruvylshikimate-3-Phospahte Syntheses').[61] The patent covered thirty-four claims, and the dispute centred on a 'DNA molecule comprising DNA encoding a kinetically efficient, glyphosate tolerant EPSP enzyme', for which there were a total of 34 claims encompassing a range of invention including DNA sequences, isolated DNA sequences, DNA molecules, methods of producing glyphosate tolerant plants and the glyphosate plants.[62] Monsanto brought an action in the Hague District Court, alleging that the import of the patented DNA sequence constituted patent infringement of their European patent based solely on the presence of the DNA molecules that encode the EPSP enzymes in the processed soy meal. In reply to Monsanto's claim, *Cefetra* argued that it was not enough for the patented DNA sequence to be present in the soy meal and that Monsanto had to show that the DNA present in the soy meal also performs its function in the soy meal by making it tolerant to glyphosate. The defendants based this contention on the argument that Article 9 of the *Biotechnology Directive* imposes a limit on the circumstances in which patent protection is extended to products containing patented genetic information. Article 9 of the Biotechnology Directive states:

> [t]he protection conferred by a patent on a product containing or consisting of genetic information shall extend to all material, same as provided in Article 5(1), in which the product is incorporated and in which the genetic information is contained and performs its function.

While the Hague District Court made a number of determinations in relation to the facts in dispute, including that 'DNA cannot perform its function in soy meal, which is dead material',[63] the Court referred a number of questions of law to the Court of Justice including whether Article 9 of the *Biotechnology Directive*:[64]

> be interpreted as meaning that the protection provided under that provision can be invoked even in a situation such as that in the present proceedings, in which the product (the DNA sequence) forms part of a material imported into the European Union (soy meal) and does not perform its function at the time of the alleged infringement, but has indeed performed its function (in the soy plant)

[61] European Patent 0 546 090, 'Glyphosate Tolerant 5-Enolpyruvylshikimate-3-Phospahte Synthases'.
[62] Ibid.
[63] *Monsanto Technology LLC* v. *Cefetra BU and others* Case (C-428/08) [2011] [2011] FSR 6, [37].
[64] For a discussion of the four questions referred to the Court of Justice see M. Kock, 'Purpose-bound protection for DNA sequences' (2010) 5 *Journal of Intellectual Property Law and Practice* 495.

or would possibly again be able to perform its function after it has been isolated from that material and inserted into the cell of an organism?[65]

The Court of Justice interpreted Article 9 of the *Biotechnology Directive*, as well as the *Biotechnology Directive* more broadly as it applied to national patent law. In so doing the Court of Justice emphasised that patent protection extends to material in which a patented DNA sequence is contained only if two conditions are satisfied: first, the patented genetic information such as a DNA sequence must be present in the product; and second, the genetic information must perform the function described in the patent. In relation to the second requirement, the court pointed out that 'neither art.9 of the Directive nor any other provision thereof accords protection to a patented DNA sequence which is not able to perform its function'.[66] The Court of Justice then applied this to the facts before them. In the court's view, the first of these requirements was satisfied as there was no disputing that the imported soy meal contained the patented DNA sequence. In answering the question of whether the DNA sequence was performing its function, the Court of Justice pointed out that the patented DNA sequence only performs its function if it makes the soy meal tolerant to glyphosate. The court concluded that the use of herbicide on soy meal is not 'foreseeable, or even conceivable', and even if it was used, it could not perform its function because the soy meal is dead material.[67] As a consequence, Monsanto's European Patent 'Glyphosate Tolerant 5-Enolpyruvylshikimate-3-Phospahte Syntheses' did not protect the imported soy meal, and Monsanto could not prevent the soy meal being imported into the European Union merely because it contained, in a residual state, a patented gene sequence.

The *Monsanto* v. *Cefetra* decision is important for numerous reasons. First, the decision creates greater certainty and consistency in relation to plant patents.[68] Had the Court of Justice allowed Monsanto's patent to extend to the DNA sequence irrespective of function, patentees would be in a position to control the market of products obtained from patented genetic information regardless of whether the products performed the claimed function. As Advocate General Mengozzi pointed out, to allow such products to be protected would contradict a number of principles

[65] *Monsanto Technology LLC* v. *Cefetra BU and others* Case (C-428/08) [2011] [2011] FSR 6, [50].

[66] Ibid., [49]. [67] Ibid., [37].

[68] Contrast with G. Van Overwalle, 'The CJEU's Monsanto soybean decision and scope – Clear as mud' (2011) 42(1) *International Review of Intellectual Property and Competition Law* 1.

that underpin and inform modern patent law.[69] The first of these principles is the notion that a patent is a social contract between the patentee and the public in which the patentee must disclose how to reproduce the invention in return for gaining a temporary monopoly. The Advocate General stressed that granting patent protection to genetic information including DNA sequences, irrespective of the function that the genetic information performs, would breach this social contract because the invention is not being disclosed by the patentee. The second principle sustained by a purposive reading of patent claims is the dichotomy between a discovery and an invention, where on the one hand, isolating a DNA sequence independent of the DNA's function is a discovery (and is not patentable), while on the other hand, indicating a function of the DNA sequence is generally accepted as an invention (which is patentable).

Second, the European Court of Justice's decision may have an effect on patent claim construction. Rather than claiming rights over the soy meal itself, Monsanto's principle claim was for the protection of the patented DNA sequence contained in the soy meal, which Monsanto alleged was protected under Article 53 of the Netherlands Law on Patents 1995. As a result of this decision, in the future patentees will be more likely to directly claim products such as plants or products such as soy meal without reference to a particular DNA sequence. A successful claim to a product per se is a claim to that product however it is made, which means that any subsequent use of the product is an infringement regardless of how the product is obtained. In addition to claiming products, patentees may construct claims in such a way as to include product-by-process claims whereby a product is defined with reference to the process by which it is made.[70] As we have seen in Chapter 5, this may depend on how national patent laws treat the concept of plant variety.

Yet another reason why *Monsanto* v. *Cefetra* is important is because it highlights the significance of the protection of harvested material under UPOV 1991. One of the most often cited advantages of plant variety rights schemes is that they provide a more targeted and balanced approach to the protection of plant varieties. On this basis alone, plant variety schemes are often a more attractive option for countries that must 'provide for the protection of plant varieties either by patents or by an

[69] Opinion of A.G. Mengozzi, *Monsanto* v. *Cefetra* (C-428/08) [2011] FSR 6, [AG31]–[AG32].

[70] See, e.g., G. Grant, 'The protection conferred by product-by-process claims' [2010] *EIPR* 635.

effective *sui generis* system or by a combination thereof' under their obligations pursuant to Article 27.3(b) of TRIPS. Indeed, as a consequence of the *sui generis* nature of plant variety rights, plant variety rights schemes may be available in circumstances where patent protection is not. This is the case in Argentina, for example, where some form of plant variety rights protection has been in place since 1973 and where the Argentine government became a member of the UPOV Union in 1994.[71] So, in the case of *Monsanto* v. *Cefetra*, it is possible that plant variety protection of the patented soybean may have entitled the rights holder to a 'reasonable opportunity' to exercise their rights in relation to the propagating material. While patents protect a wide range of inventions – covering a range of plant-related innovations including gene sequences, promoters and plant breeding methods – patent protection is not absolute and does not automatically extend to plants or other products containing the patented genetic information. As a consequence, genetic information such as a DNA sequence and products containing the DNA sequence may be left unprotected, and a patentee may not have a right against the importation of harvested goods produced outside the European Union that are produced using the genetic information.[72] In clarifying this point, the decision of *Monsanto* v. *Cefetra* serves as a useful reminder of the important role of the UPOV Convention in the protection of plant-related innovations and highlights why plant variety rights offer a viable alternative or conjunctive intellectual property protection to that granted under patent law, even for plants containing patented genetic information.

As noted earlier, the extension of protection under the UPOV Convention is based on a cascading principle and is conditional upon two factors. First, the harvested material or products obtained directly from the protected variety need to be obtained through the unauthorised use of the protected plant variety. The requirement that there is the unauthorised use of the protected plant variety is relatively straightforward in situations similar to those in *Monsanto* v. *Cefetra* (and assuming there was a valid Community plant variety right) where the owner makes a number of unsuccessful attempts to control and manage the use of their intellectual property directly with farmers and the government, the use of that material is unauthorised. The second requirement for the extension of a Community plant variety right is that the plant variety owner has not had a

[71] Argentina became a member of UPOV 1978 on 25 December 1994. For a discussion of the Argentinian plant variety protection scheme see UPOV, *UPOV Report on the Impact of Plant Variety Protection* (UPOV, 2005), pp. 35–43.

[72] Kock, 'Purpose-bound protection for DNA sequences'.

'reasonable opportunity to exercise his right' in relation to the protected plant material. Importantly, then, in order to determine whether protection extends to harvested material or products obtained from the protected variety, it is necessary to consider what constitutes a 'reasonable opportunity'. The question of whether a plant variety owner has had a 'reasonable opportunity' to exercise their right over the protected plant material was considered by the German Supreme Court in relation to the importation of the garden plant *Calluna vulgaris*. As we saw earlier, the decision provides an example of a situation where the plant variety owner has not had an opportunity, let alone a 'reasonable opportunity', to exercise their rights in relation to the protected propagating material.[73] More specifically the two *Calluna vulgaris* varieties were reproduced and grown in France (where there was no plant variety protection) and were then imported back to Germany to be sold as plants. As a consequence it was only when the plants were sent back to, and traded in, Germany as harvested material that the plant variety owner was able to exercise their rights. Based on this reasoning, it would be difficult if not impossible to successfully argue that a plant variety owner – in a similar position to Monsanto's in which the plants are reproduced, grown and harvested outside the area of protection and then imported back in – ever has a 'reasonable opportunity' to exercise their plant variety right over the protected plant material. In the case of *Monsanto v. Cefetra*, the soy plants were reproduced outside the European Union and were then imported into the European Union as soy meal to be used as animal feed. So, even though Monsanto sought commercial arrangements with the defendant companies, and with the Argentine government, there was little that they could do to manage and control the use of the soy plants containing the patented DNA sequence. If, however, there is a plant variety right over the plant containing the DNA sequence, then it is possible that a rights owner could, given they did not have a 'reasonable opportunity' to exercise their right over the propagating material, be able to exercise their rights over the harvested material once it was returned to the country in which plant variety protection was held.

8.6 Conclusion

In order to gain control over new plant varieties, UPOV 1961 granted protection over the propagating material of protected plant varieties. The

[73] The German Supreme Court also concluded that retailers specialising in trading plants have an obligation to check the existence of intellectual property rights and to ensure no infringement of such rights.

limited nature of protection largely reflected concerns over the implication of granting exclusive rights over food and 'consumer material'. However, the convergence of professional interests and the desire of UPOV Members to strengthen plant variety rights resulted in a broadening of the plant material that is protected under UPOV 1991. More specifically, Articles 14(1) to (3) of UPOV 1991 introduced a 'cascading' right in which plant variety owners can, in certain circumstances, protect harvested material and products derived from harvested material.

The extension of plant protection to harvested material and products derived directly from harvested material has a number of potential consequences for UPOV-based plant variety protection schemes. One of these is that the introduction of the term 'reasonable opportunity' to Articles 14(2) and (3) of UPOV 1991 provides a clear example of the increased use of legal language within the UPOV Convention. Determining whether a plant variety owner has had a reasonable opportunity to exercise their right (over either the propagating material or the harvested material) has introduced a level of indeterminacy to the UPOV Convention. Consequently, the use of contractual and licensing arrangements has been entrenched by UPOV 1991. It even appears that the notion of 'reasonable opportunity' places an obligation on plant variety rights owners to enter into contractual relationships with the users of their protected propagating material. This has increased the possibility of disagreement and legal dispute, as well as a reliance on legal structures and lawyers. Moreover, the extension of protected plant material beyond the propagating material of the protected plant variety has changed the nature, perception and relational connection between the plant variety owner and the users of plant varieties.

Another consequence of extending the scope of protection under the UPOV Convention is that it has given plant variety rights a new lease of life. While plant variety rights have been criticised as outdated and obsolete, in some respects at least, plant variety rights are better able to deal with the protection of plant-related developments. Though further consideration needs to be given to practical, operational and administrative issues – such as eligibility for protection, variety identification and establishing the origin of the plant material – in the push to promote agricultural research and to encourage the development of new plants, it is vital to acknowledge the role that plant variety rights can play, even in cases in which plant varieties contain specific genetic information.

9 Examining and Identifying Essentially Derived Varieties

The Place of Science, Law and Cooperation

9.1 Introduction

Plant breeding is an incremental and iterative process. Plant breeders use and build upon the work of previous plant breeders and rely on existing plant varieties for the initial source of genetic variation. Common plant breeding techniques include targeted selection (where plant breeders cross two closely related parents, each with a different desirable characteristic, and then select the progeny that has both of those characteristics) and induced mutations (where the aim is to stimulate an increase in the frequency of mutation events through radiation or chemical induction).[1] Significantly, the incremental and iterative nature of plant breeding is accommodated in the UPOV Convention, with the so-called breeder's exception allowing protected varieties to remain available to plant breeders developing new plant varieties.[2]

However, the breeder's exception has, in combination with other provisions of the UPOV Convention, caused troubles for plant variety rights holders and concerns for UPOV Members and breeder organisations. Leading in to the Diplomatic Conference of 1991, there were concerns among plant breeders and UPOV Members that when combined with the low threshold of distinctness and limited infringement provisions in UPOV 1978, the breeder's exception allowed, or even encouraged, copying and plagiarism in plant breeding.[3] The combination of which resulted in an unfair advantage to second and subsequent plant breeders and weakened plant variety protection.[4] These concerns were exacerbated by the advent of molecular plant breeding techniques

[1] J. Brown and P. Caligari, *An Introduction to Plant Breeding* (Blackwell Publishing, 2008).

[2] UPOV 1991, Article 15(i)(iii); UPOV 1978, Article 5(3); UPOV 1961, Article 5(3).

[3] For example, the ISF defines plagiarism as 'any act or use of material/technology in a breeding process that purposely makes a close imitation of an existing plant variety': ISF, *ISF View on Intellectual Property* (ISF, 2012), p. 4.

[4] UPOV, *Records on the Diplomatic Conference for the Revision of the International Convention for the Protection of New Varieties of Plants 1991*, Publication No. 346(E) (UPOV, 1992), pp. 331–349, 417–420.

in the 1970s and 1980s as well as unfavourable comparisons to patent law, which did not have a comparable breeder's exception. One of the specific reasons given for introducing the concept of EDVs was to prevent the exploitation of mutations of protected varieties and varieties that had undergone a minor or trivial change in relation to the initial variety- for example, by using biotechnology- without the first plant variety rights holder being able to share in the profits.[5] Since the introduction of EDVs, the identification and examination of EDVs has been elusive and remains one of the key issues for UPOV. In 2013, for example, UPOV held a *Seminar on Essentially Derived Varieties* in Geneva that canvassed some of the technical and legal aspects of essentially derived varieties.[6] Most telling about this gathering of experts was the fact that there was no consensus on how to examine and identify EDVs.

Taking EDVs as its focus, this chapter looks at why the concept of EDVs was introduced into the UPOV Convention and how EDVs are examined and identified. The chapter begins by exploring how derived plant varieties were not only permitted but were granted protection under UPOV 1961 and UPOV 1978 by way of the breeder's exception. The chapter then considers arguments about how the (relatively) low threshold of distinctness and the limited infringement provisions in the UPOV Convention put pressure on the breeder's exception. Further pressure to do something about potential plagiarism by plant breeders resulted from the perception that plant breeders could abuse UPOV-based schemes because subsequent plant breeders had an unrestricted ability to protect derived plant varieties. These concerns over the breeder's exception were exacerbated by the advent of molecular plant breeding techniques and the desire leading up to the Diplomatic Conference of 1991 to strengthen the UPOV Convention and in so doing bring the scope of plant variety protection closer to that found in plant law. While the breeder's exception remained substantially untouched by UPOV 1991, the concept of EDVs was introduced to mediate the breeder's exception and to discourage trivial third party plant breeding. Since the introduction of EDVs, however, the examination and identification of EDVs has remained elusive. A major part of the difficulty for UPOV is that EDVs are a hybrid concept: creating scientific, legal and pragmatic questions. So,

[5] UPOV, *Fourth Meeting with International Organizations: Revision of the Convention*, IOM/IV/2 (UPOV, 1989), pp. 10–12; See also WIPO, *Introduction to Plant Variety Protection under the UPOV Convention* (2003), WIPO/IP/BIS/GE/03/00, [53]–[57].

[6] UPOV, *Seminar on Essentially Derived Varieties*, Publication 358 (Geneva, 2013), www.upov.int/edocs/pubdocs/en/upov_pub_358.pdf.

while scientists have studied essential derivation in various plant species using molecular techniques directed to plant DNA, these have proven incomplete.[7] UPOV and breeder organisations have also contributed to the examination and identification of EDVs. Most notably, the ISF has developed position papers,[8] guidelines[9] and lists of arbitrators for disputes[10] that have attempted to reconcile scientific (quantitative) and legal (qualitative) questions around the concept. Finally, we will also see how the concept of EDVs occupies a legal space and raises distinctly legal issues such as what is the standard of proof and who has the burden of proving that a variety has been essentially derived.[11] Consequently, EDVs cannot and should not be examined and identified purely on quantitative grounds, 'cultural and practical values' are also important when examining and identifying EDVs.[12] Perhaps, then, EDVs are best viewed as 'agreed facts', in which adopted guidelines and arbitration are crucial to the examination and identification of EDVs. And while science can quantify the ways in which plant varieties are the same, it cannot tell us whether that sameness should have any meaning.

9.2 The Breeder's Exception and Fears over Genetic Engineering

A notable feature of the UPOV Convention is that it includes an explicit exception to the rights of the plant breeder known as the breeder's exception.[13] The breeder's exception takes into account the nature of

[7] See, for examples, E. Noli, M. Teriaca and S. Conti, 'Identification of a threshold level to assess essential derivation in durum wheat' (2012) 29(3) *Molecular Breeding* 687; A. Kahler, et al., 'North American Study on essential derivation in Maize: II. Selection and evaluation of a panel of simple sequence repeat loci' (2010) 50(2) *Crop Science* 486; E. Jones et al., 'Development of single nucleotide polymorphism (SNP) markers for use in commercial maize (Zea mays L.) germplasm' (2009) 24(2) *Molecular Breeding* 165.

[8] See, e.g., ISF, *ISF Viw on Intellectual Property*.

[9] See ISF, *Guidelines for Handling a Dispute on Essential Derivation in Ryegrass* (ISF, 2009); ISF, *ISF Guidelines for the Handling of a Dispute on Essential Derivation of Maize Lines* (ISF, 2008); ISF, *Guidelines for the Handling of a Dispute on Essential Derivation in Oilseed Rape* (ISF, 2007).

[10] ISF, *Regulation for the Arbitration of Disputes concerning Essential Derivation (RED)* (ISF, 2015); *Issues to be Addressed by Technical Experts to Define Molecular Marker Sets for Establishing Thresholds for ISF EDV Arbitration* (ISF, 2010); ISF, *List of International Arbitrators for Essential Derivation* (ISF, 2010).

[11] *Danziger* v. *Astée* 105.003.932/01, Court of Appeal, The Hague (2009); *Danziger* v. *Azolay* 1228/03, District Court, Tel-Aviv-Jaffa (2009).

[12] As we see later in the Chapter, this is the language used by the Netherland's Court of Appeal in The Hague: see *Danziger 'Dan' Flower Farm* v. *Astée Flowers B. V.* 105.003.932/01, Court Appeal, The Hague (2009), [21].

[13] UPOV 1991, Article 15(i)(iii); UPOV 1978, Article 5(3); UPOV 1961, Article 5(3).

plant breeding and the social importance of the continued development of new plant varieties and permits the use of protected plant varieties as an initial source of genetic variation. Central in the discussions surrounding the emergence of the UPOV Convention were the particular characteristics and practices of plant breeding. Notably, UPOV 1961 was negotiated in the context of classical plant breeding methods that involved the deliberate and incremental breeding of plant varieties.[14] While the scope of the plant variety right was limited by UPOV 1961 to 'commercial uses' of the protected plant variety, participant countries felt that it was necessary to explicitly permit the use of a protected plant variety for the breeding of subsequent new plant varieties. Thus, in order to use a protected plant variety in subsequent breeding programmes, the authorisation of the plant variety owner of the protected variety was not required. Specifically, Article 5(3) of UPOV 1961 and UPOV 1978 provides:

> Authorization by the breeder or his successor in title shall not be required either for the utilization of the new variety as an initial source of variation for the purpose of creating other new varieties or for the marketing of such varieties. Such authorization shall be required, however, when the repeated use of the new variety is necessary for the commercial production of another variety.

The effect of Article 5(3) of UPOV 1961 and UPOV 1978 was to allow plant breeders to use a protected plant variety in their breeding programmes without needing to obtain the permission of the plant variety rights owner. Furthermore, if the second plant breeder created a new plant variety, and this satisfied the criteria for grant, then they are able to obtain plant variety rights protection without obtaining the permission from the first breeder.[15] While the breeder's exception is accepted as formalising the common practice of plant breeding, since the Diplomatic Conference of 1957–1961, there have been concerns over the fact that UPOV and UPOV-based schemes protect derivative plant varieties. Indeed, the inclusion of a breeder's exception has not always garnered complete support, and questions over providing protection to derived

[14] Plant breeding is often separated into three broad phases: domestication, which is more than 10,000 years old; classical plant breeding, which began in the 1700s; and molecular genetics, which emerged in the 1970s and became a regular breeding technique in the 1990s. It has been suggested that organised farming originated for a number of reasons, including the availability of water, population growth and as a by-product of religious ceremony: see, J. Hancock, *Plant Evolution and the Origin of Crop Species* (CABI Publishing, 2004), pp. 151–153.

[15] To be protected, the second plant variety must satisfy the requirements of grant. See UPOV, *Clarification of the Breeder's Exception in the 1978 and 1991 Acts of the UPOV Convention* (UPOV, 2004).

plant varieties were raised at the Diplomatic Conference of 1957–1961.[16] Most notably, there were concerns about providing protection to ornamental plant varieties that were developed through multiplication. CIOPORA argued that preliminary authorisation by the owner of the initial plant variety should be required when that variety's main characteristics were used in the development of a new plant variety.[17] In addition, the United Kingdom stated that if:

> a new variety obtained from a protected variety is not different enough from the protected variety to justify a particular protection, it would be desirable that the owner of the protected variety is justified as having the commercialisation of this selection considered as a breach of his right.[18]

During the 1980s, concerns over the breeder's exception began to take shape. One of the reasons for this was that plant variety holders had limited infringement provisions available to them under UPOV 1961 and UPOV 1978. So, despite the fact that there was a need to obtain authorisation when 'the repeated use of the new variety is necessary for the commercial production of another variety' (in the last sentence of Article 5(3) of UPOV 1961 and UPOV 1978), this was limited to identical copies of the initial plant variety or, more specifically, to the multiplication of plant varieties when there is no plant breeding effort. The implication was that this had limited effect as there was very little scope to initiate infringement proceedings in relation to plagiarism in plant breeding. Derived plant varieties could be protected in their own right without any recourse for the plant variety owner of the initial plant variety.

In addition to the concerns over a lack of infringement provisions, there were concerns about the interaction of the breeder's exception and the distinctness requirement. As we saw in Chapter 5, Article 6(1)(a) of UPOV 1961 and UPOV 1978 provides that to satisfy the criteria of distinctness a plant variety must be 'clearly distinguishable by one or more important characteristics'. Because the assessment of distinctiveness is based on physically observable 'morphological or physiological' differences, it was felt that this was a relatively easy test to satisfy. Specifically, it was felt that the threshold of distinctness was too low and that the amount of difference required between a protected plant

[16] In the past, the breeder's exception has been questioned by countries such as the United States of America: see GRAIN, *The End of Farm-Saved Seed? Industry's Wish List for the Next Revision of UPOV* (2007), www.grain.org/article/entries/58-the-end-of-farm-saved-seed-industry-s-wish-list-for-the-next-revision-of-upov.

[17] UPOV, *Actes des Conférence Internationales pour la Protection des Obtentions Végétales* 1957–1961, 1972, UPOV Publication No. 316 (UPOV, 1972), p. 92.

[18] Ibid., p. 109.

variety and a second plant variety derived from the first was minimal. Consequently, some UPOV Members and breeder organisations were of the opinion that it was too easy to satisfy the distinctness requirement and gain protection over derived plant varieties. Taken together, the breeder's exception, the limits on what constituted an infringement and the low threshold for distinctness under the UPOV Convention resulted in the perception that plant breeders could abuse UPOV-based schemes because subsequent plant breeders had an unrestricted ability to protect derived plant varieties. This, in effect, diluted the protection granted to plant variety rights holders.

Despite these concerns, up until the 1980s, it was generally felt that the incremental (i.e. time-consuming) nature of plant breeding ameliorated the impact of the breeder's exception, providing a practical barrier to the abuse of the breeder's exception. As the term suggests, incremental plant breeding takes time, effort and resources. Take, for example, the practice of targeted selection: where plant breeders might cross two parents, each with a different desirable characteristic; the plant breeder then selects the progeny that display both of the desirable characteristics. As the initial cross combines thousands of genes and the progeny may not receive the exact combination of genetic material that the breeder intended, crosses need to be carried out repeatedly over a number of generations. Due to the time lag associated with incremental plant breeding, the breeder's exception was not seen to have a detrimental impact on the plant breeders' rights. Leading up to the Diplomatic Conference of 1991 however, this practical barrier was challenged by two factors: the advent of molecular plant breeding techniques in the 1970s and 1980s and the absence of a comparable breeder's exception under patent law.

9.2.1 *Molecular Breeding Techniques, Tipping the Balance in Favour of Subsequent Plant Breeder*

The practical barrier of incremental plant breeding to potential plagiarism was, theoretically at least, lessened by the emergence of molecular breeding techniques. By unravelling the double helix structure of deoxyribonucleic acid (DNA) in the mid-1950s, James Watson and Francis Crick laid the foundation for better understanding how inherited characteristics were reproduced. Building on these discoveries, plant breeders were given the means to directly manipulate plants at a molecular level. In the lead up to the Diplomatic Conference of 1991, the molecular technique that was most prominent was genetic engineering: the use of recombinant DNA technology marked an important shift in plant breeding methodology. Using genetic engineering, plant breeders could identify a DNA sequence for a particular characteristic and then attempt to transcribe and transpose that particular characteristic into another organism.[19] Without overstating the use of these techniques in plant breeding, a number of advantages of genetic engineering were evident. One advantage was that genetic engineering enabled plant breeders to select the desirable characteristic that they were trying to breed into the new plant variety. This allowed plant breeders to achieve with relative certainty, in one generation, what might have otherwise required years using classical incremental breeding methods. A second advantage of genetic engineering was that it enabled plant breeders to transfer genetic material between unrelated species and kingdoms: such as between bacteria and plants. An example of this can be found in 'Bt Cotton', 'Bt Canola' and 'Bt Maize' in which new plant varieties had a single insect-resistance gene from the bacterium *Bacillus thuringiensis* transferred into them to produce plants that are resistant to insects.[20]

Despite the advantages offered by genetic engineering, there were concerns at the Diplomatic Conference of 1991 about the effect of such techniques on the nature of derived plant varieties. It was felt that genetic engineering and other molecular techniques had the potential to make it easier for subsequent plant breeders (i.e. competitors) to make minor or trivial adaptations to protected plant varieties. This meant that plant breeding did not have to be incremental in the same way of classical plant breeding, as the hard work of classical plant breeding had given way to the seemingly simple process of inserting a new gene into an existing plant variety. More specifically, the fear was that new technologies such as genetic engineering made it (relatively) easy to copy existing plant varieties, and for second and subsequent plant breeders to add very little of their own labour, skill and effort while still being eligible for plant variety rights protection. On this basis it was argued that the breeder's exception was open for abuse by fellow plant breeders who, after making only slight variations to the genetic structure of a new plant variety, could claim to have bred a new plant variety.[21] A hypothetical example of this is a protected orange variety that has been used to create a new orange

[19] The two main methods of producing transgenic plants are by transporting the DNA into the plant cell via the bacterium *Agrobacterium thuringiensis* or by shooting the DNA through the cell wall using biolistics: J. Dunwell, 'Review: Intellectual property aspects of plant transformation' (2005) 3 *Plant Biotechnology Journal*, 371.

[20] See, e.g., M. Metz, *Bacillus thuringiensis: A Cornerstone of Modern Agriculture* (Food Products Press, 2003).

[21] J. Guiard, 'Essential derivation: For what?' in A. Llewelyn and M. Goode (eds.), *Conference on Plant Intellectual Property within European and the Wider Global Community* (Sheffield Academy Press, 2002).

variety which is identical in shape, colour and taste to the protected variety. Generally, even if the difference between the two varieties is that the new variety has a different leaf colour, it is eligible for plant variety rights protection because it is distinct from the protected variety. As the President of the Council of UPOV stated '[d]uring the 80s, it became obvious that the influence of biotechnology on breeders' rights, and on the UPOV Convention, would become stronger and stronger'.[22] More significantly, breeding organisations felt that the advent of genetic engineering tipped the balance of plant breeders' rights in favour of subsequent second breeders because it was possible that a new plant variety which made a slight adaptation to an existing plant variety would be granted a plant variety right.[23]

9.2.2 *The Availability of Patent Protection*

Concerns over the impact of new technologies on the existing provisions of the UPOV Convention were amplified by a threat to the legitimacy of the UPOV Convention by patent law. One of the most frequently cited factors in the shift towards the use of patents to protect plant development is that patent rights are subject to fewer exceptions than alternative forms of protection.[24] In contrast to the position under the UPOV Convention, a cornerstone of which is the broad breeder's exception, any use of a patented invention without the prior authorisation of the patentee will constitute *prima facie* infringement of the patent. That said, the patent laws of most jurisdictions contain a limited defence which exempts acts done for *bona fide* experimental purposes from infringement, with the nature and scope of permitted experimental uses differing from one jurisdiction to another. In general terms however, the research exception in patent law is narrowly construed and there tends to be two approaches to the treatment of experimental use that can be discerned among the major patent systems.[25]

[22] UPOV, Publication No. 346(E), p. 164 (Address by the President of the Council of UPOV, Mr. Duffhues).

[23] ASSINSEL, *Consolidation of ASSINSEL Position Papers on Protection of Biotechnological Inventions and Plant Varieties* (1999), www.worldseed.org/ pdf.pos2_ass.pdf.

[24] The American Seed Trade Association has stated the 'open access to germplasm allowed under UPOV for breeding immediately upon commercialisation has the effect of diminishing the developer's opportunity to earn a competitive return on research investments': American Seed Trade Association, *Position Statement on Intellectual Property for the Seed Industry* (2004), www.amseed.com/newsDetail.asp?id=97.

[25] Article 30 of TRIPS allows member countries to provide limited exceptions to the exclusive rights conferred by a patent, provided that such exceptions do not unreasonably conflict with a normal exploitation of the patent and do not unreasonably

In the United States, the experimental use defence is regarded as 'truly narrow' and has no application where the use has 'the slightest commercial implication' or where the act is done 'in furtherance of the alleged infringer's legitimate business interests'.[26] On this view only acts performed 'for amusement, to satisfy idle curiosity, or for strictly philosophical inquiry' are exempt from infringement.[27] In Europe, the *European Patent Convention 1973* stipulates the grounds upon which a European patent may be granted, leaving individual members to determine what constitutes infringement of a European patent. As Europe currently does not have a single patent system, each European country has its own rules that determine the scope of the research exemption.[28] This, however, will change once the unitary patent system and the Unified Patent Court (UPC) system come into force.[29] In accordance with Article 27(b) of the Agreement on a Unified Patent Court (UPC Agreement),[30] the rights conferred by a patent do not extend to 'acts done for experimental purposes relating to the subject-matter of the patented invention'.

While questions over the exact nature of experimental use in patent law remain, there have been a number of attempts to introduce into patent law a breeder's exception of equal or comparable scope to that enjoyed by plant breeders under UPOV-based schemes. The ISF stated in its 2003 position paper on intellectual property that:

> further clarification is needed as regards the use of transgenic varieties containing patented elements and protected by Breeder's Right for further breeding. ISF is strongly attached to the breeder's exception provided for in the UPOV

prejudice the legitimate interests of the patent owner, taking account of the legitimate interests of third parties. The three-step test from Article 9(2) of the *Berne Convention* establishes the criteria against which exceptions to the right to reproduce copyright material are to be assessed. The steps are (1) that reproductions may be permitted in special cases, but (2) are not to conflict with normal exploitation of the work, and (3) must not unreasonably prejudice the legitimate interests of the copyright owner.

26 *Madey* v. *Duke University* 307 F. 3d (Fed. Cir. 2002) 1351, 1362. Significantly, the United States Court of Appeals for the Federal Circuit held in *Madey* v. *Duke University* that the activities of universities are inherently commercial and, as such, use of patented inventions by academic scientists and researchers in the United States will generally not be exempt from infringement under United States patent law.

27 Ibid., 1362. See also *Roche Products Inc.* v. *Bolar Pharmaceutical Co Inc.* 733 F. 2d 858 (1984) 862, 863.

28 To date, there have been few decisions in which the scope of this exception has been considered. However, the scope of the experimental use exception under German patent law. See, e.g., *Clinical Trials I* [1997] RPC 623 and *Clinical Trials II* [1998] RPC 423 and W. Cornish, 'Experimental use of patented inventions in European community states' (1998) 29 *International Review of Industrial Property and Copyright Law* 735.

29 See, e.g., T. Cook, 'Update on the unitary patent court and the European patent with unitary effect' (2015) 20(3) *Journal of Intellectual Property Rights*, 185.

30 *Agreement on a Unified Patent Court* (UPC Agreement), 2013/C 175/01.

Convention and is concerned that the extension of the protection of a gene sequence to the relevant plant variety itself could extinguish this exception. Therefore ISF considers that a commercially available variety protected only by Breeder's Rights and containing patented elements should remain freely available for further breeding. If a new plant variety, not an essentially derived variety resulting from that further breeding, is outside the scope of the patent's claims, it may be freely exploitable by its developer. On the contrary, if the new developed variety is an e.d.v. or if it is inside the scope of the patent's claims, consent from the owner of the initial variety or of the patent must be obtained.[31]

To date however, only France and Germany have introduced a breeder's exception of comparable scope to that provided under the UPOV Convention into their patent laws. In both countries, patent protection on a biological product 'does not extend to acts done for the purpose of creating, or discovering and developing other plant varieties'.[32] Few other countries have indicated any interest in following in France and Germany's footsteps, and it has been suggested that an experimental use exception is inconsistent with Article 30 of TRIPS which states that:[33]

Members may provide limited exceptions to the exclusive rights conferred by a patent, provided that such exceptions do not unreasonably conflict with a normal exploitation of the patent and do not unreasonably prejudice the legitimate interests of the patent owner, taking account of the legitimate interests of third parties.

The cumulative effect of the new technologies and the possibility of patent protection (generally absent a breeder's exception) were that questions over the legitimacy of the UPOV Convention were raised. More specifically, questions were raised about the liberal approach to derived plant varieties found in UPOV 1961 and UPOV 1978. It was felt that plant variety rights owners needed an avenue to challenge subsequent plant breeders because the 'costs of deploying the new technologies and the costs of developing and producing varieties' required that plant breeders' rights be strengthened to meet the changing needs of the plant breeding industry.[34] In addressing these concerns, UPOV

[31] ISF, *ISF View on Intellectual Property*. However, the Plant Committee of the Intellectual Property Law Section of the American Bar Association has emphatically rejected this proposal: see The Plant Committee, *Annual Report: Committee No. 1002* (1997–1998), www.abanet.org/intelprop/summer2004/1002.pdf.

[32] H. Ghijsen, 'Access to germplasm for further breeding in the case of patents and after-sale conditions', (2005/2006) 2 *Bio-Science Law Review* 87, 90.

[33] See, e.g., E. Misati and K. Adachi, *The Research and Experimental Exceptions in Patent Law: Jurisdictional Variations and the WIPO Development Agenda*, UNCTAD–ICTSD Project on IPRs and Sustainable Development, Policy Brief Number 7 (2010), www.unctad.org/en/docs/iprs_in20102_en.pdf.

[34] UPOV, Publication No. 346(E), p. 165.

1991 made significant amendments to UPOV 1978 including the expansion of protection to all plant genera and species, broader rights for plant breeders and the extension of the materials covered by plant breeders' rights to include harvested materials; and removing the ban on dual patent and plant breeders' rights protection for plant varieties was removed.[35] In relation to the breeder's exception, Article 15(1) of UPOV 1991 sets out particular uses that do not require the breeder's authorisation, including:

(i) acts done privately and for non-commercial purposes;
(ii) acts done for experimental purposes; and
(iii) for the purpose of breeding other varieties.

9.3 Essential Derivation

While Article 15(1) does not substantially change the breeder's exception, the introduction of EDVs had the effect of reducing the impact of the breeder's exception by expanding the scope of protection granted to plant variety rights owners. Now, the breeder's exception needs to be read in light of EDVs. Article 14(5)(a)(i) of UPOV 1991 extends the scope of protection to 'varieties which are essentially derived from the protected variety, where the protected variety is not itself an essentially derived variety'. This means that a plant variety that is 'essentially derived' from a protected variety cannot be exploited without the authorisation of the breeder of the protected variety. According to UPOV 1991, there are two steps in assessing whether a plant variety is essentially derived. The first is administrative in nature, as the putative essentially derived variety must be registered under the relevant plant variety protection scheme. As such the putative essentially derived variety satisfies the distinct, uniform and stable criteria. The second step, of assessing whether the second variety is essentially derived, is more problematic. Article 14(5)(b) of UPOV 1991 provides three (cumulative) criteria that must be satisfied for a plant variety to be declared essentially derived. These are that the protected variety:

(i) is predominantly derived from the initial variety, or from a variety that is itself predominantly derived from the initial variety, while retaining the expression of the essential characteristics that result from the genotype or combination of genotypes of the initial variety;
(ii) is clearly distinguishable from the initial variety; and

[35] B. Greengrass, 'The 1991 Act of the UPOV convention' (1991) 12 *European Intellectual Property Review* 467.

(iii) except for the differences which result from the act of derivation, [the variety] conforms to the initial variety in the expression of the essential characteristics that result from the genotype or combination of genotypes of the initial variety.

To be essentially derived, a second plant variety must be 'predominantly derived' from the initial variety, be clearly distinguishable from the initial variety and express genetic conformity to the initial variety in the expression of the essential characteristics that result from the genotype. While all three criteria must be satisfied before a declaration of essential derivation is made, there are no clear definitions for the terms 'predominantly derived', 'essential characteristics' or 'conforms to the initial variety'. What is important, though, is being able to distinguish between two levels of 'relatedness': that is, those varieties that are 'derived' which are allowed and those that are 'essentially derived', which are not allowed.

Making a determination about whether a plant variety is essentially derived involves a comparison of the putative essentially derived plant variety with the initial plant variety. Where the two plant varieties are identical, the second breeder has clearly infringed. Problems arise, however, where there is only a slight difference between the initial plant variety and the putative essentially derived plant variety. This would be the case, for example, where the second plant breeder only adds one characteristic to the initial variety.[36] To assist in the assessment of whether a plant variety is essentially derived, Article 14(5)(c) of UPOV 1991 provides a non-exhaustive list of breeding techniques that may result in an essentially derived variety. These include new plant varieties that are obtained from the originating variety by selection of a natural or induced mutant, a somaclonal variant, the selection of a variant individual from plants of the initial variety, backcrossing or transformation by genetic engineering. Importantly, though, this list is not determinative, and it is left to member states to make declarations of essential derivation.

Since the introduction of EDVs, there have been a number of attempts to establish a framework in which to examine and identify EDVs.[37] Although it was accepted from the outset that this would be a difficult task. For example, at the Diplomatic Conference of 1991, the Japanese delegation said that 'from a technical point of view it was rather difficult to decide what was an essentially derived variety and what was not'.[38]

[36] See, e.g., R. Hunter, *Essentially Derived and Dependency: Some Examples* (1999), www.cdnseed.org/pdfs/press/hunter.pdf.

[37] The Australian government commissioned a report which deals with similar issues: Expert Panel on Breeding, *Clarification of Plant Breeding Issues under the Plant Breeder's Rights Act 1994* (2002), 5, www.anbg.gov.au/breeders/index.html 5.

[38] UPOV, Publication No. 346(E), p. 346, [1119], Mr. Hayakawa (Japan).

Due to these difficulties, the Japanese delegation proposed that standard guidelines on essentially derived varieties be established 'to enable each Contracting Party to implement the provisions relating to essentially derived varieties'.[39] To this end, a resolution was adopted by the Diplomatic Conference on Article 14(5) of the 1991 Act that said:

> The Diplomatic Conference for the Revision of the International Convention for the Protection of New Varieties of Plants held from March 4 to 19, 1991, requests the Secretary-General of UPOV to start work immediately after the Conference on the establishment of draft standard guidelines for adoption by the Council of UPOV, on essentially derived varieties.[40]

However, developing rules for the examination and identification of EDVs was far from straightforward. And while the UPOV Administrative and Legal Committee (CAJ) worked on producing standard guidelines on the subject of essential derivation between 1991 and 1993, they decided to postpone their work. One reason for the postponement of this work was that under Article 37(3) of UPOV 1991 there were various 'grace periods' for developing countries in relation to the implementation of UPOV 1991.[41] This was in consideration of the fact that many developing countries had no plant variety protection and would, therefore, face the most resistance to plant variety protection.[42] Further, the CAJ felt that it would be pertinent to wait for practical experience in relation to essential derivation.[43] In the following paragraphs, we will see that a large part of the difficulty in determining how to examine and identify EDVs is that EDVs are very much a hybrid concept and process, a concept and process that involves at least part science (quantification), law (qualification) and cooperation.

9.3.1 *Genetic Conformity, Coefficients, Indices and Percentages*

Unlike the examination of distinctness, the examination of EDVs is not confined to phenotypic differences – quite the opposite in fact. The focus of much of the work into EDVs has focused on establishing quantifiable

[39] Ibid. [40] Ibid., p. 349.

[41] UPOV 1978 was left open for accession by developing countries until 31 December 1995 and by other countries until 31 December 1993.

[42] N. Byrne, *Commentary on the Substantive Law of the 1991 UPOV Convention for the Protection of Plant Varieties* (University of London, 1996).

[43] UPOV, Personal Communication with UPOV (17 January, 2016). Despite the lack of legal precedent, the Administrative and Legal Committee has decided to develop information materials concerning the 1991 Act of the UPOV Convention. This will include material on essentially derived varieties under Article 14(5) of UPOV 1991 and gives recognition to the fact that plant breeders do, in actuality, need guidance in relation to the essential derivation concept.

scientific approaches to determining EDVs or more specifically to assess genetic conformity. In 1999, ASSINSEL called for the development of 'scientifically reliable criteria' to identify EDVs. In doing so, ASSINSEL stated that the plant industry would 'probably' have to use distance coefficients to define EDV thresholds. In this sense ASSINSEL followed the scientific literature on essential derivation that favours an approach that assesses genetic conformity (similarity) between plant varieties using molecular markers. For each of the substantive criteria of Article 14(5)(b), ASSINSEL believed that the following would have to be considered:

> (i) clear distinctiveness: decided on by the office in charge of granting a right to the breeder of the variety, according to the UPOV rule of distinctness. (ii) conformity to the initial variety: could be based on reliable phenotypic characteristics and/or on reliable molecular characteristics: either close relationship in general which could lead to a "conformity threshold" parallel to the minimum distance threshold used for distinctness or only small differences in some simply inherited characteristics. If this second criterion is considered as fulfilled, then, we have to assess the third one, which is "predominant derivation from an initial variety".
> (iii) predominant derivation: implies that the initial variety or products essentially derived therefrom have been used in the breeding process. In order to prove that use, various criteria or a combination thereof may be used: combing ability, phenotypic characteristics, and molecular characteristics.

The notion of genetic conformity came to the fore in 2003, when the ISF published its *View on Intellectual Property* and in 2005, with the publication of *Essential Derivation: Information and Guidance to Breeders*.[44] The ISF said that in determining whether a plant variety was essentially derived, it was necessary to estimate the genetic conformity between the initial plant variety and the subsequent plant variety on the basis of straightforward characteristics such as molecular markers, phenotypic traits and combing abilities. If the estimate for genetic conformity exceeds a 'generally accepted threshold', there is a *prima facie* case of essential derivation. It follows that in the case where there is a *prima facie* case of essential derivation, the owner of the putative essentially derived variety, the second breeder, must show that there has been no plagiarism by submitting, among other evidence, the breeding records of the new plant variety. The ISF's *View on Intellectual Property* shows a definite move towards the use of genetic conformity and the need to establish 'generally accepted thresholds' as the cornerstone of determining essential derivation.

[44] ISF, *ISF View on Intellectual Property*.

The focus on genetic conformity and similarity in EDVs is based on the notion that genetic differences between plant varieties find their basis in variation between DNA sequences.[45] Genetic similarities in relation to EDVs are obtained using molecular markers that allow DNA segments to be tracked from parents to their progeny. In this way, molecular markers are genetic signposts that highlight the presence of genes that control particular plant characteristics. To date, the investigation of genetic similarities has been carried out on a number of different plant species and varieties including maize,[46] roses,[47] *Calluna vulgaris*[48] and wheat.[49] For example, in 1999, the ISF established a working group to study the implementation of essential derivation for lettuce. The ISF's Vegetable and Ornamental Section conducted a study in 2001 and 2002 that evaluated the genetic similarity of three varieties of lettuce based on molecular markers. These were Butter Head Greenhouse, Butter Head Field Summer and Iceberg. Using statistical analysis, van Eeuwijk and Law proposed a trigger point to initiate discussions between the plant breeders of the initial plant variety and the putative essentially derived plant variety.[50] The researchers used a Jaccard statistical distance to assess the proportion of characters that matched between the varieties of lettuce. The study proposed a threshold of 0.96 Jaccard. That is, when a putative essentially derived variety measures a genetic distance of greater than 0.96, it is presumed to be essentially derived. If this is the case, it is assumed that the breeders of these varieties of lettuce should begin discussions with one another. These findings were set out in *Guidelines for the Handling of a Dispute on EDV in Lettuce.*[51] In another

[45] I. Roldán-Ruiz et al., 'A comparative study of molecular and morphological methods of describing relationships between perennial ryegrass (Lolium perenne L.) varieties' (2001) 103(8) *Theoretical and Applied Genetics* 1138.

[46] Y. Rouselle et al., 'Study on essential derivation in Maize: III. Selection and evaluation of a panel of single nucleotide polymorphism loci for use in European and North American germplasm' (2015) 55(3) *Crop Science* 1170.

[47] B. Vosman, D. Visser, et al., 'The establishment of "essential derivation" among rose varieties, Using AFLP' (2004) 109 (8) *Theoretical and Applied Genetics* 1718.

[48] T. Borchert, J. Krueger and A. Hohe, 'Implementation of a model for identifying essentially derived varieties in vegetatively propagated *Calluna vulgaris* varieties' (2008) 9 *BMC Genetics* 56.

[49] Noli, Teriaca and Conti, 'Identification of a threshold level to assess essential derivation in durum wheat'.

[50] F. van Eeuwijk and J. Law, 'Statistical aspects of essential derivation, with illustrations based on lettuce and barley' (2004) 137 *Euphytica* 129.

[51] ISF (ISF), *Guidelines for the Handling of a Dispute on Essential Derivation in Lettuce* (2004). The International Seed Federation has also developed guidelines for maize, perennial ryegrass, cotton and oilseed rape: see ISF, *Essential Derivation* (2010), www.worldseed.org/isf/edv.html.

study, Heckenberger et al. investigated genetic similarity in maize. The French maize seed industry adopted a code of conduct which uses the genetic thresholds found by Heckenberger et al.[52] The code of conduct stipulates three ranges to be used in disputes that involve maize. If the putative essentially derived variety is above the threshold of 90 per cent, the variety is considered an essentially derived variety; between 82 and 90 per cent, there is possible essential derivation and the parties have to negotiate; and below 82 per cent, there is no essential derivation.[53]

While the scientific community has been trying to provide a scientific method for examining and identifying EDVs, the question of essential derivation is fundamentally a legal one: that is, how different do plant varieties need to be? Because of this, scientific attempts to determine essential derivation are fraught with complications. Science, by its very nature, relies on quantitative data to make comparisons between plant varieties. This means that the nature and range of genetic thresholds may vary according to a number of factors including the particular plant species and the research method used. The use of genetic distances to determine essential derivation is not suitable for all plant varieties as there is a need for phenotypic variations (that are well known) to be based on genotypic variation.[54] Furthermore, a plant species may have a narrow genetic base and therefore have inherently high genetic similarities.[55] One plant that has small genetic diversity compared to other plants is cotton (*Gossypium*).[56] In one study, it was found that the average genetic similarity of a number of cotton varieties was 89.55 per cent.[57] Where this occurs, it is extremely difficult to establish reliable standard

[52] M. Heckenberger, M. Bohn, D. Klein and A.E. Melchinger, 'Identification of essentially derived varieties obtained from biparental crosses of homozygous lines: II Morphological distances and heterosis in comparison with simple sequence repeat and amplified fragment length polymorphism data in maize' (2005) 45 *Crop Science* 1132. See also J. Dendauwa, et al., 'Variety protection by use of molecular markers: Some case studies on ornamentals' (2000) 135(1) *Plant Biosystems* 107.

[53] ISF, *Guidelines for the Handling of a Dispute on Essential Derivation in Lettuce* (2004).

[54] F. van Eeuwijk and C. Baril, 'Conceptual and statistical issues related to the use of molecular markers for distinctness and essential derivation' (2001) 546 *Act Horticulturae* 35; J. Dendauwa et al., 'Variety protection by use of molecular markers: Some case studies on ornamentals' (2000) 135(1) *Plant Biosystems* 107.

[55] J. Staub, S. Chung and G. Fazio, 'Conformity and genetic relatedness estimation in crop species having a narrow genetic base: the case of cucumber (*Cucumis sativus* L.)' (2005) 124 *Plant Breeding* 44.

[56] C. Brubaker and J. Wendel, 'RFLP diversity in cotton' in JN Jenkins and S Saha (eds.), *Genetic Improvement of Cotton: Emerging Technologies* (Enfield, Science Publishers, 2000).

[57] M. Raham, D. Hussain and Y. Zafar, 'Estimation of genetic divergence among elite cotton cultivars – Genotype by DNA fingerprinting technology' (2002) 42 *Crop Science* 2137.

thresholds from which to assess essential derivation. Another limitation of scientifically assessing essential derivation is that absolute measures of genetic similarity are not scientifically feasible. This is because the methodology used to study genetic similarities will only be a sampling strategy. Therefore, any statistical method applied to this problem has to be able to maintain a delicate balance in order to avoid excessive identification of false positives on the one hand, as well as false negatives on the other hand.[58] Consequently, estimations of similarity or dissimilarity are often influenced by the methodology used for that particular experiment.[59] Further problems arise because plant breeding techniques evolve over time and these changes may not be reflected in the agreed-upon genetic thresholds. This means that quantitative thresholds have to be continually monitored. For example, the *Guidelines for the Handling of a Dispute on Essentially Derived Varieties in Lettuce* developed by the ISF provided for a review of the protocol and the thresholds after five years 'in the light of the experience gained and the technical and scientific evolution'.[60]

Finally, it has been suggested that utilising molecular markers to assess essential derivation could actually provide a mechanism to undermine the intention of essential derivation, which is to discourage plagiarism. Ironically, a plant breeder may use marker-assisted breeding to evade a declaration of essential derivation. This could be achieved through the selection of a molecular marker profile that is 'sufficiently different' from the initial variety.[61] For instance, if the threshold for variety Y is 90 per cent, it may be possible to ensure that subsequent varieties will show genetic thresholds 85 per cent or less, despite there only being minor changes made to the initial variety. While the new plant variety may be quantitatively outside the boundary established for essential derivation, it may still draw on the important or essential features of the existing variety for its commercial appeal to the industry.

In summary, there are a number of reasons why progress on examining and identifying EDVs has been slow. Most notably, scientific attempts

[58] Eeuwijk and Baril, Conceptual and statistical issues related to the use of molecular markers for distinctness and essential derivation'.

[59] Roldán-Ruiz et al., 'A comparative study of molecular and morphological methods of describing relationships between perennial ryegrass (Lolium perenne L.) varieties', 1139.

[60] ISF, *Guidelines for the Handling of a Dispute on Essential Derivation in Lettuce* (ISF, 2004), www.worldseed.org/cms/medias/file/Rules/EssentialDerivation/Guidelines_for_the_Handling_of_a_Dispute_on_Essential_Derivation_in_Lettuce_20040525_(En).pdf.

[61] J. Donnenwirth, J. Grace and S. Smith, 'Intellectual property rights, patents, plant variety protection and contracts: A perspective from the private sector' (2004) 9 *IP Strategy Today* 19.

are hampered by the legal basis of essential derivation, as well as the important physically observable characteristics to the UPOV Convention and plant breeding more generally.[62] The limitations of quantitative approaches are generally either technical in nature or are a consequence of the nature of science, which tends to assume that essential derivation is a scientific fact that can be quantitatively determined. There is also the fundamental issue of the juridical nature of the EDVs. Indeed, some of the problems associated with scientific (un)certainty have been overcome by a consideration of the quality of the differences between the plant varieties in dispute. This is supported by the limited case law on essential derivation.[63] It is, therefore, difficult to predict how future legal disputes in relation to essential derivation will be settled and whether the provisions of UPOV 1991 will have substantive bearing on these disputes.

9.3.2 *Legally Qualifying Essential Derivation*

Examining and identifying EDVs will not always be difficult. There are instances where the genetic information and morphological characteristics tell the same, or a very similar, story about whether a variety is essentially derived. For example, in *Van Zanten BV* v. *Hofland BV*, it was alleged that Hofland's 'Mercurius' variety was essentially derived from Van Zanten's 'Ricastor' variety.[64] Importantly in this case, the genetic/DNA test showed no genetic difference, and the varieties were almost identical morphologically. Specifically, Van Zanten proved that the genetic conformity was approximately 100 per cent - with an almost identical Amplified Fragment Length Polymorphism (AFLP) profiles and a Jaccard coefficient of 1.0 - and that 38 out of 39 morphological characteristics were the same. This was, therefore, considered a clear case of essential derivation. What happens, however, when the genetic information and morphology tell a different story, and the genetic test is inconsistent with morphological characteristics?

The case(s) of *Astée Flowers* v. *Danziger 'Dan' Flower Farm* illustrates the challenge of examining and identifying EDVs:[65] this case was heard

[62] See Chapter 6 for more details on the importance of characteristics to the assessment of distinctness, uniformity and stability.

[63] Although, disputes around EDVs are typically settled out of court and often involve confidentiality agreements: personal communication with Mr. Peter Button, Vice Secretary-General UPOV, September 2012.

[64] District Court Hague, 310918/KG KA 08-594 (2008).

[65] In the Netherlands see *Astée Flowers B.V.* v. *Danziger 'Dan' Flower Farm*, Case 198763, Court of The Hague (13 July 2005); *Danziger 'Dan' Flower Farm* v. *Astée Flowers B.V.*

in different jurisdictions (Holland and Israel) and shows how, by giving genetic tests and morphology different weight, courts treat the question of essential derivation differently. The cases also illustrate the importance of who has the burden of proving that a variety is essentially derived. The facts of *Astée Flowers* can be summarised as follows. Two rival plant breeders were involved in disputes in relation to the *Gypsophila* plant variety.[66] Danziger had plant variety protection for the variety 'Dangymini', which had the trade mark 'Million Stars'. A few years later, Astée obtained plant variety protection for their variety 'Blancanieves'. In Europe, as a result of Astée Flowers entering the *Gypsophila* market, Danziger began sending letters throughout the horticultural industry that warned potential customers not to purchase 'Blancanieves' or 'Summer Snow' from Astée Flowers because, in doing so, they would infringe the rights that Danziger held under their *EC Plant Variety Right*. Astée Flowers brought an action in respect of the alleged (mis)representations made by Danziger Flower Farm. Danziger counterclaimed that the Astée Flowers' plant varieties were in fact EDVs and sought a declaration to this effect. Based on the claim and counterclaim, at first instance the Civil Court of The Hague had to determine whether 'Blancanieves' and 'Summer Snow' were essentially derived from 'Million Stars'. The applicable law was the *European Council Regulation on Community Plant Variety Rights* (EC 2100/94 as amended).[67] Based largely on qualitative and substantial differences between 'Dangymini' and 'Blancanieves', the Civil Court found that the varieties were not essentially derived.[68] The genetic testing information was dismissed by the Civil Court largely because it was viewed as unreliable and biased.

The decision of the Civil Court was appealed by Danziger on the basis that the Civil Court incorrectly interpreted the *Council Regulation* on EDVs and improperly ignored Danziger's evidence of genetic conformity. The Court of Appeal, however, confirmed the judgment of the Civil Court. The Court of Appeal was of the opinion that the similarities found by the genetic testing could not serve as evidence of essential derivation. The Court of Appeal, like the Civil Court before it, found the genetic

105.003.932/01, Court Appeal, The Hague (2009). In Israel see *Danziger* v. *Azolay & Astée Flowers* 001228/03, District Court, Tel-Aviv-Jaffa (5 March 2009).

[66] *Gypsophila* is a flower that normally is white in colour and is commonly called Baby's breath or Chalk plant.

[67] Articles 13(5)(a) and (6) of *European Council Regulation on Community Plant Variety Rights* (EC 2100/94 as amended).

[68] *Astée Flowers B.V.* v. *Danziger 'Dan' Flower Farm*, Case 198763, Court of The Hague (13 July 2005).

Table 9.1 *Phenotypic differences between Blancanieves and Dangypmini as determined by the Community Plant Variety Office*

Characteristic	Blancanieves	Dangypmini
Plant height	(9) very tall	(5) medium
Stem thickness	(7) thick	(3) thin
Leaf apex	(2) incurved	(1) straight
Number of petals	(7) many	(5) medium
Profile of upper part of corolla	(1) flat	(2) convex

evidence to be 'open to objection',[69] preferring to examine essential derivation through the morphological characteristics of the two plants, stating that: '[c]ontrary to Danziger, the Court is of the opinion that the putative derived variety and the original variety must also be phenotypically similar to such a high degree that the one variety differs from the other variety only in one or a few inheritable characteristics'.[70] Further, the Court of Appeal believed that the language of 'essential characteristics' in the *Council Regulation* and UPOV Convention meant that contemplation of the 'cultural and practical' value of the variety and its characteristics were crucial to examining and identifying EDVs, asserting:

> Within this context, the Court notes that which characteristics are essential to a variety is closely related to the cultural and practical values of that variety. Essential to a variety are (is) those (that) unique (combination of) characteristics which determine the cultural and practical values and from which the variety derives its varietability.[71]

So, in determining whether there was a 'substantial' difference between 'Dangymini' and 'Blancanieves', the original assessment of distinctness by the plant variety authority was important. In this case, both the Civil Court and Court of Appeal relied on the Community Plant Variety Office's assessment that found 'Blancanieves' was clearly distinguishable from all other varieties: based on observable phenotypic characteristics, there were no less than 17 out of the 21 characteristics that were different between the two plant varieties (see Table 9.1 for examples).[72] On this

[69] Ibid., [16].
[70] *Danziger 'Dan' Flower Farm* v. *Astée Flowers B.V.* 105.003.932/01, Court Appeal, The Hague (2009), [20].
[71] Ibid., [21].
[72] Ibid., [21]. Also see *Astée Flowers B.V.* v. *Danziger 'Dan' Flower Farm*, Case 198763, Court of The Hague (13 July 2005), [15].

basis, the Court of Appeal dismissed Danziger's appeal and held that Astée's variety 'Blancanieves' was not an EDV.

Danziger, whose registered office is in Israel, also initiated legal proceedings in Israel against Astée Flowers.[73] In contrast to the decision in The Hague, however, Israel's District Court of Tel-Aviv-Jaffa accepted Danziger's genetic testing and information as accurate and reliable.[74] In so doing they accepted the genetic testing as evidence of genetic conformity and, ultimately, essential derivation. Accepting AFLP testing as an appropriate means of examining and identifying EDVs, the District Court of Tel-Aviv-Jaffa made it clear that when determining the question of EDVs they preferred 'genetic conformity' over comparisons of observable traits or characteristics. Significantly, in contrast to Holland (and all other UPOV Members) under Israel's *Plant Breeders Right Law*, the plaintiff has the burden of proving EDV.[75] So accepting the genetic tests as a reliable indicator of 'genetic conformity' was a particularly important decision in the context of Israel's *Plant Breeders Right Law*: in which the burden of proof for essential derivation is reversed so that it is the defendant that must prove that their variety is not an EDV. In this case, therefore, once the Court accepted the genetic tests as reliable, the onus fell on Astée to prove that their variety was not essentially derived. According to the District Court of Tel-Aviv-Jaffa, Astée was unable to prove their claims about the way in which 'Blancanieves' was created.[76] So, the differences in morphology that were crucial in the decision in Holland were outweighed by the 'genetic conformity' between the 'Blancanieves' and 'Dangymini' varieties, with the District Court of Tel-Aviv-Jaffa concluding that 'Blancanieves' was essentially derived because:

> The Defendants have not met the burden laid upon them to prove that Blancanieves is not an Essentially Derived Variety from the Registered Variety. Hence, and since at the dates pertinent to the Action the Defendants made use of Blancanieves for the purposes of reproduction, growing and marketing – I determine that they have infringed the Plaintiff's breeders' right in the Registered Variety.[77]

[73] *Plant Breeder's Right Law 5733–1973.*

[74] *Danziger* v. *Azolay & Astée Flowers* 001228/03, District Court, Tel-Aviv-Jaffa (5 March 2009).

[75] The UPOV Convention leaves the matter of reversal of the burden of proof to the UPOV Members: see, UPOV, *Sixth Meeting with International Organizations*, IOM/6, 30 October 1992 (document IOM/6/2, 'Essentially Derived Varieties').

[76] *Danziger* v. *Azolay & Astée Flowers* 001228/03, District Court, Tel-Aviv-Jaffa (5 March 2009), [21].

[77] Ibid., [23].

Further support for a qualitative, fact-based approach to assessing essential derivation can be found in the Australian approach to essential derivation.[78] As a member of UPOV, the Australian government was active in the negotiations on essential derivation. A at the Diplomatic Conference of 1991, the Australian delegation argued that the essential derivation concept was 'legally imprecise and technically flawed'.[79] The Australian government explicitly introduced qualitative considerations to the determination of essential derivation. Under Section 4 of the *Plant Breeder's Rights Act 1994* (Cth), a plant variety is taken to be an essentially derived variety of another plant variety if:

(a) it is predominantly derived from that other plant variety; and
(b) it retains the essential characteristic that result from the genotype or combination of genotypes of that other variety; and
(c) it does not exhibit any important (as distinct from cosmetic) features that differentiate it from that other variety.

The Australian legislation goes a step further than UPOV 1991 by explicitly adding a qualitative layer to the test of essential derivation in so far as it uses the phrase 'important (as distinct from cosmetic) features'.[80] When the *Plant Breeder's Right Bill* was introduced, it was stated that plain English was used to 'promote a better understanding of, and a wider interest in, the legislation on which the plant breeder's right scheme will be based'.[81] According to the Oxford Dictionary, 'important' means that differences between the initial plant variety and the putative essentially derived variety need to be of 'great significance or value'.[82] It can be assumed that in any dispute the term 'important' will be construed in terms of functional considerations such as performance and/or market value.[83] The example given by the Australian *Expert Panel on Plant Breeding* was that of a new wheat variety that was bred with purple anthers. Anthers contain the pollen, and making changes to their colour will not affect the performance or value of the wheat crop. Consequently, such a change would fall into the category of a cosmetic feature, and the variety would then be declared to be an essentially derived variety.[84]

[78] *Plant Breeder's Rights Act 1994* (Cth).
[79] UPOV, Publication No. 346(E), p. 342 [1078].
[80] The concept of essential derivation was considered in the Australian context by the Advisory Council on Intellectual Property (ACIP): ACIP, *A Review of Enforcement of Plant Breeder's Rights: Final Report*, pp. 69–70.
[81] Commonwealth, House of Representatives, (24 August 1994), 157.
[82] Oxford Dictionary (Oxford University Press, 2003).
[83] Expert Panel on Breeding, *Clarification of Plant Breeding Issues Under the Plant Breeder's Rights Act 1994*, p. 22.
[84] Ibid.

In summary, conflicts and disputes over EDVs will be increasingly solved by reference to the law. The consideration of qualitative differences, in terms of 'essential', 'substantial' or 'important', enables the concept of EDVs to meet its goal, which is to discourage plagiarism without hindering advances in plant breeding. A juridical approach also provides a number of other advantages. Because the purpose of the legal approach is to seek, and understand, all the relevant facts, it is a dynamic approach that can adapt with the evolution of plant breeding practices and variety production. Importantly, too, the requirements for a declaration of essential derivation will be similar to those of inventiveness and obviousness in patent law – a question of fact, determined by reference to the facts of the particular case and the inferences drawn from those facts. Further, the legal assessment of essential derivation can provide a more in-depth and comprehensive decision. While it has been said that the concept of essential derivation 'might have a stultifying effect on plant breeding programs as breeders remain uncertain as to what is permissible',[85] assessing 'substantial' and 'important' differences between plant varieties may increase the degree of certainty. Subsequently, plant breeders will have greater surety that their new plant varieties are not essentially derived.[86] Relevant qualitative questions that may need to be asked by those assessing whether a plant variety is essentially derived are: What are the performance differences between the putative essentially derived variety and the initial variety? Has the new variety overcome some problem or is it merely a cosmetic difference? Has the new variety satisfied a need? And what (if any) is the benefit to the public of the new variety?

9.3.3 *Cooperation, Guidelines and Arbitration*

It is clear that legal and scientific approaches to the examination and identification of EDVs are not, on their own, enough. As a result, UPOV and plant breeder organisations have established various crop specific guidelines and explanatory notes around essential derivation. These have required cooperation and consensus among stakeholders about what constitutes EDVs for a particular species or crop. While UPOV has also developed a set of guidelines for the examination and identification of essential derivation, these do little more than acknowledge the difficulty

[85] M. Llewelyn, 'From "outmoded impediment" to global player: the evolution of plant variety rights' in D. Vaver and L. Bently (eds.), *Intellectual Property in the New Millennium* (Cambridge University Press, 2005), pp. 137–156.

[86] Expert Panel on Breeding, *Clarification of Plant Breeding Issues under the Plant Breeder's Rights Act 1994*, p. 23.

of reaching agreement on the appropriate method and measure of essential derivation.[87] UPOV do, however, clearly acknowledge that cooperation, negotiation and compromise are important elements in examining and identifying EDVs. The Administrative and Legal Committee (CAL), for example, considered the nature and scope of essential derivation and developed *Explanatory Notes on Essentially Derived Varieties under the 1991 Act of the UPOV Convention.*[88] While UPOV's *Explanatory Notes* do not provide specific details the CAJ acknowledges the difficulty of agreeing on common understandings about EDVs,[89] reinforce the idea that discussion, negotiation and cooperation are essential ingredients in examining and identifying EDVs. The UPOV Convention 'does not provide clarification of terms such as "predominantly derived" or "essential characteristics"'.[90] And in relation to examining and identifying EDVs, UPOV states:

> With regard to establishing whether a variety is an essentially derived variety, a common view expressed by members of the UPOV is that the existence of a relationship of essential derivation between protected varieties is a matter for the holders of plant breeders' rights in the varieties concerned.[91]

As we saw in Section 9.3.1, the ISF adopted *Guidelines for the Handling of a Dispute on EDV in Lettuce,*[92] and the French maize seed industry adopted a code of conduct on essential derivation. While these guidelines are based on scientific tests of genetic conformity – a Jaccard coefficient of 0.96 and thresholds of >90 per cent (EDV), 82–90 (possible EDV) and 82 per cent (not an EDV), respectively – there has been negotiation and cooperation about what the genetic testing methods and measure should be.[93] That is, guidelines on EDVs must be both supported by scientific research and accepted by plant breeders and industry. Going even further, the ISF has explicitly acknowledged the contentious nature of examining and identifying EDVs and focused on establishing guidelines and procedures around negotiation and arbitration on EDVs. For

[87] UPOV, *Explanatory Notes on Essentially Derived Varieties under the 1991 Act of the UPOV Convention,* UPOV/EXN/EDV/1) (UPOV, 2009).
[88] Ibid.; Also see UPOV, Administrative and Legal Committee, *Report on Conclusions,* Sixty-Seventh Session, CAJ/67/15, (2013), [15]–[20].
[89] UPOV, Administrative and Legal Committee Advisory Group, *Report,* Seventh Session, CAJ-AG/12/7/7 (2012), [21]–[46].
[90] UPOV, *Explanatory Notes on Essentially Derived Varieties under the 1991 Act of the UPOV Convention,* p. 6.
[91] Ibid., p. 11.
[92] ISF, *Guidelines for the Handling of a Dispute on Essential Derivation in Lettuce* (2004).
[93] The ISF has developed similar guidelines for maize, perennial ryegrass, cotton and oilseed rape.

example, Article 3 (2) of the *Regulation for the Arbitration of Disputes concerning Essential Derivation* of the ISF[94] sets out:

> The phenotypic analysis or description should preferably meet the requirements of the appropriate UPOV Technical Guideline for the crop concerned and may include additional characteristics. The molecular analysis must be performed by using the agreed methods, as mentioned in the crop specific scheme of this RED. If there is no agreed-upon molecular marker method available for the crop concerned, the method to be used and the EDV threshold as mentioned in Article 2e may be decided upon between the parties.

There are various benefits to consensus and cooperation on EDVs. First, consensus and cooperation encourage the involvement of various stakeholders and interested groups and give them time and resources to prospectively consider how to best examine and identify EDVs for particular species and crops. In so doing, it allows stakeholders to consider and weigh the merits of particular scientific thresholds, percentages and coefficients of 'genetic conformity' with the 'cultural and practical' value of the morphological characteristics. Second, any agreed guidelines or rules on EDVs can be implemented subject to revision or amendment depending on the outcomes of future scientific findings or industry realisation about what is important or significant.

9.4 Conclusion

This chapter has shown that over time the way in which third party plant breeding has been dealt with by the UPOV Convention has changed. Initially, despite concerns over the low threshold of distinctness and the absence of infringement provisions, the practical barrier – in the form of a time lag – associated with incremental plant breeding meant that the impact of the breeder's exception on the plant variety rights holder was marginal. However, the advent of molecular plant breeding techniques and the availability of patent protection meant that this practical barrier was minimised. As a consequence, the concept of EDVs was introduced by UPOV 1991.

While science plays a role in assessing essential derivation, the concept of EDVs occupies a legal space and raises distinctly legal issues such as what is the standard of proof and who has the burden of proving that a variety has been essentially derived.[95] Consequently, EDVs cannot and should not be examined and identified purely on quantitative grounds;

[94] ISF, *Regulation for the Arbitration of Disputes concerning Essential Derivation (RED)*.
[95] *Danziger* v. *Astée* 105.003.932/01, Court of Appeal, The Hague (2009); *Danziger* v. *Azolay* 1228/03, District Court, Tel-Aviv-Jaffa (2009).

'cultural and practical values' are also important when examining and identifying EDVs. Perhaps, then, EDVs are best viewed as 'agreed facts', in which adopted guidelines and arbitration are crucial to the examination and identification of EDVs. And while science can quantify the ways in which plant varieties are the same, it cannot tell us whether there are 'substantial' or 'important' differences between plant varieties, or whether that sameness should have any meaning.

10 Saving and Exchanging Seeds
Licences, Levies and Speculation

10.1 Introduction

One of the most contentious aspects of the UPOV Convention is the issue of farm-saved seed.[1] Since the introduction of the UPOV Convention, the practice of saving seed has generated heated debate among farmers, civil society organisations, plant breeder organisations, NGOs, governments, UPOV and seed companies.[2] Given the customary and significant nature of saving and exchanging seeds, it is unsurprising that it is an issue. Concerns about farm-saved seed are often fuelled by an ostensible paradox between the 'ownership' of plant varieties and a farmer's (in)ability to save and exchange seeds. On the one hand, supporters of the UPOV Convention argue that strong protection, which limits farm-saved seed, is essential to adequately reward plant breeders for their efforts and to stimulate innovation and the development of new plant varieties. For proponents of strong plant variety protection, farm-saved seed has even been referred to as the 'most frequent loophole in national implementation'.[3] On the other hand, opponents of the UPOV Convention argue that placing limits on the

[1] The word 'seed' is used throughout this Chapter in a broad sense to include other propagating material such as cuttings, tubers or bulbs.

[2] For example, in 2007, GRAIN noted that stopping farm-saved seed was at the top of the seed industry's 'wish list': GRAIN, *The End of Farm-Saved Seed? Industry's Wish List for the Next Revision of UPOV* (2007), www.grain.org/article/entries/58-the-end-of-farm-saved-seed-industry-s-wish-list-for-the-next-revision-of-upov. Also see, e.g., B. Adi, 'Intellectual property rights in biotechnology and the fate of poor farmers' agriculture' (2006) 9(1) *The Journal of World Intellectual Property* 91; M. Blakeney, 'Protection of plant varieties and farmers' rights' (2002) 24(1) *European Intellectual Property Review* 9; B. Endres and P. Goldsmith, 'Alternative business strategies in weak intellectual property environments: A law and economics analysis of the agro-biotechnology firm's strategic dilemma' (2007) 14(2) *Journal of Intellectual Property Law* 237.

[3] B. Le Buanec, 'Protection of plant-related innovations: Evolution and current discussion' (2006) 28(1) *World Patent Information* 50. The ISF has stated that farm-saved seed is 'the most complex and worrying issue for the seed industry of self-pollinated crops, but increasingly also for hybrids of vegetables and field crops': ISF, *Meeting on Enforcement of Plant Breeders' Rights*, UPOV (Geneva, 2005), p. 3.

saving and exchanging of seed undermines traditional farming practices and livelihoods and has a deleterious effect of famers, food security and poverty. For example, in 2014, the Special Rapporteur on the Right to Food (Olivier De Schutter) called for changes to plant variety protection laws including the support and facilitation of local seed exchanges.[4] And, commentators often suggest that the issue of farm-saved seed is one of the key reasons developing countries – such as India, Thailand and Chile – oppose the ratification of UPOV 1991 and adopt *sui generis* plant variety schemes.[5]

There are, however, limits with these kinds of arguments.[6] One of the problems is that much of the discussion on farm-saved seed is underpinned by speculation and rhetoric. As we will see throughout this chapter, by taking sides and relying on dichotomies (e.g. farmers vs. breeders; developed countries vs. developing countries), the issue of farm-saved seed is often reduced to political, civil or other agendas. Another way to think about many of these concerns over the UPOV Convention and farm-saved seed is through what philosopher Graham Harman calls 'overmining', a process of 'reduc[ing] upward rather than downward' so that the UPOV Convention becomes part of a bigger issue surrounding farmers' rights, technological developments and patent law.[7] This makes the UPOV Convention an easy target but also misses some of nuance in and around the issue of farm-saved seed. Another problem is that the extent of the UPOV Convention's effect, detrimental or otherwise, on farmer exchanges is largely unknown. Although the exact rates of saving and exchanging seeds remain unclear, there have been some attempts to quantify the extent to which farmers save, reuse and exchange seeds.[8] In Kenya, for example, it is

[4] See, e.g., Report of the Special Rapporteur on the Right to Food, *Final Report: The Transformative Potential of the Right to Food*, UN Document HRC/A/25/57/2014 (2014), p. 22 [25]–[28]; Berne Declaration, 'Owning seeds, accessing food: A human rights impact assessment of UPOV 1991 based on case studies in Kenya, Peru and the Philippines', Berne Declaration (2014).

[5] See, e.g., D. Jefferson, 'Development, farmers' rights, and the Ley Monsanto: The struggle over the ratification of UPOV 91 in Chile' (2014) 55 *IDEA* 31.

[6] The debates over farm-saved seed have echoed claims of piracy more commonly associated with other areas of intellectual property such as copyright law. For a historical analysis of piracy in intellectual property, see A. Johns, *Piracy: The Intellectual Property Wars from Gutenberg to Gates* (The University of Chicago Press, 2009).

[7] G. Harman, *The Quadruple Object* (Zero Books, 2011), p. 10.

[8] Other attempts to quantify the practice of saving seeds include: in 1990, it was estimated that for small grain crops like wheat and barley, seed saving was 89 per cent (Australia), 85 per cent (Spain) and 70 per cent (Canada and the United States) yet there was no mention of methodology, sample sizes or response rates: see N. Byrne, *Commentary on the Substantive Law of the 1991 UPOV Convention for the Protection of Plant Varieties*

estimated that more than 80 per cent of seed is accessed from saved seed in a number of crops including bananas, beans, cassava, millet, sorghum and sweet potato.[9]

This chapter examines how the practice of saving and exchanging seed is embodied in the UPOV Convention. The chapter begins by examining how the UPOV Convention has dealt with farm-saved seed. We will see that initially the scope of the right granted by UPOV 1961 and UPOV 1978 was limited to 'commercial purposes' of the protected plant variety.[10] As a consequence, it was generally accepted that saving seeds and other propagating material for non-commercial purposes, without the breeder's prior authorisation, was permitted. However, much to the objection of civil society organisations and some governments, the implied exception was replaced with an express, albeit optional, exemption that is subject to a number of safeguards including that it must be exercised 'within reasonable limits and subject to the safeguarding of the legitimate interests of the breeder'. In order to give context to our discussion of the UPOV Convention, in part 10.4 of this chapter, 'non-UPOV' attempts to curtail saving, reusing and exchanging seeds will be considered. Far from being an issue unique to the UPOV Convention and plant variety protection, there have been various attempts to restrict farm-saved seed including technological such as hybridisation and genetic use restriction technologies (GURTs) and patent protection and associated licences.[11] In some ways these attempts to curtail saved seed affect discussions of the UPOV Convention, particularly as the concerns over farm-saved seed often get reduced to political, civil or other agendas and are subordinated to speculation and simple dichotomies. The chapter then turns to farmer exchange networks and tries to get a sense of the extent to which the UPOV Convention's provisions on farm-saved seed affect farmers who save, reuse and exchange seeds. In so doing there appear to be only a few questions about farm-saved seed that can be answered with any certainty. For example: Is the optional farm-saved seed exception under UPOV 1991 more limited than that implicit in

(University of London, 1996), p. 61. The Australian Seed Federation estimated that 'seed piracy' cost the Australian seed industry A$300 million per annum and that rates of saved seed are wheat (95 per cent), barley (90 per cent) and oats (80 per cent) although there is no mention of how these figures were obtained: Australian Seed Federation, *Submission into the Review of Enforcement of Plant Breeder's Rights* (May 2007).

[9] M. Ayieko and D. Tschirley, *Enhancing Access and Utilization of Improved Seed for Food Security in Kenya* (Tegemeo Institute of Agricultural Policy and Development, 2006).

[10] As we saw in Chapter 8, UPOV 1978 clarified the uses of protected material to include the offering for sale or marketing of such material.

[11] There have been additional policy measures aimed at limiting farm-saved seed (including phytosanitary measures and seed certification) in order to protect farmers.

UPOV 1978 and UPOV 1961? Yes.[12] Have developed countries increasingly made use of licences and levies to receive remuneration for saved seed? Yes. Are famer exchange networks important in developing countries? Definitely. However, when it comes to the UPOV Convention's relationship with farm-saved seed, and farmer exchange networks, we must be more cautious. Perhaps the UPOV Convention and farmer exchange networks are not mutually exclusive or a zero-sum proposition. Perhaps they operate side by side and rarely, if ever, come into contact. This probably means that neither civil society's direst predictions about the UPOV Convention nor UPOV's assertion that restricting farm-saved seed (within reasonable limits and safeguarding the legitimate interests of plant breeders) is essential for the creation of new plant varieties is correct. One thing is clear: before we are able to reach any conclusions about the extent of the UPOV Convention's effect, detrimental or otherwise, on farmer exchanges more ethnographic, anthropological and interdisciplinary research is necessary.

10.2 From an Implied to an Express (But Optional) Right to Save Seed

Farm-saved seed was not expressly mentioned in either UPOV 1961 or UPOV 1978. The overarching concern for those at the Diplomatic Conference of 1957–1961 was protecting commercial plant breeding. As a consequence, the scope of the right granted by UPOV 1961 and UPOV 1978 was limited to 'commercial purposes' of the protected plant variety.[13] While the term 'commercial' was not defined by UPOV 1961, it was *generally* accepted that the use of seeds and other propagating material for non-commercial purposes, without the breeder's prior authorisation, was permitted. The implication of this was that while farmers were permitted to save seed harvested from a protected plant variety, they could not commercially market and sell seed harvested from a protected variety.[14] The distinction between commercial and non-commercial uses was maintained by UPOV 1978. Article 5(1) of UPOV 1978 lists certain acts that require the plant variety rights holders authorisation, namely, for the production for the purposes of commercial

[12] Although some countries (e.g. France) interpreted UPOV 1978 as not permitting the unauthorised use of farm-saved seed.

[13] As we saw in Chapter 6, UPOV 1978 clarified the uses of protected material to include the offering for sale or marketing of such material.

[14] See, for example, Crucible Group, *Seeding Solutions: Options for National Laws Governing Control over Genetic Resources and Biological Innovation* (International Development Research Centre, 2001).

marketing, the offering for sale and the marketing of the reproductive or vegetative propagating material of the protected plant variety.[15] As a consequence, farm-saved seed was read into both UPOV 1961 and UPOV 1978.

Based on the distinction between commercial and non-commercial uses, UPOV Members of both UPOV 1961 and UPOV 1978 were left to determine if, and how, they would implement the farm-saved seed exemption. The implementation of the farm-saved seed exception under UPOV 1961 and UPOV 1978 varied widely in Member States, with some countries only permitting farmers to plant saved seed on their own farms and others allowing farmers to trade (but not commercially market) saved seed.

The liberal nature of the implied farm-saved seed exemption under UPOV 1961 and UPOV 1978 is illustrated by the United States *Plant Variety Protection Act 1970* (PVPA). Until April 1995, the PVPA permitted farmers to sell a quantity of seed that they had saved from their own holding to other farmers without any obligation to the plant variety owner, as long as the farmer's 'primary farming occupation is the growing of crops for sale for other than reproductive purposes'.[16] In *Delta & Pine Land Co* v. *Peoples Gin Co*,[17] the Fifth Circuit considered whether the involvement of a third party broker rendered otherwise exempt sales between farmers ineligible for the exemption. The facts of the case concerned the infringement liability of a farmer's cooperative which was brokering exchanges of seed between its members. The Fifth Circuit held that the farm-saved seed exemption 'only exempts sales of the protected variety from one farmer directly to another farmer accomplished without the active intervention of a third party'.[18] The Court focused on whether the exchange between members of the farmer's cooperative constituted selling the protected seed, offering it for sale or soliciting an offer to buy it. The Court concluded that sales did not reach

[15] UPOV 1978 retained the option to extend protection to 'plants or parts thereof' when they are used commercially as propagating material in the production of ornamental plants or cut flowers. Although, few countries appeared to implement the extended version of the right: see S. Hassan, 'Ornamental plant variety rights: A recent Italian judgment' (1987) 18(2) *International Review of Industrial Property* 219.

[16] *Plant Variety Protection Act 1970*. Although the United States *Plant Variety Protection Act 1970* allowed farmers to sell seed that was harvested from protected varieties, this was amended in 1994, to comply with UPOV 1991. In effect, the amendment prohibited the sale of any saved seed without the authorisation of the variety owner: see M. Janis and J. Kesan, 'US plant variety protection: Sound and fury...?' (2002) 39 *Houston Law Review* 727.

[17] *Delta & Pine Land Co.* v. *Peoples Gin Co.* 694 F.2d 1012 (5th Circuit 1983).

[18] Ibid., 1016.

the transfer of possession clause at issue in *Delta & Pine Land Co v. Peoples Gin Co*. As a consequence, if farmers wished to sell saved seed to each other, they could do so directly, and not through intermediaries such as farm cooperatives or grain elevators. This practice is commonly referred to as 'over the fence' sales or 'brown bagging'.

A second case that reaffirms the broad nature of the farm-saved seed exemption in the United States was *Asgrow Seed Co* v. *Winterboer*.[19] In this case, the United States Supreme Court considered the ability of United States farmers to save seed of protected plant varieties. In *Asgrow Seed Co*, the respondents (the Winterboers) ran a farm in Iowa and grew two of Asgrow's protected soybean varieties. The Winterboers then sold their 'brown bagged' seeds to other farmers: they sold approximately 10,000 bushels of saved seed and did not compensate Asgrow in any way. The complaint alleged infringement under the PVPA, based on the fact that the Winterboers were selling or offering to sell Asgrow's protected soybean varieties,[20] for sexually multiplying Asgrow's novel varieties as a step in marketing those varieties for growing purposes,[21] and for dispensing the novel varieties to others in a form that could be propagated without providing notice that the seeds were of a protected variety.[22] Asgrow sued to keep the Winterboers from selling saved seed.

The Winterboers argued that their sales were permitted under the saved seed exemption because they were farmers and that they were selling directly to farmers in compliance with *Delta & Pine Land Co*. The Supreme Court held that the saved seed exemption implied the following limitation: farmers could only sell seed they had grown for the purpose of replanting on their own farms but which for some reason they did not end up planting.[23]

As we have already seen, the capacity of farmers to save, reuse and exchange seeds was implied from UPOV 1961 and UPOV 1978. Because of this, the farm-saved seed exemption was interpreted liberally by Member States to include farmers planting saved seed on their own farms and, in some cases, farmers trading but not commercially marketing saved seed. The liberal nature of the implied farm-saved seed exemption was an important issue leading up to the Diplomatic Conference of 1991, with the overall intent of the revision being to strengthen

[19] *Asgrow Seeds Co* v. *Winterboer* 115 S.Ct 788 (1995).
[20] *Plant Variety Protection Act 1970*, 7 U.S.C. § 2541(1). [21] Ibid., § 2541(3).
[22] Ibid., § 2541(6).
[23] See N. Hamilton, 'Who owns dinner: Evolving legal mechanisms for ownership of plant genetic resources' (1992–1993) 28 *Tulsa Law Journal* 587, 632–643; D. Leskien and M. Flitner, *Intellectual Property Rights and Plant Genetic Resources: Options for a Sui Generis System* (International Plant Genetic Resources Institute, 1996).

the UPOV Convention. It appears that the farm-saved seed exception was, and perhaps still is, a compromise and temporary and that '[o]nce farmers were convinced of the benefits that could accrue to them by rewarding breeders for their efforts, work could be started towards reducing the scope of the [farm-saved seed] provision'.[24] In relation to farm-saved seed, a significant change made by UPOV 1991 was the removal of the distinction between commercial and non-commercial activities. This had the effect of removing the implied right to save seed based on the fact that saving seed was 'non-commercial'. While the plant variety owners' right remains primarily over the propagating material of the protected plant variety, Article 14(1)(a) of UPOV 1991 sets out a range of uses to which the plant variety right applies:

> [*Acts in respect of the propagating material*] *(a)* Subject to Articles 15 and 16, the following acts in respect of the propagating material of the protected variety shall require the authorization of the breeder:
> i. production or reproduction (multiplication),
> ii. conditioning for the purpose of propagation,
> iii. offering for sale,
> iv. selling or other marketing,
> v. exporting,
> vi. importing,
> vii. stocking for any of the purposes mentioned in (i) to (vi), above.

The aim of Article 14(1)(a) is to cover 'all the acts we could imagine which could apply to plant varieties',[25] and in doing so it enumerates the exclusive activities which owners are able to exercise control over the propagating material of protected plant varieties.[26] In the context of farm-saved seed, Article 14(1)(a) extends protection to production, reproduction and conditioning for the purpose of propagation without reference to the purpose of this production, reproduction or propagation. Thus, under UPOV 1991 protection is granted to most uses of the plant variety, including saving and reusing seed. This change effected the farm-saved seed exemption that had been read into UPOV 1961 and UPOV 1978.

[24] Diplomatic Conference of 1991, *Records on the Diplomatic Conference for the Revision of the International Convention for the Protection of New Varieties of Plants 1991*, Publication No. 346(E) (UPOV, Geneva, 1992), p. 367 [1254] (Dr. Mick Lloyd).

[25] J. Ardley, 'The 1991 UPOV Convention, Ten Years On' in M. Llewellyn, M. Adcock and M.J. Goodes (eds.), *Proceedings of the Conference on Plant Intellectual Property within Europe and the Wider Global Community* (Sheffield Academy Press, 2002).

[26] CIOPORA has suggested that the omission of 'uses' within UPOV 1991 is non-compliance with TRIPS: CIOPORA, *Green Paper on Plant Variety Protection: Policy Statement* (2002), www.ciopora.org/fileadmin/assets/pageDownloads/2004/05/CIOPORA_Greenpaper_en.pdf. See also M. Thiele-Witting and P. Claus, 'Plant variety protection – A fascinating subject' (2003) 25 *World Patent Information* 243.

Because Article 14(1)(a) of UPOV 1991 broadened the protection provided by the UPOV Convention, it was felt that it needed to be qualified by three specific exceptions. The first exception is for a compulsory experimental and plant breeding exception (Articles 15(1)(ii) and 15(1)(iii), respectively). The second exception allows for a compulsory exception for acts done privately and for non-commercial purposes, thus covering farm-saved seed produced by subsistence farmers (Article 15(1)(i)). The third exception is 'optional' and allows governments to include a farm-saved seed exception in their national or regional laws (Article 15(2)).

Participants at the Diplomatic Conference of 1991 expressed a range of views on the incorporation of an express farm-saved seed exemption in the UPOV Convention. For example, the ISF argued that the exception was tantamount to a subsidy;[27] the French delegation wanted the seed-saving provision deleted altogether so that plant breeders' rights had the same strength as patents,[28] while the Australian approach to the issue could only be categorised as one of reluctant acceptance. In supporting the saved seed provision, the Australian delegation made it clear that its support was based upon 'political necessity' and was motivated, at least partly, by fear of a backlash from farmer groups if Australia did agree to such an exception.[29]

In addition to these overarching considerations about including an express farm-saved seed concept, there was also discussion over the substance of the possible saved seed exception. One proposal was to restrict the types of species or crops that the exemption applied to, with the Dutch delegation seeking to limit the farm-saved seed exception to the areas of agriculture in which the practice was already established (e.g. in cereals, peas and potatoes) and to prevent the exemption from applying to the horticultural sector.[30] In response, the Canadian delegation argued that there were other crops for which the farm-saved seed exemption was significant including flax or linseed, buckwheat, field beans and soybeans.[31] Ultimately, so as not to restrict accessories to UPOV 1991, it was decided that the saved seed exception would be optional and would be subject to a number of safeguards.

In adopting Article 15(2) UPOV 1991, UPOV Members included the titled 'Optional Exception' so that the implementation would be left up to national or regional governments.[32] In this way the interests of plant

[27] UPOV (1991), Publication No. 346(E), p. 357. [28] Ibid., pp. 351–352 [1159.1].
[29] Ibid., p. 367. [30] Ibid., pp. 353 [1166.2]. [31] Ibid., pp. 353 [1168].
[32] There was some debate over the title of Article 15(2), including over whether the title should be 'farm-saved seed', 'farmers' privilege' or whether there should be no title at all: Ibid., pp. 372–373 [1287]–[1299].

breeders, the differing natures of industries and UPOV Members, as well as the varying political positions were given consideration, and implementation was left up to the national or regional governments. Nonetheless, even for those Member States looking to include a farm-saved seed exception, Article 15(2) of UPOV 1991 provides only a limited framework within which to allow saving seed. Article 15(2) of UPOV 1991 states that:

> [a] Contracting Party may, within reasonable limits and subject to the safeguarding of the legitimate interests of the breeder, restrict the breeder's right in relation to any variety in order to permit farmers to use for propagating purposes, on their own holdings, the product of the harvest which they have obtained planting, on their own holdings, the protected variety or a variety covered by Article 14(5)(a)(i) or (ii).[33]

Despite the uncertainty over how Article 15(2) might be implemented, it is clear that saving seed is no longer automatically permitted under the UPOV Convention. The 1991 version of farm-saved seed cannot be interpreted liberally to authorise farmers to sell or exchange seeds 'non-commercially' with other farmers for propagating purposes.[34] In addition, the exception is subject to a number of safeguards, including that any farm seed exception in national or regional laws must be exercised 'within reasonable limits and subject to the safeguarding of the legitimate interests of the breeder'. This has generally been taken to mean that farmers are now expected to compensate or remunerate plant breeders for any farm-saved seed. Furthermore, according to the ISF, the 'reasonable limits' language in Article 15(2) requires UPOV Members to restrict the acreage, quantity of seed and species subject to the farmers' privilege, while the requirement to safeguard breeders' 'legitimate interests' required farmers to pay some form of remuneration to the breeder for their privileged acts.[35]

Since the introduction of UPOV 1991, one of the main ways in which plant breeders and seed companies have tried to safeguard the 'legitimate interests' of breeders against farm-saved seed is by licensing propagating material of protected plant varieties. While the licensing of plant varieties is not new: nonetheless, Article 15(2) of UPOV 1991 requires the farm-saved seed exception to be implemented 'within reasonable limits and subject to the safeguarding of the legitimate interests of the breeder'. This provides an explicit justification for the use of contracts

[33] UPOV 1991, Article 15(2).
[34] J. Watal, *Intellectual Property Rights in the WTO and Developing Countries* (Kluwer International, 2000), p. 141.
[35] ISF, *ISF View on Intellectual Property* (ISF, 2012).

and licences over plant varieties, particularly in relation to farm-saved seed.[36] There are two ways in which plant breeders have tended to use licences to combat farm-saved seed. The first of these is through the use of an annual 'single crop licence' or a non-propagation clause. In these cases, it is often an express condition of the use of a plant variety that farmers (or other growers) do not save seed. An example of this is Australia, where a grower was sued by a commercial fruit tree nursery for growing 'Sweetheart' and 'Black Star' varieties of cherry. It was alleged that the grower had obtained the plant varieties under a non-propagation licence but had subsequently propagated the cherry trees. The parties agreed to a settlement of more than $700,000 in damages, and the defendants were forced to remove approximately 14,000 trees.[37]

Despite the (relatively) common use of non-propagation clauses, there are questions that need to be addressed. The use of restrictive covenants in licensing agreements to negate policy-based exceptions to infringement of intellectual property rights is a contentious issue. On general principles it is difficult to justify the use of such terms; however, it is often the case that contractual provisions which override statutory protections from infringement of intellectual property rights are valid and enforceable.[38]

A second way that plant breeders have used licences to 'safeguard' their legitimate interests in relation to farm-saved seed is through royalty payments. Generally speaking, because the UPOV Convention provides an exclusive right to the propagating material of the plant variety, royalties have traditionally been sought on the propagating material at the point of sale.[39] As we have seen, though, the problem for plant breeders is that

[36] While seed licences that impose express restrictions on a farmers' ability to save seed have generated controversy, courts have been willing to uphold seed use restrictions contained in licensing agreements and 'bag tags'. Therefore farmers planting seeds that were purchased subject to these licensing arrangements risk prosecution for saving seed: see K. Aoki, 'Weeds, seeds and deeds: Recent skirmishes in the seed wars' (2003) 11 *Cardozo Journal International and Comparative Law* 247, 255; D. Uchtmann, 'Can farmers save Roundup Ready® beans for seed? McFarling and Trantham cases say no' (2002) 19 *Agricultural Law Update* 4.

[37] Although the parties agreed on the terms of the settlement and the case did not go to court, the outcome is considered a judgment as the Federal Court of Australia signed off on the matter: *Fleming's Nurseries Pty Ltd & Anor* v. *Tony Hannaford* (Federal Court Proceedings No. VID 1432 of 2005); *Fleming's Nurseries Pty Ltd & Ors* v. *Portikal Thirteen Pty Ltd & Ors* (Federal Court Proceedings No. VID 1588 of 2005).

[38] See, e.g., J. Chen, 'The parable of the seeds: Interpreting the plant variety protection act in furtherance of innovation policy' (2000) 81(4) *Notre Dame Law Review* 105; M. Lemley, 'Beyond redemption: The law and policy of intellectual property licensing' (1999) 87(1) *California Law Review* 111. See also UPOV, *Symposium on Contracts in Relation to Plant Breeders' Rights*, UPOV/SYM/GE/08 (31 October 2008), www.upov.int/meetings/en/details.jsp?meeting_id=16202.

[39] For a discussion of the protected plant material, see Chapter 8.

farmers can only purchase seed once, and then rely on their own, or another farmers', saved seed for future plantings. While licences may make it clearer when infringement has occurred, difficulties of enforcement still remain, and there is also the problem of non-licensed infringement. This is particularly a problem for those seeking remuneration on subsequently derived propagating material that has been produced from the initial crop.

An alternative to seed point royalties was legitimised by UPOV 1991, where Article 14(2) introduced the possibility of placing restrictions on the use of harvested material which had been obtained from protected propagating material, and also by the use of the phrase 'within reasonable limits and subject to the safeguarding of the legitimate interests of the breeder' in Article 15(2) of UPOV 1991. The combination of these amendments has strengthened plant variety rights by providing opportunity for collecting revenue on harvested plant material post sale in certain circumstances. In Australia, for example, royalties are increasingly being imposed on the harvested material by way of end point royalties.[40] In this scenario, payment is based on the volume, quantity or weight of the harvested material sold by the farmers.[41] Strictly speaking the current Australian approach does not depend on this extension of the plant breeder's right per se (to harvested material)[42] but depends on the use of contracts.[43] The introduction of end point royalties was discussed by Kingwell and Watson,[44] who profiled a number of possible advantages and complications in the implementation of the regime. On a positive note, end point royalties were seen as a way for plant breeders to

[40] End point royalties are generally imposed on varieties of wheat, barley, canola and chickpeas.

[41] The end point royalty may be between $1 to $10 per tonne, depending on the collection and management costs, as well as the perceived 'quality' of the new variety. See, e.g. C. Lawson, 'The evolution of a workable scheme for end point royalties for plant varieties' (2013) 94 *Intellectual Property Forum: Journal of the Intellectual and Industrial Property Society of Australia and New Zealand* 36.

[42] Australia implemented this under Section 14 of the *Plant Breeder's Rights Act 1994* (Cth). It was felt that this amendment allowed for the imposition of 'breeder royalties on the delivery of grains': Explanatory Memorandum, *Plant Breeder's Rights Amendment Act 2002* (Cth), 6. The ability to place a royalty on the harvested material was also seen to be facilitated by Section 4 of the *Plant Breeder's Rights Amendment Act* 2002 (Cth). At the time Section 18 was repealed (and replaced), it was stressed that the '[a]gricultural industry is anticipating the introduction of the amendments positively as they will facilitate commercial arrangements based on plant breeder's rights, including through a system of end point royalties': see Explanatory Memorandum, *Plant Breeder's Rights Amendment Act 2002* (Cth), 11.

[43] J. Sanderson, 'Back to the future: Possible mechanisms for the management of plant varieties in Australia' (2007) 30(3) *University of New South Wales Law Journal* 686.

[44] R. Kingwell and A. Watson, 'End point royalties for plant breeding in Australia' (1998) 5 (3) *Agenda* 323.

obtain a return on their investment and at the same time reduce the cost of the seed for the farmer.[45] This would have the consequence of encouraging the adoption of new plant varieties.[46] On the other hand, problems of collection were foreshadowed, particularly relating to obtaining accurate farmer declarations and variety identification.[47] Advocates of the end point royalty regime (most notably plant breeders and the seed industry) also argue that end point royalties are the most equitable means of ensuring continued investment in plant breeding because farmers contribute proportionately, based on the success of the crop rather than on the amount of seed that they purchase.[48]

As a result of the desultory fashion in which end point royalties have been introduced, more than ten years after their implementation, there are a number of prevailing concerns about the regime. There is still suspicion that the revenue from end point royalties is not getting back to the breeder or is not being used for further breeding activities.[49] There is also concern that the current administration and collection strategies used for end point royalties are too complex, inefficient and costly. On one hand, owners and the seed industry lament the cost of collection because it dilutes their returns. On the other hand, some farmers argue that the collection and administration fees are too high and that the fees are set inconsistently and arbitrarily.[50]

The interaction between farm-saved seed provisions and the extension of plant variety rights to harvested material and products – as well as the importance of licensing for farm-saved seed – is seen in the Australian Federal Court decision of *Cultivaust* v. *Grain Pool Pty Ltd*.[51] The decision of *Cultivaust* v. *Grain Pool Pty Ltd* considered a

[45] J. Hamblin, 'Research for the pulse industry in Australia', paper presented to ABARE Outlook97 (1997) Canberra (4–6 February).

[46] A. Watson, 'The impact of plant breeder's rights and royalties on investment in public and private breeding and commercialisation of grain cultivars' (1997), a paper prepared for the Grains Research and Development Corporation; A. Lazenby et al., *Trials and Errors: A Review of Variety Testing and Release Procedures in the Australian Grains Industry* (Grains Research and Development Corporation, 1994).

[47] Kingwell and Watson, 'End point royalties for plant breeding in Australia'. Importantly though, errors in declaration could potentially be either to the detriment or to the advantage of the plant breeder as farmers can make either positive (state that they have a protected variety when they do not) or negative (state that they do not have a protected variety when they do) errors.

[48] Ibid.

[49] See, e.g., 'EPR debate defines crop breeding future', *Ground Cover* (2006) 14.

[50] For example, in 2007, the Australian Wheat Board management fees range from $0.30 for one variety of wheat (Drysdale) to $1.40 for another variety of wheat (Sentinel). 'Australian Wheat Board Seeds Variety Licence 2007', *Ground Cover* 66 (2007) 16.

[51] (2004) 62 IPR 11. For a discussion see Sanderson, 'Back to the future: Possible mechanisms for the management of plant varieties in Australia'.

range of issues under the *Plant Breeder's Rights Act 1994* (Cth), including whether the plant variety owner (Cultivaust, as licensee of the State of Tasmania) had a 'reasonable opportunity' to exercise its rights over the propagating material. Justice Mansfield interpreted and applied Australia's farm-saved seed provisions subject to the 'cascading' principle to extend protection to harvested material and products: Section 17 of the *Plant Breeder's Rights Act 1994 (Cth)*. Justice Mansfield examined the conditions under which Section 14 would operate in relation to farm-saved seed, stating that saved seed would be categorised as 'harvested material' if two factual conditions were satisfied.[52] First, Section 14(1)(a) of the *Plant Breeder's Rights Act 1994* requires that the second (or subsequent) generation crop not be authorised by the grantee of the *Plant Breeder's Rights Act 1994*. That is, plant breeders holding a plant breeder's right can give permission to farmers to harvest (and sell) crops grown from farm-saved seed. Secondly, Section 14(1)(b) of the *Plant Breeder's Rights Act 1994* requires that the grantee does not have a reasonable opportunity to exercise plant breeder's rights in relation to the propagating material.[53]

On the facts, Justice Mansfield found that the rights' owner had known that farmers were saving the seed and harvesting second and subsequent generation crops without authority and so had a reasonable opportunity to exercise their rights. However, they had not been able to reach an agreement with the respondent. The court did not go into any further detail on how the plant variety owners would have exercised their rights over the saved seed, other than to say that they had a 'reasonable opportunity' to do so. In finding that Cultivaust did have a reasonable opportunity to exercise its right in relation to the propagating material, Justice Mansfield identified possible *indicia* of what constitutes a 'reasonable opportunity'. According to Justice Mansfield, the *indicia* of what constitutes a 'reasonable opportunity' were knowing that such crops were being grown and harvested, understanding that the crops were themselves subject to the plant variety rights protection and, if relying on another entity (e.g. to obtain some form of payment or royalty),

[52] *Cultivaust Pty Ltd* v. *Grain Pool Pty Ltd* (2004) 62 IPR 11, 51.

[53] On the facts presented, Justice Mansfield held that Cultivaust and the Tasmanian government did have a reasonable opportunity to exercise their plant breeder's rights in relation to the propagating material leading to each harvest, and therefore Section 14 did not apply. According to Justice Mansfield, the *indicia* of what constitutes a 'reasonable opportunity' were knowing that such crops were being grown and harvested, understanding that the crops were themselves subject to the *Plant Breeder's Rights Act 1994* by reason of Section 14 and if relying on another body (e.g. to obtain end point royalties) knowledge that there had been no agreement: Ibid., 50–51.

knowledge that there had been no agreement in relation to payments or royalties on the harvested material.[54]

The Australian position suggests that it might only be in limited circumstances that a plant variety owner does not have a 'reasonable opportunity' to exercise their rights on the propagating or harvested material. In fact, Justice Mansfield stated that '[a]t the time of the initial sale of certified Franklin barley, Tasmania and Cultivaust could have imposed conditions upon the disposition of second and subsequent generations of crop'.[55]

Another strategy used to safeguard the interests of plant breeders for farm-saved seed is through the imposition of levies. In this case, farmers may be required to make an additional levy payment based on the quantity of seed purchased or the harvested material they produce. This payment can then be distributed proportionally to plant breeders depending on the number of farmers saving seed. One example of this approach is in Europe, where the requirement to pay a fee on farm-saved seed is embodied in statute.[56] The provision allowing farm-saved seed can be found in Article 14 of *Council Regulation (EC) 2100/94 of 27 July 1994 on Community Plant Variety Rights* and the *Implementing Rules on the Agricultural Exemption.*[57] Article 14(1) establishes a tiered system under which a plant variety owner can claim royalties over farm-saved seed. Article 14(1) states that:

> '...farmers are authorized to use for propagating purposes in the field, on their own holding the product of the harvest which they have obtained by planting, on their own holding, propagating material of a variety other than a hybrid or synthetic variety, which is covered by a Community plant variety right'.[58]

[54] *Cultivaust Pty Ltd* v. *Grain Pool Pty Ltd* (2004) 62 IPR 11, 50–51.

[55] Cultivaust appealed the decision to the Full Federal Court. The Full Federal Court expressed doubt about Justice Mansfield's construction of Section 14 due to confusion over the distinction between the primary rights under Section 11 of the *Plant Breeder's Rights Act 1994* and the secondary rights that arise by way of infringement under Section 53(1): *Cultivaust Pty Ltd* v. *Grain Pool Pty Ltd* (2005) 67 IPR 162, 174.

[56] In France, there is a 'Voluntary Contribution' – although it is, in fact, mandatory as *La Contribution Volontaire Obligatoire* translates to Mandatory Voluntary Contribution. This is consistent with the European Regulations and based on reciprocal commitments by breeders and wheat producers. The Voluntary Contribution currently only applies to all bread wheat, regardless of what seed was used. See GRAIN, *The End of Farm-Saved Seed? Industry's Wish List for the Next Revision of UPOV* (2007), p. 8.

[57] *Commission Regulation (EC) No 1768/95 of 24 July 1995 implementing rules on the agricultural exemption provided for in Article 14(3) of Council Regulation (EC) No 2100/94 on Community Plant Variety Rights*, as amended by *Commission Regulation (EC) No 2605/98 of 3 December 1998 amending Regulation (EC) No 1768/95 implementing rules on the agricultural exemption provided for in Article 14(3) of Council Regulation (EC) No 2100/94 on Community Plant Variety Rights.*

[58] This applies only to agricultural plant species listed in Article 14(2) of *Council Regulation (EC) No 2100/94 on Community Plant Variety Rights* such as barley, wheat and potatoes.

Article 14(3) sets out six criteria determining the conditions which give effect to the exemption, but generally farmers who save protected propagating material, other than small farmers,[59] are required to pay equitable remuneration that is 'sensibly lower than the amount charged for the licensed production of propagating material of the same variety in the same area; the actual level of this equitable remuneration may be subject to variation over time, taking into account the extent to which use will be made of the derogation'.

The factors taken into account when determining the amount of remuneration to be paid are elaborated in the *Implementing Rules on the Agricultural Exemption.*[60] More specifically, Article 5(1) of the *Implementing Rules* states that equitable remuneration is to be paid to the plant variety rights owner based on a contract between the holder and the farmer concerned. Where there is no contract, however, Article 5(2) states that 'the level of remuneration shall be sensibly lower than the amount charged for the licensed production of propagating material'.[61] Despite the fact that the European Commission, in 1998, amended the *Implementing Rules on the Agricultural Exemption* and imposed a rate of 50 per cent of the original payment on farm-saved seed, questions remain over the level of 'equitable remuneration' payable on farm-saved seed.[62] In the case of *Saatgut-Treuhandverwaltungs GmbH* v. *Deppe, Hennings, Lubbe,*[63] the European Court of Justice considered the criteria by which the equitable remuneration a farmer must pay for propagation of a protected plant variety under the farm-saved seed provision may be fixed

[59] A small farmer is a farmer who grows no more than ninety-two tonnes of cereals on farm: *Council Regulation (EC) No 2100/94 on Community Plant Variety Rights*, Article 14(3).

[60] *Commission Regulation (EC) No 1768/95 of 24 July 1995 implementing rules on the agricultural exemption* provided for in Article 14 (3) of *Council Regulation (EC) No 2100/94 on Community plant variety rights*, as amended by Commission Regulation (EC) No 2605/98. See M. Llewelyn and M. Adcock, *European Plant Intellectual Property* (Hart, 2006), pp. 230–233.

[61] It is also specified in these regulations that farmers and processors are obliged to supply information as to the use of farm-saved seeds and that official bodies involved in the monitoring of the production may equally provide this information to the breeders. Private collecting agencies have been set up by seed companies, who collect royalties directly. In the United Kingdom, the British Society of Plant Breeders (BSPB) collects royalties. (1996) Memorandum of Association: The British Society of Plant Breeders Limited; BSPB (2004) BSPB Strategic Plan 2004–2009. Other functions include education, enforcement activities. See The Fair Play Campaign, *Fair Play on Farm Saved Seed*, www.fairplay.org.uk.

[62] *Commission Regulation (EC) No 1768/95 of 24 July 1995 implementing rules on the agricultural exemption provided for in Article 14 (3)* of *Council Regulation (EC) No 2100/ 94 on Community plant variety rights*, as amended by Commission Regulation (EC) No 2605/98.

[63] European Court of Justice, Joined Cases C-7/05 to C-9/05 (8 June 2006).

in the absence of a contract between the parties. At first instance, the German Federal Supreme Court stated that 'sensibly lower' implies an appreciable reduction in price against the original cost.[64] On appeal, the European Court of Justice held that a rate of 80 per cent was not 'sensibly lower', having regard to Article 5(1) of the *Implementing Regulation*, which fixes the rate of 50 per cent of the regular licence fee. However, the European Court of Justice stressed that the 50 per cent mentioned in the European Commission's *Community Plant Variety Rights Regulation* is neither an upper nor a lower limit.[65]

In this part we have seen how UPOV 1991 requires governments to ensure that any farm-saved seed exception is implemented in a way that safeguards the 'legitimate interests' of breeders. In doing so, many plant variety schemes encourage the use of licences or levies that require farmers to renumerate plant breeders for any seed that they have saved. Yet far from being unique to the UPOV Convention, there have been various attempts to limit the practice of saving, reusing and exchanging seed. So, in the next part, before we look at farmer exchange networks, it is worth considering some of these attempts including hybridisation and genetic use restriction technologies (GURTs), as well as patent protection and associated licences. Before doing so, however, it is important to stress that many of these techniques, tools and strategies were not specifically used to limit farm-saved seed. Rather, they are adopted for various reasons and serve other purposes including intellectual property or business strategy, breeding outcomes associated with yield, phytosanitary or biosecurity.

10.3 Using Technology and Patents to Restrict the Saving and Exchange of Seeds

One way in which plant breeders have attempted to limit farm-saved seed is through technological means. While technical solutions to the issue of farm-saved seed emerged with deterioration techniques used at the end of the nineteenth century,[66] such techniques became more systematic with the commercial use of (F1) hybrids. Hybridisation involves the

[64] *Saatgut-Treuhandverwaltungs GmbH* v. *Ulrich Deppe and others*, Joined Cases C-7/05 to C-9/05 (European Court of Justice) 8 June 2006.

[65] For a general discussion on the European approach, see Llewelyn and Adcock, *European Plant Intellectual Property*, pp. 230–233.

[66] A technique where plant varieties 'deteriorate' in the farmer's field: J.P. Berlan and R. Lewontin, 'Cashing in on life: Operation terminator' (1998) *Le Monde Diplomatique*, December 1998, http://mondediplo.com/1998/12/02gen.

crossing of two genetically distinct plants and results, after a number of generations, in progeny that possess the most desirable (or exhibit amplification of the desired) characteristics of both plants. In the early 1900s, it was recognised that (F1) hybrid plant varieties exhibited hybrid vigour and had yields higher than open-pollinated plant varieties.[67] Since then, hybridisation has been a major tool in plant breeding and has been used successfully to develop new plant varieties of corn, cotton, canola and sorghum.[68] While the improved characteristics of the hybrid varieties were the primary justification of hybrid technique(s), it has been suggested that the proprietary character of hybridisation was also appealing to plant breeders and the seed industry.[69]

One consequence of hybridisation for farmers is that the seed of second (and later) generations of hybrid varieties lose some of their yield potential and uniformity. This means that seed saved from a hybrid plant variety, when replanted, may display a considerable decrease in the quality and quantity at the next harvest. There is little, if any, benefit in saving and replanting hybrid seed, and therefore, it becomes necessary for farmers to purchase new propagating material each year. Because of the reduced output and quality of second (and later) generations, hybridisation serves as a way of ensuring biological control over the propagating material of plant varieties.[70] In this way, hybridisation became an effective way for plant breeders to protect their innovation from unauthorised use and exploitation, as farmers could only reliably access the sought-after trait for one, or possibly a few, generation(s). Because of this protective characteristic, it was suggested that plant breeders would 'divert their efforts to the creation of hybrid varieties, with an inherent biological solution to the problem [of farm-saved seed]'.[71] A limitation

[67] Hybrid vigour refers to an increase in the level of production and is also known as heterosis. For an overview of the development of plant breeding, see M.C. Kharkwal and Roy D. Roy, 'A century of advances in plant breeding methodologies' in H.K. Jain and M.C. Kharkwal (eds.), *Plant Breeding: Mendelian to Molecular Approaches* (2004) 21; M. Koornneef & P. Stam 'Changing paradigms in plant breeding' (2001) 125 (1) *Plant Physiology* 156.

[68] B. Dhillon, A. Singh, B. Lather and G. Srinivasan, 'Advances in hybrid breeding methodology' in H.K. Jain and M.C. Kharkwal (eds.), *Plant Breeding: Mendelian to Molecular Approaches* (Kluwer Academic Publishers, 2004), p. 419.

[69] J. Kloppenburg, *First the Seed: The Political Economy of Plant Biotechnology* (Cambridge University Press, 1988), p. 125.

[70] The use of technological protection measures may also prolong the monopoly beyond the life of any plant breeder's right or patent.

[71] Byrne, *Commentary on the Substantive Law of the 1991 UPOV Convention for the Protection of Plant Varieties*, p. 61. Also see Kloppenberg, *First the Seed*; C. Fowler, *Unnatural Selection: Technology, Politics and Plant Evolution*.

for plant variety owners was that some crops (including wheat and barley) are difficult to breed using hybridisation techniques.[72]

Another example of technology being used to restrict reproduction of plants is GURTs, which are an 'innovation induced in response to the problems of appropriability and weaknesses in existing intellectual property rights (IPR) institutions'.[73] GURTs can be classified into two types: (ii) trait-related (T-GURTs), which regulate the expression of certain characteristics, for example, drought resistance; and (ii) variety-related (V-GURTs), which can render the subsequent generation sterile.[74] GURTs, and especially V-GURTs, provide a means of biologically protecting against unauthorised reproduction of the seed, making any saved seed inert.[75] Put simply, if a farmer saves seed from these plants, they will not grow. As the seeds only produce one harvest and if a farmer wants additional seed of that variety, he or she must go back to the seed company to purchase it.[76] GURTs, therefore, overcome the problem of compliance and enforcement under plant variety rights, patents and common law contract.[77] In comparison to the legal

[72] For a discussion of hybridisation in various crops, see S.S. Banga and S.K. Banga (eds.) *Hybrid Cultivar Development* (Springer,1998).

[73] C. Srinivasan and C. Thirtle, *Impact of Terminator Technologies in Developing Countries: A Famework for Economic Analysis* (The International Consortium on Agricultural Biotechnology Research, 2002). For an explanation of the early technology, see M. Crouch, *How the Terminator Terminates: An Explanation for the Non-scientist of a Remarkable Patent for Killing Second Generation Seeds of Crop Plants* (1998) www.edmonds-institute.org/crouch.html. For more detail see R. Jefferson, et al. (1999), *Genetic Use Restriction Technologies: Technical Assessment of the Set of New Technologies which Sterilize or Reduce the Agronomic Value of Second Generation Seed, as Exemplified by US Patent No 5,723,765 and WO 94/03619*, UNEP/CBD/SBSTTA4/9/ Rev.1, 25–35.

[74] V-GURTs are also known as 'terminator genes', a term coined by the Rural Advancement Foundation International (RAFI) (now the action group on Erosion, Technology and Concentration (ETC)) in 1998: ETC Group 'Terminator and Traitor' (2006), www.etcgroup.org/en/issues/terminator_traitor.html.

[75] While the first relevant patent was awarded in 1998, to Delta and Pine Land Company and the United States Department of Agriculture for the 'Control of Plant Gene Expression' (US Patent 5,723,765), there have been a number of patents granted since for various methods of producing 'controlled sterility' (for example, US Patents 5,808,034 and 6,700,039).

[76] J. Oczek, 'In the aftermath of the "Terminator" technology controversy: IP protections for genetically engineered seeds and the right to save and replant seed' (2000) 41 *Boston College Law Review* 627, 628. In addition to this, some commentators argue that GURTs are more a 'strategic bargaining device' that can be used to coerce governments into implementing some form of intellectual property protection for plants: see S. Hubicki and B. Sherman, 'The killing fields: Intellectual property and genetic use restriction technologies' (2005) 28(3) *University of New South Wales Law Journal* 740, 746.

[77] These seeds are different from hybrid seeds because hybrid seeds are not sterile and offer, at least in theory, the benefits of 'improved agronomic performance'. On a commercial level, GURTs will not be successful until the technology is paired with a

mechanisms for prohibiting farm-saved seed, it was felt that GURTs were 'broader, more effective and less limited by time constraints, than the protection conferred by intellectual property rights'.[78] According to one of the pioneers of GURTs, Melvin Oliver, there was a 'need to come up with a system that allowed you to self police your technology, other than trying to put on laws and legal barriers to farmers saving seeds, and to try and stop foreign interests from stealing the technology'.[79] Despite the potential for limiting the impact of farm-saved seed, concerns the sterilisation of plants (and other biological material) and the potential ramifications for farming practices have meant that GURTs have been controversial.[80]

Discussions regarding GURTs have involved a number of forums, most notably, the Conference of the Parties to the CBD.[81] The CBD focused on the potential consequences of GURTs, including the impact of GURTs on agricultural diversity, biosafety, ethics, intellectual property protection and the practice of farm-saved seed.[82] Initially, the Conference of the Parties to the CBD expressed concern in relation to GURTs, asking governments to act cautiously and requesting that the Subsidiary Body on Scientific, Technical and Technological Advice (SBTTA) consider, and assess, possible consequences of the commercialisation of GURTs.[83] In 2000, the Conference of the Parties recommended that governments should be prohibited from field testing and commercialising GURTs. In effect, this established a 'de facto' moratorium on the field testing and commercial use of GURTs until transparent

desirable characteristic such as drought tolerance or salinity resistance so that farmers are given an incentive to purchase the seeds containing GURTs over those that do not.

[78] Jefferson, et al., UNEP/CBD/SBSTTA4/9/Rev.1, 13.

[79] Quoted in V. Shiva, *Stolen Harvest: The Hijacking of the Global Food Supply* (2000) 82. See also R. Edwards, 'End of the germ line: Farmers may soon be entirely reliant on seed companies' (28 March 1998) *New Scientist* .

[80] See, e.g., A. Segarra and J. Rawson, *The 'Terminator Gene' and Other Genetic Use Restriction Technologies (GURTs) in Crops*, CRS Report for Congress (1999) www.ncseonline.org/NLE/CRSreports/Agriculture/ag-83.cfm. As a result, in 1999, Monsanto (one of the major companies involved) announced that '[u]ntil a thorough, independent examination of gene protection systems has been conducted and all points of view considered, we will not commercialize these technologies': Statement from Monsanto, 'Gene Protection Technologies; a Monsanto Background Statement' (May 1999).

[81] At 25 March 2016, there were 196 parties to the CBD.

[82] *Consequences of the Use of the New Technology for the Control of Plant Gene Expression for the Conservation and Sustainable Use of Biological Diversity*, UNEP/CBD/SBSTTA/4/Inf.3 Supplementary information to UNEP/CBD/SBSTTA/4/9/Rev.1 (17 May 1999).

[83] *Report of the Fourth Meeting of the Conference of the Parties to the Convention on Biological Diversity*, UNEP/CBD/COP/4/27, www.biodiv.org/doc/meetings/cop/cop-04/official/cop-04–27-en.pdf.

scientific assessments of the potential impacts were made and the socio-economic impacts of GURTs were validated.[84] The Conference of the Parties adopted Decision V/5 (Agricultural biological diversity) Section 3, paragraph 23, which recommended that:

in the current absence of reliable data on genetic use restriction technologies, without which there is an inadequate basis on which to assess their potential risks, and in accordance with the precautionary approach, products incorporating such technologies should not be approved by Parties for field testing until appropriate scientific data can justify such testing, and for commercial use until appropriate, authorized and strictly controlled scientific assessments with regard to, inter alia, their ecological and socio-economic impacts and any adverse effects for biological diversity, food security and human health have been carried out in a transparent manner and the conditions for their safe and beneficial use validated.[85]

At the Eighth Meeting of the Conference of the Parties to the CBD, held in 2006, the issue of GURTs was raised again. Australia, Canada, New Zealand and the biotechnology industry initiated discussion regarding the field testing of GURTs, arguing for a 'case-by-case' approach which could open the door for countries to start commercialising GURTs and other similar technologies.[86] However, the Conference of the Parties encouraged parties to '[c]ontinue to undertake further research, within the mandate of decision V/5 section III, on the impacts of genetic use restriction technologies'.[87] While the 'case-by-case' recommendation failed to gather the necessary support, the issue is likely to be raised again, although it was absent from Ninth Meeting of the Conference of the Parties to the CBD,

[84] Some national legislatures, notably India and Brazil, have passed laws prohibiting the use, sale, registration and patenting of GURTs. More recently, a private members Bill was introduced into the Canadian Parliament that proposes to 'prohibit the release, sale, importation and use of organisms incorporating, or altered by, variety-genetic use restriction technologies', Bill C-448, House of Commons of Canada (31 May 2007). However, in spite of the moratorium, companies continued to file patents in the field of GURTs: see C. Pendleton, 'The Peculiar Case of 'Terminator' Technology: Agricultural Biotechnology and Intellectual Property Protection at the Crossroads of the Third Green Revolution' (2004) 23 *Biotechnology Law Report* 1; Report by Greenpeace and Kein Patent auf Leben, *Terminator Technology: Patents and Patent Applications: No Patents on Life* (2005), www.keinpatent.de/doc/Terminator_en.pdf; ETC Group, *Terminator: The Sequel* (2007), www.etcgroup.org/en/materials/publications.html?pub_id=635.

[85] The *Convention on Biological Diversity* adopted Decision V/5 (Agricultural biological diversity) Section III, paragraph 23 (2000), www.biodiv.org/decisions/default.aspx?m=COP-05&id=7147&lg=0.

[86] *Report of the Eighth Meeting of the Conference of the Parties to the Convention on Biological Diversity*, UNEP/CBD/COP/8/31 (15 June 2006), www.cbd.int/doc/meetings/cop/cop-08/official/cop-08–31-en.pdf.

[87] The *Convention on Biological Diversity* adopted Decision VIII/23 (Agricultural biological diversity) Section A (2006) 216, www.biodiv.org/doc/decisions/COP-08-dec-en.pdf.

held from 19–30 May 2008 (in Germany), and it is highly unlikely that GURTs will gain the necessary support in the foreseeable future.[88]

While the commercial use of GURTs is unlikely, Hubicki and Sherman have argued that GURTs act as an important tool in the development of national plant variety rights laws by placing pressure on governments to introduce some form of intellectual property over plant varieties.[89] In addition, GURTs may influence the UPOV Convention, or other legal schemes, by forcing the contemplation of technological strategies such as GURTs on national and regional legislators. Indeed, intellectual property laws may directly or indirectly prohibit technological strategies for farm-saved seed such as GURTs. One example of this can be found in India, where the Indian government contemplated and explicitly safeguarded against GURTs and similar technologies. The *Protection of Plant Varieties and Farmers' Rights Act 2001 (PVFRA)* requires plant variety applications to be accompanied by an affidavit stating that the plant variety does not contain 'any gene or gene sequence involving terminator technologies'.[90]

Another strategy used by plant breeders to limit the effect of farm-saved seed is to seek protection under patent law. Generally, there is no saved seed exemption under patent law, which means that the seed industry views patents as providing a greater level of protection.[91] In practice though, protecting plant varieties using patent law is not without its problems. In particular, there have been questions over the possibility of an implied licence under patent law that may encompass saving seed. In the United States, for example, when a farmer purchases propagating material from a plant breeder under patent law, they receive an implied licence to use the propagating material for its normal purpose: that is, growing a crop from the propagating material.[92] This does not envision a

[88] The United States is not a party to the CBD, and as the country with the bulk of research and development in plant-related science and technology, it was felt that this could be significant to the commercial adoption of GURTs. See J. Barron, 'Genetic use restriction technologies: Do the potential environmental harms outweigh the economic benefits' (2008) 20 *Georgetown International Environmental Law Review* 271.

[89] Hubicki and Sherman, 'The killing fields: Intellectual property and genetic use restriction technologies'.

[90] *Protection of Plant Varieties and Farmers' Rights Act* (2001), s 18(1)(c).

[91] For a discussion of the differences between plant breeder's rights and patents: see Llewelyn and Adcock, *European Plant Intellectual Property*, pp. 29–37; M. Janis and J. Kesan, "US plant variety protection: Sound and fury...?', 745–52; WIPO-UPOV, *Symposium on the Co-Existence of Patents and Plant Breeder's Rights in the Promotion of Biotechnological Developments* (2002), www.upon.int/en/documents/syposium2002/index1.htm.

[92] *Patent Act* 35 U.S.C., § 271. This is known as the doctrine of 'first sale' in the United States: see B. Endres, 'State authorized seed saving: Political pressures and

determination of the intention of the grower, so that the implied licence may extend to selling the patented material as food or seed. A limitation upon the doctrine of implied licence is that it does not permit the purchaser to make an article which embodies the patented invention. That is, an implied licence may only apply to the propagating material actually sold. This has important ramifications for the operation of the doctrine of implied licence in relation to animate subject matter, such as plants, which can be reproduced without direct human intervention. Relying on the existence of an implied licence, it has been suggested that when a farmer purchases patented seed in an authorised sale, without any contrary contractual terms from the patent owner, it is possible that the farmer is deemed to have received an implied patent licence to use, and dispose of, the seed in 'ordinary' ways.[93]

Despite questions over the possibility of an implied licence, the stronger view is that the patentee's rights in relation to subsequent generations of plants produced from patented seed that has been legitimately purchased by the purchaser are not exhausted, nor does the purchaser obtain any implied licence to use or deal with second-generation seed even in the absence of any restrictive term or condition in the licence limiting the purchaser's ability to save and use seed produced from seed legitimately purchased by them. Therefore, the use of farm-saved seed without the authorisation of the patent holder is likely to constitute an infringement of the patent.

This point was made clear in the United States context by the United States Court of Appeals for the Federal Circuit in *Monsanto Co* v. *McFarling*.[94] This was a case involving a North American farmer (McFarling) who was found to have infringed Monsanto's patents for glyphosate-resistant plant cells when he saved and replanted soybean seeds from a previous year's crop. McFarling argued that as he had initially purchased the soybean seeds used to produce the initial crop, Monsanto's rights in relation to the seeds saved from this initial crop had been exhausted. The United States Court of Appeals for the Federal Circuit rejected this argument. The Court held that the first sale doctrine of exhaustion did not apply to the seeds saved by McFarling because not only did '[t]he original sale of seeds . . . not confer a licence to construct new seeds', but also 'since the new seeds grown from the original batch had never been sold they entailed no principle of patent

constitutional restraints' (2004) 9 *Drake Journal of Agricultural Law* 323; M. Janis, 'Supplemented forms of intellectual property protection for plants' (2004) 6 *Minnesota Journal Law Science and Technology* 305.

[93] See, e.g. M. Janis, 'Supplemented forms of intellectual property protection for plants'.

[94] *Monsanto Co.* v. *McFarling* 302 F. 2d 1291 (2002).

exhaustion'.[95] This reasoning was upheld by the United States Court of Appeals for the Federal Circuit in *Monsanto Co* v. *Scruggs*.[96] In addition, the Court of Appeals stated that: '[w]ithout the actual sale of the second generation seed to Scruggs, there can be no patent exhaustion. The fact that a patented technology can replicate itself does not give the purchaser the right to use replicated copies of the technology. Applying the first sale doctrine to subsequent generations of self-replicating technology would eviscerate the rights of the patent holder'.[97] According to this reasoning, any use of second-generation seed will amount to infringement of the patent irrespective of whether the initial sale of seed is subject to this prohibition or not, at least in respect of asexually produced plants and genetically modified plants and animals which contain a patented gene.[98] This leads to the potentially anomalous situation whereby any use of second-generation seed, for example, the sale of seed from wheat crops as grain, will amount to infringement of the patent even though the purpose of the initial sale or licence was the production of a commercial crop. In the *Monsanto* cases, this anomaly was avoided by the terms of the contract which permitted the use of the purchased seeds for the production of a single commercial crop.

While not limited to patent law, and it is possible to have a restrictive licensing agreements on plant varieties protected using UPOV-based schemes or those with no intellectual property protection, the use of licence agreements is more prevalent for plant innovations protected under patent law. Under licence agreements, companies may pursue contractual remedies if a farmer saves seed, uses the harvested seed for replanting on her own field or sells the saved seed to another farmer. For example, a farmer purchasing protected propagating material may sign a licence, sometimes known as a 'Technology Agreement' (for patented material) or a Seed Licence (for plants protected by plant variety rights), with the seed provider, which may contain a clause whereby the farmer relinquishes rights to the seeds produced by the harvested crop.[99] These types of clauses are aimed at preventing farmers from saving, reusing or reselling protected propagating material and have been described as 'terminator clauses'.[100]

[95] Ibid, 1299. [96] *Monsanto Co.* v. *Scruggs* 459 F.3d 1328 (2006). [97] Ibid., 1336.

[98] For sexually produced plants and animals, however, this will depend on the scope of the claims, in particular whether they may validly extend to progeny of patented varieties since these will rarely breed true-to-type.

[99] For the United States experience, see S. Muscati, 'Terminator technology: Protection of patents or a threat to the patent system?' (2005) 45 *IDEA: The Journal of Law and Technology* 477.

[100] S. Ohlgart, 'The terminator gene: IP rights vs. the farmers' common law right to save seed' (2002) 7 *Drake Journal of Agricultural Law* 473.

Farmers who buy Monsanto's patented seeds must generally sign a contract promising not to save seeds from the resulting crop, which means they must buy new seeds every year. The seeds are valuable because they are resistant to the herbicide Roundup, itself a Monsanto product. For example, a sample 2008 Monsanto Technology Agreement stipulates that farmers agree to use seed containing 'Monsanto Technologies' solely for planting a single commercial crop and:

> Not to save any crop produced from Seed for planting and not to supply Seed produced from Seed to anyone for planting other than to a Monsanto licensed seed company; Not to save or clean any crop produced from Seed for planting and not to supply Seed produced from Seed; Not to transfer any Seed containing patented Monsanto Technologies to any other person or entity for planting.[101]

The relationship between patents, licence agreements and farm-saved seed is considered in a number of legal disputes involving agrochemical and agricultural biotechnology company, Monsanto. In one such case,[102] in 2013, the Supreme Court of the United States unanimously ruled that a farmer could not use Monsanto's patented genetically altered Roundup Ready soybeans to create new seeds without paying Monsanto a fee.[103] The farmer, Vernon Bowman, had purchased Roundup Ready seeds for several years, using them for his first crop of the season. This was consistent with Monsanto's licence agreement that permitted farmers to plant the purchased seed in only one season and allowed farmers to consume or sell the soybean crops but not to save any of the harvested soybeans for replanting. However, for the second crop of the season, Bowman purchased soybeans from a grain elevator; these soybeans were intended for human or animal consumption and were outside the terms of the licensing agreement. But Bowman had correctly anticipated that, since most of the local farmers also used Roundup Ready seeds, many of the purchased soybeans would contain Monsanto's patented technology and, therefore, be glyphosate-resistant. Once planted, Bowman treated the plants from the saved soybeans with glyphosate and in so doing killed all plants without the Roundup Ready trait. He then harvested the resulting soybeans that contained that trait and saved some of these harvested seeds to use in next season's planting.

[101] Monsanto, *Sample Monsanto Technology/Stewardship Agreement* (2008), www.monsanto.com/sitecollectiondocuments/tug_sample.pdf.

[102] Other disputes include the Canadian case of *Monsanto Canada Inc.* v. *Schmeiser* (C.A.) [2003] 2 F.C. 165 and the United States case of *Monsanto Co.* v. *Geertson Seed Farms* 561 US (2010).

[103] *Bowman* v. *Monsanto Co. et al* 569 U. S._ (2013). See A. Berg, 'Understanding the relation between the doctrine of patent exhaustion and self-replicating technologies after *Bowman v. Monsanto Co*' (2014) 59 *Loius University Law Journal* 59.

He did this for eight years. After discovering what Bowman was doing, Monsanto sued him for infringing its patents on Roundup Ready seed. In response, Bowman argued that his actions were protected by the doctrine of patent exhaustion, in which the purchaser of a patented article, or any subsequent owner, has the right to use or resell that article.

On appeal, the Supreme Court unanimously ruled that 'patent exhaustion does not permit a farmer to reproduce patented seeds through planting and harvesting without the patent holder's permission'.[104] In so doing, the Supreme Court affirmed the decisions handed down by the Southern District Court and Court of Appeals for the Federal Circuit.[105] Bowman's main argument was that the patent exhaustion doctrine allowed him to do what he liked with the soybeans that he purchased from the grain elevator legally. However, Justice Kagan said the exhaustion doctrine did not apply to the way Bowman used the seeds, stating: '[u]nder the patent exhaustion doctrine, Bowman could resell the patented soybeans he purchased from the grain elevator; so too he could consume the beans himself or feed them to his animals'; however, the exhaustion doctrine 'does not enable Bowman to make additional patented soybeans without Monsanto's permission (either express or implied)'.[106] Further, Justice Kagan said that permitting farmers to do what Bowman had done would extinguish the worth of Monsanto's patent, observing that, if purchasers were allowed to replicate the invention, there would be a 'mismatch between invention and reward' and the patent would give little protection to the inventor.[107]

The positon on patents and farm-saved seed is not the same throughout the world. In Europe, for example, the joint operation of Articles 8, 9 and 10 of the *European Biotechnology Directive 1998* means that a farmer does not infringe a patent simply by virtue of sowing a crop from patented seed or seed containing a patented gene and producing further seed or propagating material containing a patented gene.[108] Article 8(1) provides that the protection conferred by a patent on a biological material possessing

[104] Ibid., 4.

[105] At trial the Southern District Court of Indiana rejected Bowman's argument and awarded damages of $84,456 to Monsanto. On appeal, the Federal Circuit affirmed the trial decision: citing *Monsanto Co.* v. *Scruggs* 459 F.3d 1328 (2006), where the Federal Circuit held that the purchaser of a patented technologies which can replicate itself is not authorised to use replicated copies of it, as this practice 'would eviscerate the rights of the patent holder'.

[106] *Bowman v. Monsanto Co. et al* 569 U. S._ (2013), 5. [107] Ibid., 7.

[108] A new draft European regulation on Plant Reproductive Material has been proposed. For an overview, see Community Plant Variety Office, *Draft New EU Plant Reproductive Material Law,* www.cpvo.europa.eu/main/en/home/news/press-releases-and-communications/228-draft-new-eu-plant-reproductive-material-law.

specific characteristics shall extend to any biological material derived from the patented biological material through propagation or multiplication in an identical or divergent form and possessing those same characteristics. Further, Article 9 provides that the protection conferred by a patent on a product containing or consisting of genetic information shall extend to all material in which the product is incorporated and in which the genetic information is contained and performs its function. Article 10, however, provides that the protection conferred by Articles 8 and 9 shall not extend to biological material obtained from the propagation or multiplication of biological material placed on the market by the holder of the patent or with her consent, where the multiplication of propagation necessarily results from the application for which the biological material was marketed, provided that the material obtained is not subsequently used for other propagation or multiplication. So, when taken together, the relevant Articles of the *European Biotechnology Directive 1998* mean that a farmer does not infringe a patent simply by virtue of sowing a crop from patented seed or seed containing a patented gene and producing further seed or propagating material containing a patented gene.

Having looked at how the UPOV Convention, technology and patents are used to restrict farmers saving and exchanging seed, we can see that far from being an issue unique to the UPOV Convention, there have been various attempts to limit farm-saved seed including technological such as hybridisation and genetic use restriction technologies (GURTs) and patent protection and associated licences. Unfortunately, these attempts to limit saved seed have affected discussions over the UPOV Convention and farmers' capacity to save and exchange seeds. And issues over the saving and exchanging of seed sometimes get reduced to political, civil or other agendas. This means that contextual and thorough analysis can be subordinated to theory, assumption and speculation. Or that issues around farm-saved seed and the UPOV Convention get conflated with all attempts - whether technological, patent or contractual - to limit farm-saved seed. In hoping to add context to farm-saved seed discussion it is now necessary to turn our attention to the practice of saving, reusing and exchanging seed.

10.4 Farming and Farmer Exchange Networks

There is no denying that many famers, particularly small-scale farmers in developing countries, are doing it tough. Farming is often hard, physical work. And many farmers struggle to feed their own families. In some circumstances farmers have been 'forced' off their land, struggle with increasing debts, have crops destroyed by drought or monsoon and fight to maintain adequate water supplies, infrastructure and other resources.[109] Perhaps the most alarming measure of just how tough it is for some farmers is the rate of suicide among farmers in some countries.[110] Although statistics are hard to come by, it is estimated that between 2001 and 2011, thousands of Indian farmers committed suicide each year.[111]

The reasons why small-scale farmers are struggling are varied and complex. And while there is little doubt that improving access to quality seeds will facilitate agriculture productivity and rural well-being, it is just part of the solution. It is, therefore, important to not overemphasise the potential impact of the UPOV Convention on farmers' practices and livelihoods. It is likely that there are much more important and pressing issues. In a book titled *Feeding the World: An Economic History of Agriculture, 1800–2000*, Giovanni Federico examines the factors that led to successful agriculture and food production between 1800 and 2000.[112] In doing so Federico identifies a broad range of factors including agricultural policies (e.g. tariffs, subsidies), technical progress, land, labour, credit, institutional change, commercialisation and the environment. What this means is that while we should not trivialise the issue of farm-saved seed, neither should we speculate about how the UPOV Convention's farm-saved seed provisions hamper farmers nor its ability to encourage the development of new plant varieties.

To begin to understand how the UPOV Convention effects the saving and exchange of seed, it is crucial to understand more about the nature of farmer exchange networks and the place of protected plant variety varieties in these networks.[113] In many developing countries, farmers retain and use plant varieties independently of the more 'formal' or 'commercial'

[109] See, e.g, S. Liberti, *Land Grabbing: Journeys in the New Colonialism* (Verso, 2013), translated by E. Flannelly; L. Stringer, C. Twyman and L. Gibbs, 'Learning from the South: common challenges and solutions for small-scale farming' (2008) 174(3) *The Geographical Journal* 235; M.A. Jouanjean, *Targeting Infrastructure Development to Foster Agricultural Trade and Market Integration in Developing Countries: An Analytical Review* (Overseas Development Institute, 2013).

[110] Suicide among farmers is not limited to developing countries. See, e.g., L. Bryant and B. Garnham, 'The fallen hero: masculinity, shame and farmer suicide in Australia' (2015) 22(1) *Gender, Place & Culture* 67.

[111] See, e.g. S. Anderson and G. Genicot, 'Suicide and property rights in India' (2015) 114 *Journal of Development Economics* 64; P. Sainath, 'In 16 years, farm suicides cross a quarter million' (2011) 29 *The Hindu* 84–110; 1 Manoranjitham et al., 'Risk factors for suicide in rural south India' (2010) 196(1) *The British Journal of Psychiatry* 26.

[112] G. Federico, *Feeding the World: An Economic History of Agriculture*, 1800–2000 (Princeton University Press, 2005).

[113] Farmer exchange networks have been referred to interchangeably as seed exchange networks, farmer seed systems, traditional seed systems and informal seed systems'. In this Chapter I use the term 'farmer exchange networks' to signify the key role played by farmers and the relational aspects of such exchanges.

breeding and seed industry. Broadly defined, a farmer exchange network involves the transfer of seeds and other propagating material from farmer to farmer, either through 'gifting, swapping, bartering, or purchase, and also via trading or sale which occurs outside of the commercial seed sector and formal regulation'.[114] In this way, seeds are often exchanged with relatives, neighbours and strangers.[115] While the exact figures on the use of farmer exchange networks are difficult to come by, it appears that the rate is high among small-scale farmers. It has been estimated, for example, that in Kenya, approximately 80 per cent of seed comes from farmer exchange networks.[116] In a case study on farmer saving and exchanging of barley in rural communities of Northern Morocco, researchers used structured interviews trying to quantify seed exchange.[117] The researchers found that in the rural communes of North Morocco seed exchanges were maintained almost exclusively through 'informal' seed networks and that nearly 90 per cent of farmers cultivated traditional varieties, with 70–90 per cent of seeds coming from their own holdings and the remainder from local markets. Interestingly, rather than plan their seed saving, farmers 'simply plant the new crop using the seed remaining from the previous year'.[118] Further, farmers gave the following reasons for saving and reusing seeds:

> The reasons cited by most farmers for obtaining seed from outside sources reflect problems with their own seed source, such as having insufficient seeds, a seed stock that is contaminated with too many weed seeds, or poor germination. There does not appear to be a conscious choice to renew their seed from outside sources in the hopes of obtaining new adaptive traits or to increase the diversity of their own seed stock.[119]

In another study, farmers in Syria received new barley varieties from the ICARDA barley breeding programme in return for collaborating in barley breeding research.[120] Each farmer received 100–200kg of seed,

[114] O. Coomes et al., 'Farmer seed networks make a limited contribution to agriculture? Four common misconceptions' (2015) 25 *Food Policy* 41, 42.

[115] See, e.g., L. Samberg, C. Shennan and E. Zavaleta, 'Farmer seed exchange and crop diversity in a changing agricultural landscape in the southern highlands of Ethiopia' (2013) 41(3) *Human Ecology* 477; S. McGuire, 'Securing access to seed: Social relations and sorghum seed exchange in eastern Ethiopia' (2008) 36(2) *Human Ecology* 217; A. Sirabanchongkran et al., 'Varietal turnover and seed exchange: Implications for conservation of rice genetic diversity on-farm' (2004) 29(2) *International Rice Research Notes* 18–20.

[116] Ayieko and Tschirley, *Enhancing Access and Utilization of Improved Seed for Food Security in Kenya*.

[117] H. Jensen et al., 'A case study of seed exchange networks and gene flow for barley (Hordeum vulgare subsp. vulgare) in Morocco' (2013) 60(3) *Genetic Resources and Crop Evolution* 1119.

[118] Ibid., 1129. [119] Ibid.

[120] A. Aw-Hassan, A. Mazid and H. Salahieh, 'The role of informal farmer-to-farmer seed distribution in diffusion of new barley varieties in Syria' (2008) 44 *Experimental Agriculture* 413.

and over a five-year research period, the 52 farmers who initially received seeds of the new barley varieties transferred these to another 156 farmers. In the fifth year of the study, the new varieties accounted for 26 per cent of the seed cultivated by the monitored farmers so that:

> About half the farmers (49%) used their own seeds saved from the previous season and 37% of them purchased seed from their neighbours. Most of the new growers reported purchased seeds of the new variety from farmers who had grown it in the previous year. Neighbours were the most important source of seeds.[121]

There are a number of advantages of farmer exchange networks. For many small-scale famers obtaining seeds from nearby famers is likely to be common practice, with many seed exchanges occurring within a close radius.[122] Because of the nature of farmer exchange networks, there are social advantages of farmer exchange networks such as prestige, power and status.[123] Further, the seeds exchanged in farmer networks are particularly valuable because they are adapted to local conditions such as rainfall, soil composition and traditional farming practices. Yet proximity is not the only reason for the development of exchange networks, and interactions between farmers may be the result of many factors including agroclimatic regions, altitude, ethnolinguistic regions, access to markets and agricultural extension services and marriage networks.[124] Indeed, farmer exchanges occur 'not only because of physical proximity, social relationships and availability of information about seeds, but also because seeds from distant places are less likely to be adapted to the environment where they are to be planted'.[125] Because farmers commonly exchange seeds with neighbours, relatives and strangers, other benefits of farmer exchange networks relate to agrobiodiversity.[126] For these reasons some NGOs and farmer associations facilitate seed exchanges.[127]

121 Ibid., 417.

122 M. Bellon, D. Hodson and J. Hellin, 'Assessing the vulnerability of traditional maize seed systems in Mexico to climate change' (2011) 108 *Proc Nat Academy Science* 13432.

123 See, e.g. P. Richards et al., 'Seed systems for African food security: Linking molecular genetic analysis and cultivar knowledge in West Africa' (2009) 45 *International Journal of Technology Management* 196.

124 Samberg, Shennan and Zavaleta, 'Farmer seed exchange and crop diversity in a changing agricultural landscape in the southern highlands of Ethiopia'.

125 M. Pautasso et al., 'Seed exchange networks for agrobiodiversity conservation. A review' (2013) 33(1) *Agronomy for Sustainable Development* 151, 163.

126 See, e.g., K. Chambers and S. Brush, 'Geographic influences on maize seed exchange in the Bajio Mexico' (2010) 62 *The Professional Geographer* 305; O. Coomes, 'Of stakes, stems, and cuttings: The importance of local seed systems in traditional Amazonian societies' (2010) 62 *The Professional Geographer* 323.

127 See, e.g. M. Thomas, E. Demeulenaere, C. Bonneuil and I. Goldringer, 'On-farm conservation in industrialized countries: A way to promote dynamic management of biodiversity within agroecosystems' in N. Maxted et al. (eds.), *Agrobiodiversity*

To what extent does the UPOV Convention effect farmer exchange networks? Opponents of the UPOV Convention often argue that placing limits on the saving and exchanging of seed undermines traditional farming practices and livelihoods and, thus, has a deleterious effect of farmers, food security and poverty. Indeed, the issue of farm-saved seed is one of the key reasons developing countries – such as India, Thailand, Chile and African countries – oppose the ratification of UPOV 1991 and instead adopt *sui generis* plant variety schemes. These concerns are largely based on the rationale that the farm-saved seed provisions found in Article 15(2) UPOV 1991 are defined narrowly and are 'subject to the safeguarding of the legitimate interest of the breeder' and, as a consequence, merely permit the saving and use of seed on a farmer's 'own holding'. It does not allow farmer exchanges. Furthermore, Article 15(1)(i) of UPOV 1991, which holds that the breeder's right shall not extend to acts done privately and for non-commercial purposes, does not permit the exchange of protected varieties among smallholder farmers.

However, too little is known about the extent to which the UPOV Convention influences farmer exchange networks. And opposition to the UPOV Convention based on deleterious effects on farmer exchange networks tends to be based more on theory, assumption and speculation.[128] Part of the difficulty is that national seed systems, that involve plant variety protection, are rarely used by small-scale farmers. There are a number of reasons for this including the cost of purchasing seed,[129] the unsuitability of registered varieties for farmers[130] and the strength of farmer exchange networks.[131] The situation is further complicated because many of the countries that are the centre of these concerns (e.g. Sub-Saharan African countries) are not yet members of UPOV and, therefore, do not have national plant variety protection schemes.

10.5 Tentative Observations and Future Research

When considering the extent to which the UPOV Convention influences farmer exchange networks, we can make only tentative observations. To this end, and to conclude this Chapter, I will make only tentative

Conservation: Securing the Diversity of Wild Relatives and Landraces (CABI, 2012), pp. 173–180.

[128] See B. De Jonge, 'Plant variety protection in Sub-Saharan Africa: Balancing commercial and smallholder farmers' interests' (2014) 7 *Journal Politics and Law* 100.

[129] R. Tripp, *Seed Provision and Agricultural Development* (James Currey, 2001).

[130] S. Ceccarelli and S. Grando, 'Participatory plant breeding in cereals' in M. Carena (ed.), *Cereals* (Springer, 2009), pp. 395–494.

[131] D. Bazile, *State-Farmer Partnerships for Seed Diversity in Mali* (IIED, 2006).

observations. First, it is likely that only a very small proportion of the seeds exchanged in farmer exchange networks would be protected by plant variety rights schemes. Even in countries that have both plant variety protection and smallholder farmers, only a small portion of seeds exchanged would be protected by plant variety rights. On this point we can look to Kenya, who, as a member of UPOV since 1991, has had plant variety protection since 1972 and also a large portion of smallholder farmers. In discussing Kenya's *Seed and Plant Varieties Act*,[132] Munyi observes:

> smallholder farmers are not substantial participants in formal seed markets, which are characterized by among others certified seeds or seeds protected with some form of PBR. Indeed, with a few exceptions, mainly maize and wheat, significant portions of seed and planting material in Kenya are accessed from farmer-based sources.[133]

Secondly, farmer exchange networks and plant variety protection are not mutually exclusive. They are not necessarily in competition with one another. The consequence of this is that farmer exchange networks may operate side by side with 'commercial' seed networks that involve plant varieties that have plant variety protection.[134] Thirdly, because a country has farm-saved seed provisions consistent with the UPOV Convention does not mean that it will necessarily effect or be enforced in a way that effects farmer exchange networks. One way of protecting farmers against potential adverse effects of UPOV's farm-saved seed provisions is through the provision of compulsory licensing. To assure public access to protected varieties, the granting of compulsory licences is recognised in Article 17(2) of UPOV 1991 so long as the 'Contracting Party concerned shall take all measures necessary to ensure that the breeder receives equitable remuneration'. Two examples of different national approaches can be found in Europe and India. Under Article 29 of *European Regulation No. 2100/94 on Community Plant Variety Rights*, a compulsory licence can be granted if it is shown that the grant of a licence would be in the public interest. This is a broadly written compulsory licence, and an applicant does not need to show that the variety is not being exploited adequately or has not been available for a period of time. Further, although it is contemplated, there is no explicit

[132] *Seeds and Plant Varieties Act 1972*, as amended by the *Seeds and Plant Varieties (Amendment) Act 2012.*

[133] P. Munyi, 'Plant variety protection regime in relation to relevant international obligations: Implications for smallholder farmers in Kenya' (2015) 8(1–2) *The Journal of World Intellectual Property* 65, 66.

[134] See, e.g., N. Louwaars and W. de Boef, 'Integrated seed sector development in Africa: A conceptual framework for creating coherence between practices, programs and policies' (2012) 26 *Journal Crop Improvement* 39.

requirement that the compulsory licence involve the payment of a royalty fee.[135] In India, the *PVFRA* also permits the granting of compulsory licences, although there are a number of requirements that need to be satisfied first. These include a time period of three years,[136] and that: the reasonable requirements of the public for seed or other propagating material of the variety have not been satisfied or that the seed or other propagating material of the variety is not available to the public at a reasonable price and pray for the grant of a compulsory licence to undertake production, distribution and sale of the seed or other propagating material of that variety.[137] Alternatively, it is possible that farmers will implicitly be protected against any adverse effects of UPOV's farm-saved seed provisions. Jefferson has, for example, suggested that the perception that farmer seed exchange is detrimental is misguided.[138] Citing an implied research exception for universities and research institutes in patent law, Jefferson has suggested that it is possible for a similar implied farm-saved seed exception to operate under plant variety protection laws so that plant variety rights holders may observe a de facto exception, electing to forgo legal actions for infringement.[139] Finally, and less tentatively, more research is needed into the extent to which the UPOV Convention may affect farmer exchange networks. Further research is needed that embraces the complexity of the issues around farm-saved seed: research that is interdisciplinary, anthropological or ethnographic.

[135] *European Regulation No. 2100/94 of 27 July 1994 on Community Plant Variety Rights.*
[136] PVFPA, Section 47(1). [137] Ibid.
[138] D. Jefferson, 'Towards a balanced regime of intellectual property rights for agricultural innovations' (2014) 19 *Journal of Intellectual Property Rights* 395.
[139] Ibid., 398–399.

11 The Nature of UPOV and the UPOV Convention

This book examined the history and nature of the UPOV Convention. More specifically, it examined the key concepts and practices contained in the Convention. In so doing this book sheds light on an open-ended and contingent engagement with scientific, legal, technical, political, social and institutional actors. As Chapter 2 demonstrated, the emergence of the UPOV Convention as the peak international treaty that regulates plant development was not an inevitable response to advances in science and technology. Rather, it was the result of an ongoing dialectic of resistance and accommodation, involving many influences and actors including the growing economic importance of plants, the intentions and will of breeder and intellectual property organisations, the nature and characteristics of plants as biological entities and the (un) suitability of existing protection mechanisms – including hybridisation, patent, trade mark and certification laws – to protect the work of plant breeders. Importantly, too, despite the fact that the emergence of a distinct legal regime to regulate plant varieties marked a radical change in the international intellectual property landscape, the UPOV Convention was not the creation of a completely new legal regime, with new concepts and legal rights. It was informed by, drew from and distinguished itself from existing non-legal and legal protection of plant varieties.

Acknowledging that the UPOV Convention is not isolated but is modelled on its 'forebears' is not only important to situate the UPOV Convention within broader legal, political and scientific frameworks but also for the way in which specific principles and concepts – such as plant breeder, plant variety, the scope of protection, farm-saved seed and EDVs – are considered. By doing so we can begin to avoid the trap of overmining or undermining. This is the tendency to either reduce discussion of the UPOV Convention down to a narrow frame (e.g. DNA or molecular techniques) or up to a broad one (e.g. farmers' rights, biodiversity or conservation). In order to take the UPOV Convention seriously, we must, therefore, take the time to identify and analyse the various heterogeneous elements or actors that have been assembled to establish, revise and maintain the UPOV Convention. And it is up to academics, policy makers and advocates to avoid simple explanations, answers and rhetoric about UPOV and the UPOV Convention. Instead, we must embrace complexity and contingency. Chapter 6, for example, showed that while there is no denying the importance of molecular techniques to plant science and breeding, this does not mean that this is suitable for the assessment of distinctness, uniformity and stability. Rather than be beholden to advances in science and technology, the way in which plant varieties are described and distinguished in the UPOV Convention is contingent on the utilitarian nature of plant variety development and end-users (such as farmers, gardeners and nurserymen). The way in which plant varieties are currently described and distinguished in the UPOV Convention embodies a level of richness, complexity and context that technological determinism ignores: end-users of new plant varieties are not particularly interested in the DNA sequence of their new wheat, rose or lettuce variety.

A number of more specific observations about the nature of UPOV and the UPOV Convention can be made. Most importantly, UPOV and the UPOV Convention have technical, social and normative legitimacy around plant variety protection. As UPOV's membership has expanded, the Convention's legitimacy and authority have strengthened. While the reasons for seeking UPOV Membership, and the implementation of UPOV-based national or regional plant variety protection laws, are varied – and include a mixture of obligations to comply with TRIPS and trade agreements, the realisation that there are regional and global benefits associated with joining UPOV and even as a way to reassert or express sovereignty – UPOV membership is an effective way of encouraging plant breeding and facilitating agricultural development. ARIPO, for example, joined UPOV in large part because it believes that regional laws consistent with UPOV 1991 are the best way to sustainably promote plant breeding and encourage the developed of improved plant varieties.

Of course not all UPOV Members want the same thing. And not all plant breeders, civil society organisations or governments will agree with UPOV's direction and approaches. For example, many civil society groups and some African countries expressed reservations over ARIPO's decision to join UPOV. Such reservations are often based largely on the grounds that implementing national laws that are compliant with UPOV 1991 would have deleterious effects on the rights of farmers to save,

[1] See, e.g., C. Oguamanam, 'Breeding apples for oranges: Africa's misplaced priority over plant breeders' rights' (2015) 18(5) *The Journal of World Intellectual Property* 165.

exchange or sell farm-saved seeds.[1] Yet while issues related to farmers' rights are undeniably important, they tend to disregard very important details about the nature of UPOV and the UPOV Convention. Since its inception in 1961, UPOV and its committees and working parties have worked passionately, uniformly and effectively towards providing and promoting plant breeding through UPOV-based plant variety protection and 'an effective system of plant variety protection, with the aim of encouraging the development of new varieties of plants, for the benefit of society'.[2] This means that the UPOV Convention tends to operate in a distinct space from the concerns of around farmers' rights biodiversity and conservation. This has been explicitly acknowledged by UPOV for many years. For instance, in UPOV's communications with the CBD Secretariat, it was noted the UPOV Convention is not an instrument for access and benefit sharing; its objective is to encourage plant breeding.[3]

Perhaps, then, a key aspect of the nature of UPOV and the UPOV Convention is that – while it is not entirely indifferent to the broader social issues related to famers' rights, conservation and biodiversity – it has its own clearly articulated objectives and hierarchy of offices, which have delineated spheres of objectives, competence, functions and technical abilities. Further, too little is known about the overall impact of the UPOV Convention and UPOV-based plant variety protection. And support for, or opposition to, the UPOV Convention tends to be based more on principle, assumption and speculation.

Another notable feature of the nature of UPOV is that it is membership-driven and goal-oriented. In this way UPOV represents the interests of its member and plant breeders with passion, commitment and a clear sense of purpose. It is representative of all of its members, not one member or interest group. Importantly, too, the UPOV Convention reflects the collective interests of its members. This book has shown that as much as an international organisation such as UPOV can, it has shown itself to be dynamic and increasingly democratic and transparent. Over the years, UPOV has actively sought new members and has made itself more open to observers. Furthermore, UPOV and the UPOV Convention have developed and facilitated a system of cooperation. One area where cooperation is evident is in relation to variety denominations. Indeed, the impact of the UPOV Convention on variety denominations is strengthened through cooperation between UPOV Members and the UPOV Council, Office of the UPOV, national plant variety offices and

[2] UPOV, *Mission Statement*, www.upov.int/about/en/mission.html.

[3] UPOV, *UPOV Reply of 23 January 2009, to the Letter of the Executive Secretary of the Secretariat of the Convention on Biological Diversity (CBD) of 19 December 2008*, www.upov.int/about/en/key_issues.html.

the ICRAs. As we have seen, often national plant variety offices accept denominations submitted in other UPOV Members, so that there is a single denomination for each variety worldwide. Further, UPOV's requirements, rules and practices for variety denominations are recognised under the *Cultivated Plant Code* as providing de facto registration of a plant name by horticultural authorities; once a plant variety is registered for plant variety rights protection under national UPOV-based schemes, the designated name is recognised in the relevant ICRA as the name of the plant and often details are shared and the examination of variety denomination under UPOV is carried out in conjunction with the lists of the ICRA.[4] Another concept that illustrates the cooperative nature of UPOV and the UPOV Convention is EDVs. In Chapter 9, I argued that EDVs are a hybrid concept and are best viewed as 'agreed facts', in which negotiated and agreed-upon guidelines and arbitration are crucial.

Perhaps, most importantly though, UPOV and the UPOV Convention have standardised and normalised several concepts and practices around plant variety protection. In so doing the UPOV Convention provides the legal framework around which plant variety protection is structured and sets the normative framework around plant breeding and plant variety protection. Furthermore, UPOV has designed, published and maintained various guidelines, explanatory notes, information documents, databases and standards. Many of these activities clarify aspects of the Convention and maintain and improve the effectiveness of the UPOV system.

In this way UPOV informs and underpins the work of many actors including plant breeders, taxonomists, researchers, lawyers, administrator and advocates. Most notably, UPOV's normative influence comes from the way in which plant varieties are described, distinguished and named. In Chapter 7 we saw how that notwithstanding advances in science and technology physically observable characteristics have played and continue to play a key role in the identification, description and comparison of plants. In fact, UPOV 1991 and UPOV's committees and working groups have reinforced the pivotal role of physically observable characteristics in the assessment of distinctness, uniformity and stability. Chapter 6 showed how, since its introduction in 1961, the UPOV Convention has played an increasingly significant role in the naming of plants and, to some extent at least, has ameliorated some of the concerns and confusion over plant names stemming from the Plant Codes. By developing rules, practices and repositories related to variety

[4] U. Loscher, 'Variety denomination according to plant breeders' rights' (1985) 182 *International Symposium on Taxonomy of Cultivated Plants*, 59.

denomination, UPOV established a legal framework that has facilitated the consistent and effective naming of plants.

Yet another aspect of the nature of UPOV and the UPOV Convention worth pointing out is that they have become more legal and political. This is particularly evident in the concepts and practices around plant breeder, plant variety, protected material and EDVs. Take, for example, the concept of plant variety. There has been an expansion and proliferation in the kinds of plant varieties referred to in both the UPOV Convention and *sui generis* plant variety protection schemes. While some of these varieties (most notably EDVs) strengthen plant variety protection, others have been introduced to recognise the role of farmers and local communities in the development of new plant varieties (e.g. farmers' varieties and local domestic varieties). Another example of the way in which the UPOV Convention has been rendered more legal and political is through the extension of plant protection to harvested material and products derived directly from harvested material. As we saw in Chapter 8, determining whether a plant variety owner has had a 'reasonable opportunity' to exercise their right over either the propagating material or the harvested material has introduced a level of indeterminacy to the UPOV Convention. It appears, though, that the notion of 'reasonable opportunity' places an obligation on plant variety rights owners to licence their plant varieties in ways that contemplate possible uses and abuses of their plant variety right. This has increased the possibility of disagreement and legal dispute, bringing with it challenges of implementation, application and interpretation.

To finish the book, I would like to provide a comment on the future. As we have seen throughout this book, UPOV is a membership-driven and goal-oriented institution that represents the interests of its members. In so doing UPOV has achieved a great deal for plant variety protection. Because UPOV and for that matter the UPOV Convention are representatives of all of UPOV Members, it is conceivable that as more countries and organisations join UPOV and begin the processes of implementing regional or national plant variety protection laws, these countries will inform, shape and sustain future revisions to the UPOV Convention. Similar to the emergence of the UPOV Convention during the 1950s and 1960s, future revisions to the UPOV Convention are not inevitable nor will they be sudden. Future revisions of the UPOV Convention will emerge from ongoing, open-ended and contingent engagement with scientific, legal, technical, political, social and institutional actors. In this way it is up to UPOV Members to set the agenda for UPOV, as well as the parameters around which future iterations of the UPOV Convention and its key concepts and practices will be shaped.

Appendix 1 UPOV 1961/1972

ACT OF 1961/1972
INTERNATIONAL CONVENTION FOR THE PROTECTION OF NEW VARIETIES OF PLANTS

adopted by the Diplomatic Conference
on December 1, 1961

and

ADDITIONAL ACT OF NOVEMBER 10, 1972, AMENDING THE INTERNATIONAL CONVENTION FOR THE PROTECTION OF NEW VARIETIES OF PLANTS

adopted by the Diplomatic Conference
on November 10, 1972

International Convention for the Protection of New Varieties of Plants*

of December 2, 1961

Table of Contents**

* Official English translation.
** This Table of Contents is added for the convenience of the read. It does not appear in the original (French) text of the Convention.

The Contracting States, Convinced of the importance attaching to the protection of new varieties of plants not only for the development of agriculture in their territory but also for safeguarding the interests of breeders,

Conscious of the special problems arising from the recognition and protection of the right of the creator in this field and particularly of the limitations that the requirements of the public interest may impose on the free exercise of such a right,

Deeming it highly desirable that these problems to which very many States rightly attach importance should be resolved by each of them in accordance with uniform and clearly defined principles,

Anxious to reach an agreement on these principles to which other States having the same interests may be able to adhere,

Have agreed as follows:

Article 1 [Purpose of the Convention; Constitution of a Union; Seat of the Union][1]

(1) The purpose of this Convention is to recognise and to ensure to the breeder of a new plant variety, or to his successor in title, a right the content and the conditions of exercise of which are defined hereinafter.

(2) The States parties to this Convention, hereinafter referred to as member States of the Union, constitute a Union for the Protection of New Varieties of Plants.

(3) The seat of the Union and its permanent organs shall be at Geneva.

Article 2 [Forms of Protection; Meaning of "Variety"]

(1) Each member State of the Union may recognise the right of the breeder provided for in this Convention by the grant either of a special title of protection or of a patent. Nevertheless, a member State of the Union whose national law admits of protection under both these forms may provide only one of them for one and the same botanical genus or species.

(2) For the purposes of this Convention, the word "variety" applies to any cultivar, clone, line, stock or hybrid which is capable of cultivation and which satisfies the provisions of subparagraphs (1)*(c)* and *(d)* of Article 6.

Article 3 [National Treatment]

(1) Without prejudice to the rights specially provided for in this Convention, natural and legal persons resident or having their headquarters in one of the member States of the Union shall, in so far as the recognition and protection of the breeder's right are concerned, enjoy in the other member States of the Union the same treatment as is accorded or may hereafter be accorded by the respective laws of such States to their own nationals, provided that such persons comply with the conditions and formalities imposed on such nationals.

(2) Nationals of member States of the Union not resident or having their headquarters in one of those States shall likewise enjoy the same rights provided that they fulfil such obligations as may be imposed on them for the purpose of enabling the new varieties which they have bred to be examined and the multiplication of such varieties to be controlled.

[1] Articles have been given titles to facilitate their identification. There are no titles in the signed (French) text.

Article 4 [Botanical Genera and Species Which Must or May Be Protected; Reciprocity; Possibility of Declaring that Articles 2 and 3 of the Paris Convention for the Protection of Industrial Property Are Applicable]

(1) This Convention may be applied to all botanical genera and species.

(2) The member States of the Union undertake to adopt all measures necessary for the progressive application of the provisions of this Convention to the largest possible number of botanical genera and species.

(3) Each member State of the Union shall, on the entry into force of this Convention in its territory, apply the provisions of the Convention to at least five of the genera named in the list annexed to the Convention.

Each member State further undertakes to apply the said provisions to the other genera in the list, within the following periods from the date of the entry into force of the Convention in its territory:

(a) within three years, to at least two genera;
(b) within six years, to at least four genera;
(c) within eight years, to all the genera named in the list.

(4) Any member State of the Union protecting a genus or species not included in the list shall be entitled either to limit the benefit of such protection to the nationals of member States of the Union protecting the same genus or species and to natural and legal persons resident or having their headquarters in any of those States, or to extend the benefit of such protection to the nationals of other member States of the Union or of the member States of the Paris Union for the Protection of Industrial Property and to natural and legal persons resident or having their headquarters in any of those States.

(5) Any member State of the Union may, on signing this Convention or on depositing its instrument of ratification or accession, declare that, with regard to the protection of new varieties of plants, it will apply Articles 2 and 3 of the Paris Convention for the Protection of Industrial Property.

Article 5 [Rights Protected; Scope of Protection]

(1) The effect of the right granted to the breeder of a new plant variety or his successor in title is that his prior authorization shall be required for the production, for purposes of commercial marketing, of the reproductive or vegetative propagating material, as such, of the new variety, and for the offering for sale or marketing of such material. Vegetative propagating material shall be deemed to include whole plants. The breeder's right shall extend to ornamental plants or parts thereof normally marketed for purposes other than propagation

when they are used commercially as propagating material in the production of ornamental plants or cut flowers.

(2) The authorization given by the breeder or his successor in title may be made subject to such conditions as he may specify.

(3) Authorization by the breeder or his successor in title shall not be required either for the utilization of the new variety as an initial source of variation for the purpose of creating other new varieties or for the marketing of such varieties. Such authorization shall be required, however, when the repeated use of the new variety is necessary for the commercial production of another variety.

(4) Any member State of the Union may, either under its own law or by means of special agreements under Article 29, grant to breeders, in respect of certain botanical genera or species, a more extensive right than that set out in paragraph (1) of this Article, extending in particular to the marketed product. A member State of the Union which grants such a right may limit the benefit of it to the nationals of member States of the Union which grant an identical right and to natural and legal persons resident or having their headquarters in any of those States.

Article 6 [Conditions Required for Protection]

(1) The breeder of a new variety or his successor in title shall benefit from the protection provided for in this Convention when the following conditions are satisfied:

(a) Whatever may be the origin, artificial or natural, of the initial variation from which it has resulted, the new variety must be clearly distinguishable by one or more important characteristics from any other variety whose existence is a matter of common knowledge at the time when protection is applied for. Common knowledge may be established by reference to various factors such as: cultivation or marketing already in progress, entry in an official register of varieties already made or in the course of being made, inclusion in a reference collection or precise description in a publication.

A new variety may be defined and distinguished by morphological or physiological characteristics. In all cases, such characteristics must be capable of precise description and recognition.

(b) The fact that a variety has been entered in trials, or has been submitted for registration or entered in an official register, shall not prejudice the breeder of such variety or his successor in title.

At the time of the application for protection in a member State of the Union, the new variety must not have been offered for sale or marketed, with the agreement of the breeder or his successor in title, in the territory of that State, or for longer than four years in the territory of any other State.

(c) The new variety must be sufficiently homogeneous, having regard to the particular features of its sexual reproduction or vegetative propagation.

(d) The new variety must be stable in its essential characteristics, that is to say, it must remain true to its description after repeated reproduction or propagation or, where the breeder has defined a particular cycle of reproduction or multiplication, at the end of each cycle.

(e) The new variety shall be given a denomination in accordance with the provisions of Article 13.

(2) Provided that the breeder or his successor in title shall have complied with the formalities provided for by the national law of each country, including the payment of fees, the grant of protection in respect of a new variety may not be made subject to conditions other than those set forth above.

Article 7 [Official Examination of New Varieties; Provisional Protection]

(1) Protection shall be granted only after examination of the new plant variety in the light of the criteria defined in Article 6. Such examination shall be adapted to each botanical genus or species having regard to its normal manner of reproduction or multiplication.

(2) For the purposes of such examination, the competent authorities of each country may require the breeder or his successor in title to furnish all the necessary information, documents, propagating material or seeds.

(3) During the period between the filing of the application for protection of a new plant variety and the decision thereon, any member State of the Union may take measures to protect the breeder or his successor in title against wrongful acts by third parties.

Article 8 [Period of Protection]

(1) The right conferred on the breeder of a new plant variety or his successor in title shall be granted for a limited period. This period may not be less than fifteen years. For plants such as vines, fruit trees and their rootstocks, forest trees and ornamental trees, the minimum period shall be eighteen years.

(2) The period of protection in a member State of the Union shall run from the date of the issue of the title of protection.

(3) Each member State of the Union may adopt longer periods than those indicated above and may fix different periods for some classes

of plants, in order to take account, in particular, of the requirements of regulations concerning the production and marketing of seeds and propagating material.

Article 9 [Restrictions in the Exercise of Rights Protected] The free exercise of the exclusive right accorded to the breeder or his successor in title may not be restricted otherwise than for reasons of public interest.

When any such restriction is made in order to ensure the widespread distribution of new varieties, the member State of the Union concerned shall take all measures necessary to ensure that the breeder or his successor in title receives equitable remuneration.

Article 10 [Nullity and Forfeiture of the Rights Protected]

(1) The right of the breeder shall be declared null and void, in accordance with the provisions of the national law of each member State of the Union, if it is established that the conditions laid down in subparagraphs *(a)* and *(b)* of paragraph (1) of Article 6 were not effectively complied with at the time when the title of protection was issued.

(2) The breeder or his successor in title shall forfeit his right when he is no longer in a position to provide the competent authority with reproductive or propagating material capable of producing the new variety with its morphological and physiological characteristics as defined when the right was granted.

(3) The right of the breeder or his successor in title may become forfeit if:

  (a) after being requested to do so and within a prescribed period, he does not provide the competent authority with the reproductive or propagating material, the documents and the information deemed necessary for checking the new variety, or he does not allow inspection of the measures which have been taken for the maintenance of the variety; or

  (b) he has failed to pay within the prescribed period such fees as may be payable to keep his rights in force.

(4) The right of the breeder may not be annulled and the right of the breeder or his successor in title may not become forfeit except on the grounds set out in this Article.

Article 11 [Free Choice of the Member State in Which the First Application is Filed; Application in Other Member States; Independence of Protection in Different Member States)

(1) The breeder or his successor in title may choose the member State of the Union in which he wishes to make his first application for protection of his right in respect of a new variety.
(2) The breeder or his successor in title may apply to other member States of the Union for protection of his right without waiting for the issue to him of a title of protection by the member State of the Union in which he made his first application.
(3) The protection applied for in different member States of the Union by natural or legal persons entitled to benefit under this Convention shall be independent of the protection obtained for the same new variety in other States whether or not such States are members of the Union.

Article 12 [Right of Priority]

(1) Any breeder or his successor in title who has duly filed an application for protection of a new variety in one of the member States of the Union shall, for the purposes of filing in the other member States of the Union, enjoy a right of priority for a period of twelve months. This period shall run from the date of filing of the first application. The day of filing shall not be included in such period.
(2) To benefit from the provisions of the preceding paragraph, the further filing must include an application for protection of the new variety, a claim in respect of the priority of the first application and, within a period of three months, a copy of the documents which constitute that application, certified to be a true copy by the authority which received it.
(3) The breeder or his successor in title shall be allowed a period of four years after the expiration of the period of priority in which to furnish, to the member State of the Union with which he has filed an application for protection in accordance with the terms of paragraph (2), the additional documents and material required by the laws and regulations of that State.
(4) Such matters as the filing of another application or the publication or use of the subject of the application, occurring within the period provided for in paragraph (1), shall not constitute grounds for objection to an application filed in accordance with the foregoing conditions. Such matters may not give rise to any right in favour of a third party or to any right of personal possession.

Article 13 [Denomination of New Varieties of Plants]

(1) A new variety shall be given a denomination.

(2) Such denomination must enable the new variety to be identified; in particular, it may not consist solely of figures.

The denomination must not be liable to mislead or to cause confusion concerning the characteristics, value or identity of the new variety or the identity of the breeder. In particular, it must be different from every denomination which designates, in any member State of the Union, existing varieties of the same or a closely related botanical species.

(3) The breeder or his successor in title may not submit as the denomination of a new variety either a designation in respect of which he enjoys the protection, in a member State of the Union, accorded to trade marks, and which applies to products which are identical or similar within the meaning of trade mark law, or a designation liable to cause confusion with such a mark, unless he undertakes to renounce his right to the mark as from the registration of the denomination of the new variety.

If the breeder or his successor in title nevertheless submits such a denomination, he may not, as from the time when it is registered, continue to assert his right to the trade mark in respect of the above-mentioned products.

(4) The denomination of the new variety shall be submitted by the breeder or his successor in title to the authority referred to in Article 30. If it is found that such denomination does not satisfy the requirements of the preceding paragraphs, the authority shall refuse to register it and shall require the breeder or his successor in title to propose another denomination within a prescribed period. The denomination shall be registered at the same time as the title of protection is issued in accordance with the provisions of Article 7.

(5) A new variety must be submitted in member States of the Union under the same denomination. The competent authority for the issue of the title of protection in each member State of the Union shall register the denomination so submitted, unless it considers that denomination unsuitable in that State. In this case, it may require the breeder or his successor in title to submit a translation of the original denomination or another suitable denomination.

(6) When the denomination of a new variety is submitted to the competent authority of a member State of the Union, the latter shall communicate it to the Office of the Union referred to in Article 15, which shall notify it to the competent authorities of the other member States of the Union. Any member State of the Union may address its objections, if any, through the said Office, to the State which communicated the denomination.

The competent authority of each member State of the Union shall notify each registration of the denomination of a new variety and each refusal of registration to the Office of the Union, which shall inform the competent authorities of the other member States of the Union. Registrations shall also be communicated by the Office to the member States of the Paris Union for the Protection of Industrial Property.

(7) Any person in a member State of the Union who offers for sale or markets reproductive or vegetative propagating material of a new variety shall be obliged to use the denomination of that new variety, even after the expiration of the protection of that variety, in so far as, in accordance with the provisions of paragraph (10), prior rights do not prevent such use.

(8) From the date of issue of a title of protection to a breeder or his successor in title in a member State of the Union:
   (a) the denomination of the new variety may not be used, in any member State of the Union, as the denomination of another variety of the same or a closely related botanical species;
   (b) the denomination of the new variety shall be regarded as the generic name for that variety. Consequently, subject to the provisions of paragraph (10), no person may, in any member State of the Union, apply for the registration of, or obtain protection as a trade mark for, a denomination identical to or liable to cause confusion with such denomination, in respect of identical or similar products within the meaning of trade mark law.

(9) It shall be permitted, in respect of the same product, to add a trade mark to the denomination of the new variety.

(10) Prior rights of third parties in respect of signs used to distinguish their products or enterprises shall not be affected. If, by reason of a prior right, the use of the denomination of a new variety is forbidden to a person who, in accordance with the provisions of paragraph (7), is obliged to use it, the competent authority shall, if need be, require the breeder or his successor in title to submit another denomination for the new variety.

Article 14 [Protection Independent of Measures Regulating Production, Certification and Marketing]

(1) The right accorded to the breeder in pursuance of the provisions of this Convention shall be independent of the measures taken by each member State of the Union to regulate the production, certification and marketing of seeds and propagating material.

(2) However, such measures shall, as far as possible, avoid hindering the application of the provisions of this Convention.

Article 15 [Organs of the Union] The permanent organs of the Union shall be:

(a) the Council;
(b) the Secretariat General, entitled the Office of the International Union for the Protection of New Varieties of Plants. That Office shall be under the high authority of the Swiss Confederation.

Article 16 [Composition of the Council; Votes]

(1) The Council shall consist of representatives of the member States of the Union. Each member State of the Union shall appoint one representative to the Council and an alternate.
(2) Representatives or alternates may be accompanied by assistants or advisers.
(3) Each member State of the Union shall have one vote in the Council.

Article 17 [Observers in Meetings of the Council]

(1) States which have signed but not yet ratified this Convention shall be invited as observers to meetings of the Council. Their representatives shall be entitled to speak in a consultative capacity.
(2) Other observers or experts may also be invited to such meetings.

Article 18 [Officers of the Council]

(1) The Council shall elect a President and a first Vice-President from among its members. It may elect other Vice-Presidents. The first Vice-President shall take the place of the President if the latter is unable to officiate.
(2) The President shall hold office for three years.

Article 19 [Meetings of the Council]

(1) Meetings of the Council shall be convened by its President.
(2) A regular session of the Council shall be held annually. In addition, the President may convene the Council at his discretion; he shall convene it, within a period of three months, if a third of the member States of the Union so request.

Article 20 [Rules of Procedure of the Council; Administrative and Financial Regulations of the Union]

(1) The Council shall lay down its rules of procedure.
(2) The Council shall adopt the administrative and financial regulations of the Union, after having consulted the Government of the Swiss Confederation. The Government of the Swiss Confederation shall be responsible for ensuring that the regulations are carried out.

(3) A majority of three-quarters of the member States of the Union shall be required for the adoption of such rules and regulations and any amendments to them.

Article 21 [Duties of the Council] The duties of the Council shall be to:

(a) study appropriate measures to safeguard the interests and to encourage the development of the Union;

(b) examine the annual report on the activities of the Union and lay down the programme for its future work;

(c) give to the Secretary-General, whose functions are set out in Article 23, all necessary directions, including those concerning relations with national authorities;

(d) examine and approve the budget of the Union and fix the contribution of each member State in accordance with the provisions of Article 26;

(e) examine and approve the accounts presented by the Secretary-General;

(f) fix, in accordance with the provisions of Article 27, the date and place of the conferences referred to in that Article and take the measures necessary for their preparation;

(g) make proposals to the Government of the Swiss Confederation concerning the appointment of the Secretary-General and senior officials; and

(h) in general, take all necessary decisions to ensure the efficient functioning of the Union.

Article 22 [Majorities Required for Decisions of the Council] The Council's decisions shall be taken by a simple majority of the members present, except in the cases provided for in Articles 20, 27, 28 and 32, and for the vote on the budget and the fixing of the contributions of each member State. In these last two cases, the majority required shall be three quarters of the members present.

Article 23 [Tasks of the Office of the Union; Responsibilities of the Secretary-General: Appointment of Staff]

(1) The Office of the Union shall have the task of carrying out all the duties and tasks entrusted to it by the Council. It shall be under the direction of the Secretary-General.

(2) The Secretary-General shall be responsible to the Council; he shall be responsible for carrying out the decisions of the Council.

He shall submit the budget for the approval of the Council and shall be responsible for its implementation.

He shall make an annual report to the Council on his administration and a report on the activities and financial position of the Union.

(3) The Secretary-General and the senior officials shall be appointed, on the proposal of the Council, by the Government of the Swiss Confederation, which shall determine the terms of their appointment.

The terms of service and the remuneration of other grades in the Office of the Union shall be determined by the administrative and financial regulations.

Article 24 [Supervisory Function of the Swiss Government] The Government of the Swiss Confederation shall supervise the expenditure and accounts of the Office of the International Union for the Protection of New Varieties of Plants. It shall submit an annual report on its supervisory function to the Council.

Article 25 [Cooperation with the Unions Administered by BIRPI] The procedures for technical and administrative cooperation between the Union for the Protection of New Varieties of Plants and the Unions administered by the United International Bureaux for the Protection of Industrial, Literary and Artistic Property shall be governed by rules established by the Government of the Swiss Confederation in agreement with the Unions concerned.

Article 26 [Finances]

(1) The expenses of the Union shall be met from:
   (a) annual contributions of member States of the Union;
   (b) payments received for services rendered; and
   (c) miscellaneous receipts.

(2) For the purpose of determining the amount of their annual contributions, the member States of the Union shall be divided into three classes:

First class	five units
Second class	three units
Third class	one unit

Each member State of the Union shall contribute in proportion to the number of units of the class to which it belongs.

(3) For each budgetary period, the value of the unit of contribution shall be obtained by dividing the total expenditure to be met from the contributions of member States by the total number of units.

(4) Each member State of the Union shall indicate, on joining the Union, the class in which it wishes to be placed. Any member State of the Union may, however, subsequently declare that it wishes to be placed in another class.

Such declaration must be made at least six months before the end of the financial year preceding that in which the change of class is to take effect.

Article 27 [Revision of the Convention]

(1) This Convention shall be reviewed periodically with a view to the introduction of amendments designed to improve the working of the Union.

(2) For this purpose, conferences shall be held every five years, unless the Council, by a majority of five-sixths of the members present, considers that the convening of such a conference should be brought forward or postponed.

(3) The proceedings of a conference shall be effective only if at least half of the member States of the Union are represented at it.

A majority of five-sixths of the member States of the Union represented at the conference shall be required for the adoption of a revised text of the Convention.

(4) The revised text shall enter into force, in respect of member States of the Union which have ratified it, when it has been ratified by five-sixths of the member States of the Union. It shall enter into force thirty days after the deposit of the last of the instruments of ratification. If, however, a majority of five-sixths of the member States of the Union represented at the conference considers that the revised text includes amendments of such a kind as to preclude, for member States of the Union which do not ratify the revised text, the possibility of continuing to be bound by the former text in respect of the other member States of the Union, the revised text shall enter into force two years after the deposit of the last of the instruments of ratification. In such case, the former text shall, from the date of such entry into force, cease to bind the States which have ratified the revised text.

Article 28 [Languages To Be Used by the Office and in the Council]

(1) The English, French and German languages shall be used by the Office of the Union in carrying out its duties.

(2) Meetings of the Council and of revision conferences shall be held in the three languages.

(3) If the need arises, the Council may decide, by a majority of three-quarters of the members present, that further languages shall be used.

Article 29 [Special Agreements for the Protection of New Varieties of Plants]
Member States of the Union reserve the right to conclude among themselves special agreements for the protection of new varieties of plants, in so far as such agreements do not contravene the provisions of this Convention.

Member States of the Union which have not taken part in making such agreements shall be allowed to accede to them at their request.

Article 30 [Implementation of the Convention on the Domestic Level; Special Agreements on the Joint Utilization of Examination Services]

(1) Each member State of the Union shall undertake to adopt all measures necessary for the application of this Convention.

In particular, each member State shall undertake to:

(a) ensure to nationals of the other member States of the Union appropriate legal remedies for the effective defence of the rights provided for in this Convention,

(b) set up a special authority for the protection of new varieties of plants or to entrust their protection to an existing authority; and

(c) ensure that the public is informed of matters concerning such protection, including as a minimum the periodical publication of the list of titles of protection issued.

(2) Special agreements may also be concluded between member States of the Union, with a view to the joint utilization of the services of the authorities entrusted with the examination of new varieties in accordance with the provisions of Article 7 and with assembling the necessary reference collections and documents.

(3) It shall be understood that, on depositing its instrument of ratification or accession, each member State must be in a position, under its own domestic law, to give effect to the provisions of this Convention.

Article 31 [Signature and Ratification; Entry Into Force]

(1) This Convention shall be open for signature until December 2, 1962, by States represented at the Paris Conference for the Protection of New Varieties of Plants.

(2) This Convention shall be subject to ratification; instruments of ratification shall be deposited with the Government of the French Republic, which shall notify such deposit to the other signatory States.

(3) When the Convention has been ratified by at least three States, it shall enter into force in respect of those States thirty days after the deposit of the third instrument of ratification. It shall enter into force, in respect of each State which ratifies thereafter, thirty days after the deposit of its instrument of ratification

Article 32 [Accession; Entry Into Force]

(1) This Convention shall be open to accession by non-signatory States in accordance with the provisions of paragraphs (3) and (4) of this Article.

(2) Applications for accession shall be addressed to the Government of the Swiss Confederation, which shall notify them to the member States of the Union.

(3) Applications for accession shall be considered by the Council having particular regard to the provisions of Article 30.

Having regard to the nature of the decision to be taken and to the difference in the rule adopted for revision conferences, accession by a non-signatory State shall be accepted if a majority of four-fifths of the members present vote in favor of its application.

Three-quarters of the member States of the Union must be represented when the vote is taken.

(4) In the case of a favorable decision, the instrument of accession shall be deposited with the Government of the Swiss Confederation, which shall notify the member States of the Union of such deposit.

Accession shall take effect thirty days after the deposit of such instrument.

Article 33 [Communications Indicating the Genera and Species Eligible for Protection]

(1) When ratifying this Convention, in the case of a signatory State, or when submitting an application for accession, in the case of any other State, each State shall give, in the first case to the Government of the French Republic and in the second case to the Government of the Swiss Confederation, the list of genera or species in respect of which it undertakes to apply the provisions of the Convention in accordance with the requirements of Article 4. In addition, it shall specify, in the case of genera or species referred to in paragraph (4) of that Article, whether it intends to avail itself of the option of limitation available under that provision.

(2) Each member State of the Union which subsequently decides to apply the provisions of this Convention to other genera or species shall communicate the same information as is required under paragraph (1) of this Article to the Government of the Swiss Confederation and to the Office of the Union, at least thirty days before its decision takes effect.

(3) The Government of the French Republic or the Government of the Swiss Confederation, as the case may be, shall immediately communicate to all the member States of the Union the information referred to in paragraphs (1) and (2) of this Article.

Article 34 [Territories]

(1) Every member State of the Union, either on signing or on ratifying or acceding to this Convention, shall declare whether the Convention applies to all or to a part of its territories or to one or more or to all of the States or territories for which it is responsible.

This declaration may be supplemented at any time thereafter by notification to the Government of the Swiss Confederation. Such notification shall take effect thirty days after it has been received by that Government.

(2) The Government which has received the declarations or notifications referred to in paragraph (1) of this Article shall communicate them to all member States of the Union.

Article 35 [Transitional Limitation of the Requirement of Novelty] Notwithstanding the provisions of Article 6, any member State of the Union may, without thereby creating an obligation for other member States of the Union, limit the requirement of novelty laid down in that Article, with regard to varieties of recent creation existing at the date of entry into force of this Convention in respect of such State.

Article 36 [Transitional Rules Concerning the Relationship Between Variety Denominations and Trade Marks)

(1) If, at the date of entry into force of this Convention in respect of a member State of the Union, the breeder of a new variety protected in that State, or his successor in title, enjoys in that State the protection of the denomination of that variety as a trade mark for identical or similar products within the meaning of trade mark law, he may either renounce the protection in respect of the trade mark or submit a new denomination for the variety in the place of the previous denomination. If a new denomination has not been submitted within a period of six months, the breeder or his successor in title may not continue to assert his right to the trade mark for the above-mentioned products.

(2) If a new denomination is registered for the variety, the breeder or his successor in title may not prohibit the use of the previous denomination by persons obliged to use it before the entry into force of this Convention, until a period of one year has expired from the publication of the registration of the new denomination.

Article 37 [Preservation of Existing Rights] This Convention shall not affect existing rights under the national laws of member States of the Union or under agreements concluded between such States.

Article 38 [Settlement of Disputes]

(1) Any dispute between two or more member States of the Union concerning the interpretation or application of this Convention which is not settled by negotiation shall, at the request of one of the States concerned, be submitted to the Council, which shall endeavour to bring about agreement between the member States concerned.

(2) If such agreement is not reached within six months from the date when the dispute was submitted to the Council, the dispute shall be referred to an arbitration tribunal at the request of one of the parties concerned.

(3) The tribunal shall consist of three arbitrators.

Where two member States are parties to a dispute, each of those States shall appoint an arbitrator.

Where more than two member States are parties to a dispute, two of the arbitrators shall be appointed by agreement among the States concerned.

If the States concerned have not appointed the arbitrators within a period of two months from the date on which the request for convening the tribunal was notified to them by the Office of the Union, any of the member States concerned may request the President of the International Court of Justice to make the necessary appointments.

In all cases, the third arbitrator shall be appointed by the President of the International Court of Justice.

If the President is a national of one of the member States parties to the dispute, the Vice-President shall make the appointments referred to above, unless he is himself also a national of one of the member States parties to the dispute. In this last case, the appointments shall be made by the member of the Court who is not a national of one of the member States parties to the dispute and who has been selected by the President to make the appointments.

(4) The award of the tribunal shall be final and binding on the member States concerned.

(5) The tribunal shall determine its own procedure, unless the member States concerned agree otherwise.

(6) Each of the member States parties to the dispute shall bear the costs of its representation before the arbitration tribunal; other costs shall be borne in equal parts by each of the States.

Article 39 [Reservations] Signature and ratification of and accession to this Convention shall not be subject to any reservation.

Article 40 (Duration and Denunciation of the Convention; Discontinuation of the Application of the Convention to Territories]

(1) This Convention shall be of unlimited duration.

(2) Subject to the provisions of paragraph (4) of Article 27, if a member State of the Union denounces this Convention, such denunciation shall take effect one year after the date on which notification of denunciation is made by the Government of the Swiss Confederation to the other member States of the Union.

(3) Any member State may at any time declare that the Convention shall cease to apply to certain of its territories or to States or territories in respect of which it has made a declaration in accordance with the provisions of Article 34. Such declaration shall take effect one year after the date on which notification thereof is made by the Government of the Swiss Confederation to the other member States of the Union.

(4) Such denunciations and declarations shall not affect rights acquired by reason of this Convention prior to the expiration of the time limit laid down in paragraphs (2) and (3) of this Article.

Article 41 [Copies of the Convention; Language and Official Translations of the Convention]

(1) This Convention is drawn up in a single copy in the French language. That copy is deposited in the archives of the Government of the French Republic.

(2) A certified true copy shall be forwarded by that Government to the Governments of all signatory States.

(3) Official translations of this Convention shall be made in the Dutch, English, German, Italian and Spanish languages.

Annex List referred to in Article 4, paragraph (3)

Species to be protected in each genus

1.	Wheat	*Triticum aestivum* L. ssp. *vulgare* (Vill., Host) Mac Kay *Triticum durum* Desf.
2.	Barley	*Hordeum vulgare* L. s. lat.
3.	Oats	*Avena sativa* L. *Avena byzantina* C. Koch
	or Rice	*Oryza sativa* L.
4.	Maize	*Zea mays* L.
5.	Potato	*Solanum tuberosum* L.
6.	Peas	*Pisum sativum* L.
7.	Beans	*Phaseolus vulgaris* L. *Phaseolus coccineus* L.

8.	Lucerne	*Medicago sativa* L. *Medicago varia* Martyn
9.	Red Clover	*Trifolium pratense* L.
10.	Ryegrass	*Lolium* sp.
11.	Lettuce	*Lactuca sativa* L.
12.	Apples	*Malus domestica* Borkh
13.	Roses	*Rosa* hort.
	or Carnations	*Dianthus caryophyllus* L.

If two optional genera are chosen – numbers 3 or 13 above – they shall be counted as one genus only.

ADDITIONAL ACT OF NOVEMBER 10, 1972, AMENDING THE INTERNATIONAL CONVENTION FOR THE PROTECTION OF NEW VARIETIES OF PLANTS

adopted by the Diplomatic Conference on November 10, 1972

Additional Act of November 10, 1972, Amending the International Convention for the Protection of New Varieties of Plants*

Table of Contents**

* Official English translation.
** This Table of Contents is added for the convenience of the reader. It does not appear in the original (French) text of the Additional Act.

The Contracting States, Considering that in the light of the experience gained since the entry into force of the International Convention for the Protection of New Varieties of Plants, of December 2, 1961, the system of contributions of member States of the Union provided for by that Convention does not allow for sufficient differentiation among the member States of the Union as to the share in the total of the contributions that should be allotted to each of them,

Considering further that it is desirable to amend the provisions of that Convention on the contributions of member States of the Union and, in the event of arrears in the payment of such contributions, on the right to vote,

Having regard to the provisions of Article 27 of the said Convention.

Have agreed as follows:

Article I [Amended Version of Article 22 of the Convention (Majorities Required for Decisions of the Council)] [1] Article 22 of the International Convention for the Protection of New Varieties of Plants, of 2 December, 1961 (hereinafter referred to as the Convention), shall be replaced by the following text:

Decisions of the Council shall be taken by a simple majority of the members present, except in the cases provided for in Articles 20, 27, 28 and 32, for the vote on the budget, for the fixing of the contributions of each member State of the Union, for the faculty provided for in paragraph (5) of Article 26 concerning payment of one-half of the contribution corresponding to Class V and for any decision regarding voting rights under paragraph (6) of Article 26. In these last four cases, the majority required shall be three-quarters of the members present.

Article II [Amended Version of Article 26 of the Convention (Finances)] Article 26 of the Convention shall be replaced by the following text:

"(1) The expenses of the Union shall be met from:

(a) annual contributions of member States of the Union;

(b) payments received for services rendered;

(c) miscellaneous receipts".

"(2) For the purpose of determining the amounts of their annual contributions, the member States of the Union shall be divided into five classes:

Class I	5 units
Class II	4 units
Class III	3 units

[1] Articles have been given titles to facilitate their identification. There are no titles in the signed (French) text.

Class IV 2 units
Class V 1 unit"
"Each member State of the Union shall contribute in proportion to the number of units of the class to which it belongs".
"(3) For each budgetary period, the value of the unit of contribution shall be obtained by dividing the total expenditure to be met from the contributions of member States of the Union by the total number of units'.
"(4) Each member State of the Union shall indicate, on joining the Union, the class in which it wishes to be placed. Any member State of the Union may, however, subsequently declare that it wishes to be placed in another class".

"Such declaration must be addressed to the Secretary-General of the Union at least six months before the end of the financial year preceding that in which the change of class is to take effect".
"(5) At the request of a member State of the Union or of a State applying for accession to the Convention according to Article 32 and indicating the wish to be placed in Class V, the Council may, in order to take account of exceptional circumstances, decide to allow such State to pay only one-half of the contribution corresponding to Class V. Such decision will stand until the State concerned waives the faculty granted or declares that it wishes to be placed in another class or until the Council revokes its decision".
"(6) A member State of the Union which is in arrears in the payment of its contributions may not exercise its right to vote in the Council if the amount of its arrears equals or exceeds the amount of the contributions due from it for the preceding two full years, but it shall not be relieved of its obligations under this Convention, nor shall it be deprived of any other rights thereunder. However, the Council may allow such a State to continue to exercise its right to vote if, and as long as, the Council is satisfied that the delay in payment is due to exceptional and unavoidable circumstances."

Article III [Applicability of Paragraph (6) of the Amended Version of Article 26 of the Convention] The provisions of paragraph (6) of Article 26 shall apply only if all member States of the Union have ratified or acceded to this Additional Act.

Article IV [Contribution Classes of Member States] Member States of the Union shall be placed in the class under this Additional Act which contains the same number of units as the class they have chosen under the Convention, unless, at the moment of depositing their instrument of ratification or accession, they express the wish to be placed in another class under this Additional Act.

Article V [Signature; Ratification; Accession]

(1) This Additional Act shall be open for signature until April 1, 1973, by member States of the Union and by signatory States of the Convention.
(2) This Additional Act shall be subject to ratification.
(3) This Additional Act shall be open to accession by non-signatory States in accordance with the provisions of paragraphs (2) and (3) of Article 32 of the Convention.
(4) After the entry into force of this Additional Act, a State may accede to the Convention only if it accedes to this Additional Act at the same time.
(5) Instruments of ratification of or accession to this Additional Act by States which have ratified the Convention or which ratify it at the same time as they ratify or accede to this Additional Act shall be deposited with the Government of the French Republic. Instruments of ratification of or accession to this Additional Act by States which have acceded to the Convention or which accede to it at the same time as they ratify or accede to this Additional Act shall be deposited with the Government of the Swiss Confederation.

Article VI [Entry Into Force]

(1) This Additional Act shall enter into force in accordance with the first and second sentences of paragraph (4) of Article 27 of the Convention.
(2) With respect to any State which deposits its instrument of ratification of or accession to this Additional Act after the date of its entry into force, this Additional Act shall enter into force thirty days after the deposit of such instrument.

Article VII [Reservations] No reservations to this Additional Act are permitted.

Article VIII [Original Copy of the Additional Act; Language and Official Translations of the Additional Act; Notifications; Registration of the Additional Act]

(1) This Additional Act shall be signed in a single original in the French language, which shall be deposited in the archives of the Government of the French Republic.

(2) Official translations of this Additional Act shall be established by the Secretary-General of the Union, after consultation with the interested Governments, in Dutch, English, German, Italian and Spanish, and in such other languages as the Council of the Union may designate. In the latter event, the Secretary-General of the Union shall also establish an official translation of the Convention in the language so designated.

(3) The Secretary-General of the Union shall transmit two copies, certified by the Government of the French Republic, of the signed text of this Additional Act to the Governments of the States referred to in paragraph (1) of Article V, and on request to the Government of any other State.

(4) The Secretary-General of the Union shall register this Additional Act with the Secretariat of the United Nations.

(5) The Government of the French Republic shall notify the Secretary-General of the Union of the signatures of this Additional Act and of the deposit with that Government of instruments of ratification or accession. The Government of the Swiss Confederation shall notify the Secretary-General of the Union of the deposit with that Government of instruments of ratification or accession.

(6) The Secretary-General of the Union shall inform the member States of the Union and the signatory States of the Convention of the notifications received pursuant to the preceding paragraph and of the entry into force of this Additional Act.

Appendix 2 UPOV 1978

ACT OF 1978 INTERNATIONAL CONVENTION FOR THE PROTECTION OF NEW VARIETIES OF PLANTS

of December 2, 1961, as Revised at Geneva on November 10, 1972, and on October 23, 1978

Table of Contents

The Contracting Parties,

Considering that the International Convention for the Protection of New Varieties of Plants of December 2, 1961, amended by the Additional Act of November 10, 1972, has proved a valuable instrument for international cooperation in the field of the protection of the rights of the breeders,

Reaffirming the principles contained in the Preamble to the Convention to the effect that:

(a) they are convinced of the importance attaching to the protection of new varieties of plants not only for the development of agriculture in their territory but also for safeguarding the interests of breeders,

(b) they are conscious of the special problems arising from the recognition and protection of the rights of breeders and particularly of the limitations that the requirements of the public interest may impose on the free exercise of such a right,
(c) they deem it highly desirable that these problems, to which very many States rightly attach importance, should be resolved by each of them in accordance with uniform and clearly defined principles,

Considering that the idea of protecting the rights of breeders has gained general acceptance in many States which have not yet acceded to the Convention,

Considering that certain amendments in the Convention are necessary in order to facilitate the joining of the Union by these States,

Considering that some provisions concerning the administration of the Union created by the Convention require amendment in the light of experience,

Considering that these objectives may be best achieved by a new revision of the Convention,

Have agreed as follows:

Article 1 Purpose of the Convention; Constitution of a Union; Seat of the Union

(1) The purpose of this Convention is to recognise and to ensure to the breeder of a new plant variety or to his successor in title (both hereinafter referred to as "the breeder") a right under the conditions hereinafter defined.
(2) The States parties to this Convention (hereinafter referred to as "the member States of the Union") constitute a Union for the Protection of New Varieties of Plants.
(3) The seat of the Union and its permanent organs shall be at Geneva.

Article 2 Forms of Protection

(1) Each member State of the Union may recognise the right of the breeder provided for in this Convention by the grant either of a special title of protection or of a patent. Nevertheless, a member State of the Union whose national law admits of protection under both these forms may provide only one of them for one and the same botanical genus or species.
(2) Each member State of the Union may limit the application of this Convention within a genus or species to varieties with a particular manner of reproduction or multiplication, or a certain end-use.

Article 3 National Treatment; Reciprocity

(1) Without prejudice to the rights specially provided for in this Convention, natural and legal persons resident or having their registered office in one of the member States of the Union shall, in so far as the recognition and protection of the right of the breeder are concerned, enjoy in the other member States of the Union the same treatment as is accorded or may hereafter be accorded by the respective laws of such States to their own nationals, provided that such persons comply with the conditions and formalities imposed on such nationals.

(2) Nationals of member States of the Union not resident or having their registered office in one of those States shall likewise enjoy the same rights provided that they fulfil such obligations as may be imposed on them for the purpose of enabling the varieties which they have bred to be examined and the multiplication of such varieties to be checked.

(3) Notwithstanding the provisions of paragraph (1) and paragraph (2), any member State of the Union applying this Convention to a given genus or species shall be entitled to limit the benefit of the protection to the nationals of those member States of the Union which apply this Convention to that genus or species and to natural and legal persons resident or having their registered office in any of those States.

Article 4 Botanical Genera and Species Which Must or May be Protected

(1) This Convention may be applied to all botanical genera and species.

(2) The member States of the Union undertake to adopt all measures necessary for the progressive application of the provisions of this Convention to the largest possible number of botanical genera and species.

(3) (a) Each member State of the Union shall, on the entry into force of this Convention in its territory, apply the provisions of this Convention to at least five genera or species.

(b) Subsequently, each member State of the Union shall apply the said provisions to additional genera or species within the following periods from the date of the entry into force of this Convention in its territory:

(i) within three years, to at least ten genera or species in all;

(ii) within six years, to at least eighteen genera or species in all;

(iii) within eight years, to at least twenty-four genera or species in all.

(c) If a member State of the Union has limited the application of this Convention within a genus or species in accordance with the provisions of Article 2(2), that genus or species shall nevertheless, for the purposes of subparagraph *(a)* and subparagraph *(b)*, be considered as one genus or species.

(4) At the request of any State intending to ratify, accept, approve or accede to this Convention, the Council may, in order to take account of special economic or ecological conditions prevailing in that State, decide, for the purpose of that State, to reduce the minimum numbers referred to in paragraph (3), or to extend the periods referred to in that paragraph, or to do both.

(5) At the request of any member State of the Union, the Council may, in order to take account of special difficulties encountered by that State in the fulfilment of the obligations under paragraph (3)*(b)*, decide, for the purposes of that State, to extend the periods referred to in paragraph (3)*(b)*.

Article 5 Rights Protected; Scope of Protection

(1) The effect of the right granted to the breeder is that his prior authorisation shall be required for

- the production for purposes of commercial marketing
- the offering for sale
- the marketing

of the reproductive or vegetative propagating material, as such, of the variety.

Vegetative propagating material shall be deemed to include whole plants. The right of the breeder shall extend to ornamental plants or parts thereof normally marketed for purposes other than propagation when they are used commercially as propagating material in the production of ornamental plants or cut flowers.

(2) The authorisation given by the breeder may be made subject to such conditions as he may specify.

(3) Authorisation by the breeder shall not be required either for the utilisation of the variety as an initial source of variation for the purpose of creating other varieties or for the marketing of such varieties. Such authorisation shall be required, however, when the repeated use of the variety is necessary for the commercial production of another variety.

(4) Any member State of the Union may, either under its own law or by means of special agreements under Article 29, grant to breeders, in respect of certain botanical genera or species, a more extensive right

than that set out in paragraph (1), extending in particular to the marketed product. A member State of the Union which grants such a right may limit the benefit of it to the nationals of member States of the Union which grant an identical right and to natural and legal persons resident or having their registered office in any of those States.

Article 6 Conditions Required for Protection

(1) The breeder shall benefit from the protection provided for in this Convention when the following conditions are satisfied:

(a) Whatever may be the origin, artificial or natural, of the initial variation from which it has resulted, the variety must be clearly distinguishable by one or more important characteristics from any other variety whose existence is a matter of common knowledge at the time when protection is applied for. Common knowledge may be established by reference to various factors such as: cultivation or marketing already in progress, entry in an official register of varieties already made or in the course of being made, inclusion in a reference collection, or precise description in a publication. The characteristics which permit a variety to be defined and distinguished must be capable of precise recognition and description.

(b) At the date on which the application for protection in a member State of the Union is filed, the variety

(i) must not—or, where the law of that State so provides, must not for longer than one year—have been offered for sale or marketed, with the agreement of the breeder, in the territory of that State, and

(ii) not have been offered for sale or marketed, with the agreement of the breeder, in the territory of any other State for longer than six years in the case of vines, forest trees, fruit trees and ornamental trees, including, in each case, their rootstocks, or for longer than four years in the case of all other plants.

Trials of the variety not involving offering for sale or marketing shall not affect the right to protection. The fact that the variety has become a matter of common knowledge in ways other than through offering for sale or marketing shall also not affect the right of the breeder to protection.

(c) The variety must be sufficiently homogeneous, having regard to the particular features of its sexual reproduction or vegetative propagation.

(d) The variety must be stable in its essential characteristics, that is to say, it must remain true to its description after repeated reproduction or propagation or, where the breeder has defined a particular cycle of reproduction or multiplication, at the end of each cycle.
(e) The variety shall be given a denomination as provided in Article 13.

(2) Provided that the breeder shall have complied with the formalities provided for by the national law of the member State of the Union in which the application for protection was filed, including the payment of fees, the grant of protection may not be made subject to conditions other than those set forth above.

Article 7 Official Examination of Varieties; Provisional Protection

(1) Protection shall be granted after examination of the variety in the light of the criteria defined in Article 6. Such examination shall be appropriate to each botanical genus or species.
(2) For the purposes of such examination, the competent authorities of each member State of the Union may require the breeder to furnish all the necessary information, documents, propagating material or seeds.
(3) Any member State of the Union may provide measures to protect the breeder against abusive acts of third parties committed during the period between the filing of the application for protection and the decision thereon.

Article 8 Period of Protection

The right conferred on the breeder shall be granted for a limited period. This period may not be less than fifteen years, computed from the date of issue of the title of protection. For vines, forest trees, fruit trees and ornamental trees, including, in each case, their rootstocks, the period of protection may not be less than eighteen years, computed from the said date.

Article 9 Restrictions in the Exercise of Rights Protected

(1) The free exercise of the exclusive right accorded to the breeder may not be restricted otherwise than for reasons of public interest.
(2) When any such restriction is made in order to ensure the widespread distribution of the variety, the member State of the Union concerned shall take all measures necessary to ensure that the breeder receives equitable remuneration.

Article 10 Nullity and Forfeiture of the Rights Protected

(1) The right of the breeder shall be declared null and void, in accordance with the provisions of the national law of each member State of the Union, if it is established that the conditions laid down in Article 6(1)*(a)* and Article 6(1)*(b)* were not effectively complied with at the time when the title of protection was issued.
(2) The right of the breeder shall become forfeit when he is no longer in a position to provide the competent authority with reproductive or propagating material capable of producing the variety with its characteristics as defined when the protection was granted.
(3) The right of the breeder may become forfeit if:
 (a) after being requested to do so and within a prescribed period, he does not provide the competent authority with the reproductive or propagating material, the documents and the information deemed necessary for checking the variety, or he does not allow inspection of the measures which have been taken for the maintenance of the variety; or
 (b) he has failed to pay within the prescribed period such fees as may be payable to keep his rights in force.
(4) The right of the breeder may not be annulled or become forfeit except on the grounds set out in this Article.

Article 11 Free Choice of the Member State in Which the First Application is Filed; Application in Other Member States; Independence of Protection in Different Member States

(1) The breeder may choose the member State of the Union in which he wishes to file his first application for protection.
(2) The breeder may apply to other member States of the Union for protection of his right without waiting for the issue to him of a title of protection by the member State of the Union in which he filed his first application.
(3) The protection applied for in different member States of the Union by natural or legal persons entitled to benefit under this Convention shall be independent of the protection obtained for the same variety in other States whether or not such States are members of the Union.

Article 12 Right of Priority

(1) Any breeder who has duly filed an application for protection in one of the member States of the Union shall, for the purpose of filing in

the other member States of the Union, enjoy a right of priority for a period of twelve months. This period shall be computed from the date of filing of the first application. The day of filing shall not be included in such period.

(2) To benefit from the provisions of paragraph (1), the further filing must include an application for protection, a claim in respect of the priority of the first application and, within a period of three months, a copy of the documents which constitute that application, certified to be a true copy by the authority which received it.

(3) The breeder shall be allowed a period of four years after the expiration of the period of priority in which to furnish, to the member State of the Union with which he has filed an application for protection in accordance with the terms of paragraph (2), the additional documents and material required by the laws and regulations of that State. Nevertheless, that State may require the additional documents and material to be furnished within an adequate period in the case where the application whose priority is claimed is rejected or withdrawn.

(4) Such matters as the filing of another application of the publication or use of the subject of the application, occurring within the period provided for in paragraph (1), shall not constitute grounds for objection to an application filed in accordance with the foregoing conditions. Such matters may not give rise to any right in favour of a third party or to any right of personal possession.

Article 13 Variety Denomination

(1) The variety shall be designated by a denomination destined to be its generic designation. Each member State of the Union shall ensure that subject to paragraph (4) no rights in the designation registered as the denomination of the variety shall hamper the free use of the denomination in connection with the variety, even after the expiration of the protection.

(2) The denomination must enable the variety to be identified. It may not consist solely of figures except where this is an established practice for designating varieties. It must not be liable to mislead or to cause confusion concerning the characteristics, value or identity of the variety or the identity of the breeder. In particular, it must be different from every denomination which designates, in any member State of the Union, an existing variety of the same botanical species or of a closely related species.

(3) The denomination of the variety shall be submitted by the breeder to the authority referred to in Article 30(1)*(b)*. If it is found that

such denomination does not satisfy the requirements of paragraph (2), that authority shall refuse to register it and shall require the breeder to propose another denomination within a prescribed period. The denomination shall be registered at the same time as the title of protection is issued in accordance with the provisions of Article 7.

(4) Prior rights of third parties shall not be affected. If, by reason of a prior right, the use of the denomination of a variety is forbidden to a person who, in accordance with the provisions of paragraph (7), is obliged to use it, the authority referred to in Article 30(1)*(b)* shall require the breeder to submit another denomination for the variety.

(5) A variety must be submitted in member States of the Union under the same denomination. The authority referred to in Article 30(1)*(b)* shall register the denomination so submitted, unless it considers that denomination unsuitable in its State. In the latter case, it may require the breeder to submit another denomination.

(6) The authority referred to in Article 30(1)*(b)* shall ensure that all the other such authorities are informed of matters concerning variety denominations, in particular the submission, registration and cancellation of denominations. Any authority referred to in Article 30(1)*(b)* may address its observations, if any, on the registration of a denomination to the authority which communicated that denomination.

(7) Any person who, in a member State of the Union, offers for sale or markets reproductive or vegetative propagating material of a variety protected in that State shall be obliged to use the denomination of that variety, even after the expiration of the protection of that variety, in so far as, in accordance with the provisions of paragraph (4), prior rights do not prevent such use.

(8) When the variety is offered for sale or marketed, it shall be permitted to associate a trade mark, trade name or other similar identification with a registered variety denomination. If such an indication is so associated, the denomination must nevertheless be easily recognizable.

Article 14 Protection Independent of Measures Regulating Production, Certification and Marketing

(1) The right accorded to the breeder in pursuance of the provisions of this Convention shall be independent of the measures taken by each member State of the Union to regulate the production, certification and marketing of seeds and propagating material.

(2) However, such measures shall, as far as possible, avoid hindering the application of the provisions of this Convention.

Article 15 Organs of the Union

The permanent organs of the Union shall be:
(a) the Council;
(b) the Secretariat General, entitled the Office of the International Union for the Protection of New Varieties of Plants.

Article 16 Composition of the Council; Votes

(1) The Council shall consist of the representatives of the member States of the Union. Each member State of the Union shall appoint one representative to the Council and one alternate.
(2) Representatives or alternates may be accompanied by assistants or advisers.
(3) Each member State of the Union shall have one vote in the Council.

Article 17 Observers in Meetings of the Council

(1) States not members of the Union which have signed this Act shall be invited as observers to meetings of the Council.
(2) Other observers or experts may also be invited to such meetings.

Article 18 President and Vice-Presidents of the Council

(1) The Council shall elect a President and a first Vice-President from among its members. It may elect other Vice-Presidents. The first Vice-President shall take the place of the President if the latter is unable to officiate.
(2) The President shall hold office for three years.

Article 19 Sessions of the Council

(1) The Council shall meet upon convocation by its President.
(2) An ordinary session of the Council shall be held annually. In addition, the President may convene the Council at his discretion; he shall convene it, within a period of three months, if one-third of the member States of the Union so request.

Article 20 Rules of Procedure of the Council; Administrative and Financial Regulations of the Union

The Council shall establish its rules of procedure and the administrative and financial regulations of the Union.

Article 21 Tasks of the Council

The tasks of the Council shall be to:

(a) study appropriate measures to safeguard the interests and to encourage the development of the Union;
(b) appoint the Secretary-General and, if it finds it necessary, a Vice Secretary-General and determine the terms of appointment of each;
(c) examine the annual report on the activities of the Union and lay down the programme for its future work;
(d) give to the Secretary-General, whose functions are set out in Article 23, all necessary directions for the accomplishment of the tasks of the Union;
(e) examine and approve the budget of the Union and fix the contribution of each member State of the Union in accordance with the provisions of Article 26;
(f) examine and approve the accounts presented by the Secretary-General;
(g) fix, in accordance with the provisions of Article 27, the date and place of the conferences referred to in that Article and take the measures necessary for their preparation; and
(h) in general, take all necessary decisions to ensure the efficient functioning of the Union.

Article 22 Majorities Required for Decisions of the Council

Any decision of the Council shall require a simple majority of the votes of the members present and voting, provided that any decision of the Council under Article 4(4), Article 20, Article 21*(e)*, Article 26(5)*(b)*, Article 27(1), Article 28(3) or Article 32(3) shall require three-fourths of the votes of the members present and voting. Abstentions shall not be considered as votes.

Article 23 Tasks of the Office of the Union; Responsibilities of the Secretary-General; Appointment of Staff

(1) The Office of the Union shall carry out all the duties and tasks entrusted to it by the Council. It shall be under the direction of the Secretary-General.
(2) The Secretary-General shall be responsible to the Council; he shall be responsible for carrying out the decisions of the Council. He shall submit the budget for the approval of the Council and shall be responsible for its implementation. He shall make an annual report

to the Council on his administration and a report on the activities and financial position of the Union.

(3) Subject to the provisions of Article 21 *(b)*, the conditions of appointment and employment of the staff necessary for the efficient performance of the tasks of the Office of the Union shall be fixed in the administrative and financial regulations referred to in Article 20.

Article 24 Legal Status

(1) The Union shall have legal personality.

(2) The Union shall enjoy on the territory of each member State of the Union, in conformity with the laws of that State, such legal capacity as may be necessary for the fulfilment of the objectives of the Union and for the exercise of its functions.

(3) The Union shall conclude a headquarters agreement with the Swiss Confederation.

Article 25 Auditing of the Accounts

The auditing of the accounts of the Union shall be effected by a member State of the Union as provided in the administrative and financial regulations referred to in Article 20. Such State shall be designated, with its agreement, by the Council.

Article 26 Finances

(1) The expenses of the Union shall be met from:
- the annual contributions of the member States of the Union;
- payments received for services rendered;
- miscellaneous receipts.

(2) (a) The share of each member State of the Union in the total amount of the annual contributions shall be determined by reference to the total expenditure to be met from the contributions of the member States of the Union and to the number of contribution units applicable to it under paragraph (3). The said share shall be computed according to paragraph (4).

(b) The number of contribution units shall be expressed in whole numbers or fractions thereof, provided that such number shall not be less than one-fifth.

(3) (a) As far as any State is concerned which is a member State of the Union on the date on which this Act enters into force with respect to that State, the number of contribution units applicable

to it shall be the same as was applicable to it, immediately before the said date, according to the Convention of 1961 as amended by the Additional Act of 1972.

(b) As far as any other State is concerned, that State shall, on joining the Union, indicate, in a declaration addressed to the Secretary-General, the number of contribution units applicable to it.

(c) Any member State of the Union may, at any time, indicate, in a declaration addressed to the Secretary-General, a number of contribution units different from the number applicable to it under subparagraph *(a)* or subparagraph *(b)*. Such declaration, if made during the first six months of a calendar year, shall take effect from the beginning of the subsequent calendar year; otherwise it shall take effect from the beginning of the second calendar year which follows the year in which the declaration was made.

(4) (a) For each budgetary period, the amount corresponding to one contribution unit shall be obtained by dividing the total amount of the expenditure to be met in that period from the contributions of the member States of the Union by the total number of units applicable to those States.

(b) The amount of the contribution of each member State of the Union shall be obtained by multiplying the amount corresponding to one contribution unit by the number of contribution units applicable to that State.

(5) (a) A member State of the Union which is in arrears in the payment of its contributions may not, subject to paragraph *(b)*, exercise its right to vote in the Council if the amount of its arrears equals or exceeds the amount of the contributions due from it for the preceding two full years. The suspension of the right to vote does not relieve such State of its obligations under this Convention and does not deprive it of any other rights thereunder.

(b) The Council may allow the said State to continue to exercise its right to vote if, and as long as, the Council is satisfied that the delay in payment is due to exceptional and unavoidable circumstances.

Article 27 Revision of the Convention

(1) This Convention may be revised by a conference of the member States of the Union. The convocation of such conference shall be decided by the Council.

(2) The proceedings of a conference shall be effective only if at least half of the member States of the Union are represented at it. A majority of

five-sixths of the member States of the Union represented at the conference shall be required for the adoption of a revised text of the Convention.

Article 28 Languages Used by the Office and in Meetings of the Council

(1) The English, French and German languages shall be used by the Office of the Union in carrying out its duties.
(2) Meetings of the Council and of revision conferences shall be held in the three languages.
(3) If the need arises, the Council may decide that further languages shall be used.

Article 29 Special Agreements for the Protection of New Varieties of Plants

Member States of the Union reserve the right to conclude among themselves special agreements for the protection of new varieties of plants, in so far as such agreements do not contravene the provisions of this Convention.

Article 30 Implementation of the Convention on the Domestic Level; Contracts on the Joint Utilisation of Examination Services

(1) Each member State of the Union shall adopt all measures necessary for the application of this Convention; in particular, it shall:
 (a) provide for appropriate legal remedies for the effective defence of the rights provided for in this Convention;
 (b) set up a special authority for the protection of new varieties of plants or entrust such protection to an existing authority;
 (c) ensure that the public is informed of matters concerning such protection, including as a minimum the periodical publication of the list of titles of protection issued.
(2) Contracts may be concluded between the competent authorities of the member States of the Union, with a view to the joint utilisation of the services of the authorities entrusted with the examination of varieties in accordance with the provisions of Article 7 and with assembling the necessary reference collections and documents.
(3) It shall be understood that, on depositing its instrument of ratification, acceptance, approval or accession, each State must be in a position, under its own domestic law, to give effect to the provisions of this Convention.

Article 31 Signature

This Act shall be open for signature by any member State of the Union and any other State which was represented in the Diplomatic Conference adopting this Act. It shall remain open for signature until October 31, 1979.

Article 32 Ratification, Acceptance or Approval; Accession

(1) Any State shall express its consent to be bound by this Act by the deposit of:
   (a) its instrument of ratification, acceptance or approval, if it has signed this Act; or
   (b) its instrument of accession, if it has not signed this Act.
(2) Instruments of ratification, acceptance, approval or accession shall be deposited with the Secretary-General.
(3) Any State which is not a member of the Union and which has not signed this Act shall, before depositing its instrument of accession, ask the Council to advise it in respect of the conformity of its laws with the provisions of this Act. If the decision embodying the advice is positive, the instrument of accession may be deposited.

Article 33 Entry into Force; Closing of Earlier Texts

(1) This Act shall enter into force one month after the following two conditions are fulfilled:
   (a) the number of instruments of ratification, acceptance, approval or accession deposited is not less than five; and
   (b) at least three of the said instruments are instruments deposited by States parties to the Convention of 1961.
(2) With respect to any State which deposits its instrument of ratification, acceptance, approval or accession after the conditions referred to in paragraph (1)*(a)* and paragraph (1)*(b)* have been fulfilled, this Act shall enter into force one month after the deposit of the instrument of the said State.
(3) Once this Act enters into force according to paragraph (1), no State may accede to the Convention of 1961 as amended by the Additional Act of 1972.

Article 34 Relations Between States Bound by Different Texts

(1) Any member State of the Union which, on the day on which this Act enters into force with respect to that State, is bound by the

Convention of 1961 as amended by the Additional Act of 1972 shall, in its relations with any other member State of the Union which is not bound by this Act, continue to apply, until the present Act enters into force also with respect to that other State, the said Convention as amended by the said Additional Act.

(2) Any member State of the Union not bound by this Act ("the former State") may declare, in a notification addressed to the Secretary-General, that it will apply the Convention of 1961 as amended by the Additional Act of 1972 in its relations with any State bound by this Act which becomes a member of the Union through ratification, acceptance or approval of or accession to this Act ("the latter State"). As from the beginning of one month after the date of any such notification and until the entry into force of this Act with respect to the former State, the former State shall apply the Convention of 1961 as amended by the Additional Act of 1972 in its relations with any such latter State, whereas any such latter State shall apply this Act in its relations with the former State.

Article 35 Communications Concerning the Genera and Species Protected; Information to be Published

(1) When depositing its instrument of ratification, acceptance or approval of or accession to this Act, each State which is not a member of the Union shall notify the Secretary-General of the list of genera and species to which, on the entry into force of this Act with respect to that State, it will apply the provisions of this Convention.

(2) The Secretary-General shall, on the basis of communications received from each member State of the Union concerned, publish information on:

(a) the extension of the application of the provisions of this Convention to additional genera and species after the entry into force of this Act with respect to that State;

(b) any use of the faculty provided for in Article 3(3);

(c) the use of any faculty granted by the Council pursuant to Article 4(4) or Article 4(5);

(d) any use of the faculty provided for in Article 5(4), first sentence, with an indication of the nature of the more extensive rights and with a specification of the genera and species to which such rights apply;

(e) any use of the faculty provided for in Article 5(4), second sentence;

(f) the fact that the law of the said State contains a provision as permitted under Article 6(1)*(b)*(i), and the length of the period permitted;

(g) the length of the period referred to in Article 8 if such period is longer than the fifteen years and the eighteen years, respectively, referred to in that Article.

Article 36 Territories

(1) Any State may declare in its instrument of ratification, acceptance, approval or accession, or may inform the Secretary-General by written notification any time thereafter, that this Act shall be applicable to all or part of the territories designated in the declaration or notification.

(2) Any State which has made such a declaration or given such a notification may, at any time, notify the Secretary-General that this Act shall cease to be applicable to all or part of such territories.

(3) (a) Any declaration made under paragraph (1) shall take effect on the same date as the ratification, acceptance, approval or accession in the instrument of which it was included, and any notification given under that paragraph shall take effect three months after its notification by the Secretary-General.

(b) Any notification given under paragraph (2) shall take effect twelve months after its receipt by the Secretary-General.

Article 37 Exceptional Rules for Protection Under Two Forms

(1) Notwithstanding the provisions of Article 2(1), any State which, prior to the end of the period during which this Act is open for signature, provides for protection under the different forms referred to in Article 2(1) for one and the same genus or species, may continue to do so if, at the time of signing this Act or of depositing its instrument of ratification, acceptance or approval of or accession to this Act, it notifies the Secretary-General of that fact.

(2) Where, in a member State of the Union to which paragraph (1) applies, protection is sought under patent legislation, the said State may apply the patentability criteria and the period of protection of the patent legislation to the varieties protected thereunder, notwithstanding the provisions of Article 6(1)*(a)*, Article 6(1)*(b)* and Article 8.

(3) The said State may, at any time, notify the Secretary-General of the withdrawal of the notification it has given under paragraph (1). Such withdrawal shall take effect on the date which the State shall indicate in its notification of withdrawal.

Article 38 Transitional Limitation of the Requirement of Novelty

Notwithstanding the provisions of Article 6, any member State of the Union may, without thereby creating an obligation for other member States of the Union, limit the requirement of novelty laid down in that Article, with regard to varieties of recent creation existing at the date on which such State applies the provisions of this Convention for the first time to the genus or species to which such varieties belong.

Article 39 Preservation of Existing Rights

This Convention shall not affect existing rights under the national laws of member States of the Union or under agreements concluded between such States.

Article 40 Reservations

No reservations to this Convention are permitted.

Article 41 Duration and Denunciation of the Convention

(1) This Convention is of unlimited duration.
(2) Any member State of the Union may denounce this Convention by notification addressed to the Secretary-General. The Secretary-General shall promptly notify all member States of the Union of the receipt of that notification.
(3) The denunciation shall take effect at the end of the calendar year following the year in which the notification was received by the Secretary-General.
(4) The denunciation shall not affect any rights acquired in a variety by reason of this Convention prior to the date on which the denunciation becomes effective.

Article 42 Languages; Depositary Functions

(1) This Act shall be signed in a single original in the French, English and German languages, the French text prevailing in case of any discrepancy among the various texts. The original shall be deposited with the Secretary-General.
(2) The Secretary-General shall transmit two certified copies of this Act to the Governments of all States which were represented in the Diplomatic Conference that adopted it and, on request, to the Government of any other State.

(3) The Secretary-General shall, after consultation with the Governments of the interested States which were represented in the said Conference, establish official texts in the Arabic, Dutch, Italian, Japanese and Spanish languages and such other languages as the Council may designate.
(4) The Secretary-General shall register this Act with the Secretariat of the United Nations.
(5) The Secretary-General shall notify the Governments of the member States of the Union and of the States which, without being members of the Union, were represented in the Diplomatic Conference that adopted it of the signatures of this Act, the deposit of instruments of ratification, acceptance, approval and accession, any notification received under Article 34(2), Article 36(1) and (2), Article 37(1) and Article 37(3) or Article 41(2) and any declaration made under Article 36(1).

Appendix 3 UPOV 1991

ACT OF 1991
INTERNATIONAL CONVENTION FOR THE PROTECTION OF NEW VARIETIES OF PLANTS

of December 2, 1961
as Revised at Geneva on November 10, 1972,
on October 23, 1978, and on March 19, 1991

List of Articles

Chapter I Definitions

Article 1 Definitions For the purposes of this Act:

(i) "this Convention" means the present (1991) Act of the International Convention for the Protection of New Varieties of Plants;

(ii) "Act of 1961/1972" means the International Convention for the Protection of New Varieties of Plants of December 2, 1961, as amended by the Additional Act of November 10, 1972;

(iii) "Act of 1978" means the Act of October 23, 1978, of the International Convention for the Protection of New Varieties of Plants;

(iv) "breeder" means
   - the person who bred, or discovered and developed, a variety,
   - the person who is the employer of the aforementioned person or who has commissioned the latter's work, where the laws of the relevant Contracting Party so provide, or
   - the successor in title of the first or second aforementioned person, as the case may be;

(v) "breeder's right" means the right of the breeder provided for in this Convention;

(vi) "variety" means a plant grouping within a single botanical taxon of the lowest known rank, which grouping, irrespective of whether the conditions for the grant of a breeder's right are fully met, can be
   - defined by the expression of the characteristics resulting from a given genotype or combination of genotypes,
   - distinguished from any other plant grouping by the expression of at least one of the said characteristics and
   - considered as a unit with regard to its suitability for being propagated unchanged;

(vii) "Contracting Party" means a State or an intergovernmental organization party to this Convention;

(viii) "territory," in relation to a Contracting Party, means, where the Contracting Party is a State, the territory of that State and, where the Contracting Party is an intergovernmental organization, the territory in which the constituting treaty of that intergovernmental organization applies;

(ix) "authority" means the authority referred to in Article 30(1)(ii);

(x) "Union" means the Union for the Protection of New Varieties of Plants founded by the Act of 1961 and further mentioned in the Act of 1972, the Act of 1978 and in this Convention;

(xi) "member of the Union" means a State party to the Act of 1961/1972 or the Act of 1978, or a Contracting Party.

Chapter II General Obligations of the Contracting Parties

Article 2 Basic Obligation of the Contracting Parties Each Contracting Party shall grant and protect breeders' rights.

Article 3 Genera and Species to be Protected

(1) [*States already members of the Union*] Each Contracting Party which is bound by the Act of 1961/1972 or the Act of 1978 shall apply the provisions of this Convention,
   (i) at the date on which it becomes bound by this Convention, to all plant genera and species to which it applies, on the said date, the provisions of the Act of 1961/1972 or the Act of 1978 and,
   (ii) at the latest by the expiration of a period of five years after the said date, to all plant genera and species.

(2) [*New members of the Union*] Each Contracting Party which is not bound by the Act of 1961/1972 or the Act of 1978 shall apply the provisions of this Convention,
   (i) at the date on which it becomes bound by this Convention, to at least 15 plant genera or species and,
   (ii) at the latest by the expiration of a period of 10 years from the said date, to all plant genera and species.

Article 4 National Treatment

(1) [*Treatment*] Without prejudice to the rights specified in this Convention, nationals of a Contracting Party as well as natural persons resident and legal entities having their registered offices within the territory of a Contracting Party shall, insofar as the grant and protection of breeders' rights are concerned, enjoy within the territory of each other Contracting Party the same treatment as is accorded or may hereafter be accorded by the laws of each such other Contracting Party to its own nationals, provided that the said nationals, natural persons or legal entities comply with the conditions and formalities imposed on the nationals of the said other Contracting Party.

(2) [*"Nationals"*] For the purposes of the preceding paragraph, "nationals" means, where the Contracting Party is a State, the nationals of that State and, where the Contracting Party is an intergovernmental organization, the nationals of the States which are members of that organization.

Chapter III Conditions for the Grant of the Breeder's Right

Article 5 Conditions of Protection

(1) [*Criteria to be satisfied*] The breeder's right shall be granted where the variety is
   (i) new,
   (ii) distinct,

(iii) uniform and
(iv) stable.

(2) [*Other conditions*] The grant of the breeder's right shall not be subject to any further or different conditions, provided that the variety is designated by a denomination in accordance with the provisions of Article 20, that the applicant complies with the formalities provided for by the law of the Contracting Party with whose authority the application has been filed and that he pays the required fees.

Article 6 Novelty

(1) [*Criteria*] The variety shall be deemed to be new if, at the date of filing of the application for a breeder's right, propagating or harvested material of the variety has not been sold or otherwise disposed of to others, by or with the consent of the breeder, for purposes of exploitation of the variety
  (i) in the territory of the Contracting Party in which the application has been filed earlier than one year before that date and
  (ii) in a territory other than that of the Contracting Party in which the application has been filed earlier than four years or, in the case of trees or of vines, earlier than six years before the said date.

(2) [*Varieties of recent creation*] Where a Contracting Party applies this Convention to a plant genus or species to which it did not previously apply this Convention or an earlier Act, it may consider a variety of recent creation existing at the date of such extension of protection to satisfy the condition of novelty defined in paragraph (1) even where the sale or disposal to others described in that paragraph took place earlier than the time limits defined in that paragraph.

(3) [*"Territory" in certain cases*] For the purposes of paragraph (1), all the Contracting Parties which are member States of one and the same intergovernmental organization may act jointly, where the regulations of that organization so require, to assimilate acts done on the territories of the States members of that organization to acts done on their own territories and, should they do so, shall notify the Secretary-General accordingly.

Article 7 Distinctness The variety shall be deemed to be distinct if it is clearly distinguishable from any other variety whose existence is a matter of common knowledge at the time of the filing of the application. In particular, the filing of an application for the granting of a breeder's right or for the entering of another variety in an official register of varieties, in any country, shall be deemed to render that other variety a matter of common knowledge from the date of the application, provided that the application leads to the granting of a breeder's right or to the entering of the said other variety in the official register of varieties, as the case may be.

Article 8 Uniformity The variety shall be deemed to be uniform if, subject to the variation that may be expected from the particular features of its propagation, it is sufficiently uniform in its relevant characteristics.

Article 9 Stability The variety shall be deemed to be stable if its relevant characteristics remain unchanged after repeated propagation or, in the case of a particular cycle of propagation, at the end of each such cycle.

Chapter IV Application for the Grant of the Breeder's Right

Article 10 Filing of Applications

(1) [*Place of first application*] The breeder may choose the Contracting Party with whose authority he wishes to file his first application for a breeder's right.
(2) [*Time of subsequent applications*] The breeder may apply to the authorities of other Contracting Parties for the grant of breeders' rights without waiting for the grant to him of a breeder's right by the authority of the Contracting Party with which the first application was filed.
(3) [*Independence of protection*] No Contracting Party shall refuse to grant a breeder's right or limit its duration on the ground that protection for the same variety has not been applied for, has been refused or has expired in any other State or intergovernmental organization.

Article 11 Right of Priority

(1) [*The right; its period*] Any breeder who has duly filed an application for the protection of a variety in one of the Contracting Parties (the "first application") shall, for the purpose of filing an application for the grant of a breeder's right for the same variety with the authority of any other Contracting Party (the "subsequent application"), enjoy a right of priority for a period of 12 months. This period shall be computed from the date of filing of the first application. The day of filing shall not be included in the latter period.
(2) [*Claiming the right*] In order to benefit from the right of priority, the breeder shall, in the subsequent application, claim the priority of the first application. The authority with which the subsequent application has been filed may require the breeder to furnish, within a

period of not less than three months from the filing date of the subsequent application, a copy of the documents which constitute the first application, certified to be a true copy by the authority with which that application was filed, and samples or other evidence that the variety which is the subject matter of both applications is the same.

(3) [*Documents and material*] The breeder shall be allowed a period of two years after the expiration of the period of priority or, where the first application is rejected or withdrawn, an appropriate time after such rejection or withdrawal, in which to furnish, to the authority of the Contracting Party with which he has filed the subsequent application, any necessary information, document or material required for the purpose of the examination under Article 12, as required by the laws of that Contracting Party.

(4) [*Events occurring during the period*] Events occurring within the period provided for in paragraph (1), such as the filing of another application or the publication or use of the variety that is the subject of the first application, shall not constitute a ground for rejecting the subsequent application. Such events shall also not give rise to any third-party right.

Article 12 Examination of the Application Any decision to grant a breeder's right shall require an examination for compliance with the conditions under Articles 5 to 9. In the course of the examination, the authority may grow the variety or carry out other necessary tests, cause the growing of the variety or the carrying out of other necessary tests, or take into account the results of growing tests or other trials which have already been carried out. For the purposes of examination, the authority may require the breeder to furnish all the necessary information, documents or material.

Article 13 Provisional Protection Each Contracting Party shall provide measures designed to safeguard the interests of the breeder during the period between the filing or the publication of the application for the grant of a breeder's right and the grant of that right. Such measures shall have the effect that the holder of a breeder's right shall at least be entitled to equitable remuneration from any person who, during the said period, has carried out acts which, once the right is granted, require the breeder's authorization as provided in Article 14. A Contracting Party may provide that the said measures shall only take effect in relation to persons whom the breeder has notified of the filing of the application.

Chapter V The Rights of the Breeder

Article 14 Scope of the Breeder's Right

(1) [*Acts in respect of the propagating material*] *(a)* Subject to Articles 15 and 16, the following acts in respect of the propagating material of the protected variety shall require the authorization of the breeder:

(i) production or reproduction (multiplication),
(ii) conditioning for the purpose of propagation,
(iii) offering for sale,
(iv) selling or other marketing,
(v) exporting,
(vi) importing,
(vii) stocking for any of the purposes mentioned in (i) to (vi), above.

(b) The breeder may make his authorization subject to conditions and limitations.

(2) [*Acts in respect of the harvested material*] Subject to Articles 15 and 16, the acts referred to in items (i) to (vii) of paragraph (1)*(a)* in respect of harvested material, including entire plants and parts of plants, obtained through the unauthorized use of propagating material of the protected variety shall require the authorization of the breeder, unless the breeder has had reasonable opportunity to exercise his right in relation to the said propagating material.

(3) [*Acts in respect of certain products*] Each Contracting Party may provide that, subject to Articles 15 and 16, the acts referred to in items (i) to (vii) of paragraph (1)*(a)* in respect of products made directly from harvested material of the protected variety falling within the provisions of paragraph (2) through the unauthorized use of the said harvested material shall require the authorization of the breeder, unless the breeder has had reasonable opportunity to exercise his right in relation to the said harvested material.

(4) [*Possible additional acts*] Each Contracting Party may provide that, subject to Articles 15 and 16, acts other than those referred to in items (i) to (vii) of paragraph (1)*(a)* shall also require the authorization of the breeder.

(5) [*Essentially derived and certain other varieties*] *(a)* The provisions of paragraphs (1) to (4) shall also apply in relation to

(i) varieties which are essentially derived from the protected variety, where the protected variety is not itself an essentially derived variety,
(ii) varieties which are not clearly distinguishable in accordance with Article 7 from the protected variety and

(iii) varieties whose production requires the repeated use of the protected variety.

(b) For the purposes of subparagraph *(a)*(i), a variety shall be deemed to be essentially derived from another variety ("the initial variety") when

(i) it is predominantly derived from the initial variety, or from a variety that is itself predominantly derived from the initial variety, while retaining the expression of the essential characteristics that result from the genotype or combination of genotypes of the initial variety,

(ii) it is clearly distinguishable from the initial variety and

(iii) except for the differences which result from the act of derivation, it conforms to the initial variety in the expression of the essential characteristics that result from the genotype or combination of genotypes of the initial variety.

(c) Essentially derived varieties may be obtained for example by the selection of a natural or induced mutant, or of a somaclonal variant, the selection of a variant individual from plants of the initial variety, backcrossing, or transformation by genetic engineering.

Article 15 Exceptions to the Breeder's Right

(1) [*Compulsory exceptions*] The breeder's right shall not extend to

(i) acts done privately and for non-commercial purposes,

(ii) acts done for experimental purposes and

(iii) acts done for the purpose of breeding other varieties, and, except where the provisions of Article 14(5) apply, acts referred to in Article 14(1) to (4) in respect of such other varieties.

(2) [*Optional exception*] Notwithstanding Article 14, each Contracting Party may, within reasonable limits and subject to the safeguarding of the legitimate interests of the breeder, restrict the breeder's right in relation to any variety in order to permit farmers to use for propagating purposes, on their own holdings, the product of the harvest which they have obtained by planting, on their own holdings, the protected variety or a variety covered by Article 14(5)*(a)*(i) or (ii).

Article 16 Exhaustion of the Breeder's Right

(1) [*Exhaustion of right*] The breeder's right shall not extend to acts concerning any material of the protected variety, or of a variety covered by the provisions of Article 14(5), which has been sold or otherwise marketed by the breeder or with his consent in the territory of the Contracting Party concerned, or any material derived from the said material, unless such acts

(i) involve further propagation of the variety in question or
(ii) involve an export of material of the variety, which enables the propagation of the variety, into a country which does not protect varieties of the plant genus or species to which the variety belongs, except where the exported material is for final consumption purposes.

(2) [*Meaning of "material"*] For the purposes of paragraph (1), "material" means, in relation to a variety,
(i) propagating material of any kind,
(ii) harvested material, including entire plants and parts of plants, and
(iii) any product made directly from the harvested material.

(3) [*"Territory" in certain cases*] For the purposes of paragraph (1), all the Contracting Parties which are member States of one and the same intergovernmental organization may act jointly, where the regulations of that organization so require, to assimilate acts done on the territories of the States members of that organization to acts done on their own territories and, should they do so, shall notify the Secretary-General accordingly.

Article 17 Restrictions on the Exercise of the Breeder's Right

(1) [*Public interest*] Except where expressly provided in this Convention, no Contracting Party may restrict the free exercise of a breeder's right for reasons other than of public interest.
(2) [*Equitable remuneration*] When any such restriction has the effect of authorizing a third party to perform any act for which the breeder's authorization is required, the Contracting Party concerned shall take all measures necessary to ensure that the breeder receives equitable remuneration.

Article 18 Measures Regulating Commerce The breeder's right shall be independent of any measure taken by a Contracting Party to regulate within its territory the production, certification and marketing of material of varieties or the importing or exporting of such material. In any case, such measures shall not affect the application of the provisions of this Convention.

Article 19 Duration of the Breeder's Right

(1) [*Period of protection*] The breeder's right shall be granted for a fixed period.
(2) [*Minimum period*] The said period shall not be shorter than 20 years from the date of the grant of the breeder's right. For

trees and vines, the said period shall not be shorter than 25 years from the said date.

Chapter VI Variety Denomination

Article 20 Variety Denomination

(1) [*Designation of varieties by denominations; use of the denomination*]

(a) The variety shall be designated by a denomination which will be its generic designation.

(b) Each Contracting Party shall ensure that, subject to paragraph (4), no rights in the designation registered as the denomination of the variety shall hamper the free use of the denomination in connection with the variety, even after the expiration of the breeder's right.

(2) [*Characteristics of the denomination*] The denomination must enable the variety to be identified. It may not consist solely of figures except where this is an established practice for designating varieties. It must not be liable to mislead or to cause confusion concerning the characteristics, value or identity of the variety or the identity of the breeder. In particular, it must be different from every denomination which designates, in the territory of any Contracting Party, an existing variety of the same plant species or of a closely related species.

(3) [*Registration of the denomination*] The denomination of the variety shall be submitted by the breeder to the authority. If it is found that the denomination does not satisfy the requirements of paragraph (2), the authority shall refuse to register it and shall require the breeder to propose another denomination within a prescribed period. The denomination shall be registered by the authority at the same time as the breeder's right is granted.

(4) [*Prior rights of third persons*] Prior rights of third persons shall not be affected. If, by reason of a prior right, the use of the denomination of a variety is forbidden to a person who, in accordance with the provisions of paragraph (7), is obliged to use it, the authority shall require the breeder to submit another denomination for the variety.

(5) [*Same denomination in all Contracting Parties*] A variety must be submitted to all Contracting Parties under the same denomination. The authority of each Contracting Party shall register the denomination so submitted, unless it considers the denomination unsuitable within its territory. In the latter case, it shall require the breeder to submit another denomination.

(6) [*Information among the authorities of Contracting Parties*] The authority of a Contracting Party shall ensure that the authorities of all the other

Contracting Parties are informed of matters concerning variety denominations, in particular the submission, registration and cancellation of denominations. Any authority may address its observations, if any, on the registration of a denomination to the authority which communicated that denomination.

(7) [*Obligation to use the denomination*] Any person who, within the territory of one of the Contracting Parties, offers for sale or markets propagating material of a variety protected within the said territory shall be obliged to use the denomination of that variety, even after the expiration of the breeder's right in that variety, except where, in accordance with the provisions of paragraph (4), prior rights prevent such use.

(8) [*Indications used in association with denominations*] When a variety is offered for sale or marketed, it shall be permitted to associate a trademark, trade name or other similar indication with a registered variety denomination. If such an indication is so associated, the denomination must nevertheless be easily recognizable.

Chapter VII Nullity and Cancellation of the Breeder's Right

Article 21 Nullity of the Breeder's Right

(1) [*Reasons of nullity*] Each Contracting Party shall declare a breeder's right granted by it null and void when it is established

(i) that the conditions laid down in Articles 6 or 7 were not complied with at the time of the grant of the breeder's right,

(ii) that, where the grant of the breeder's right has been essentially based upon information and documents furnished by the breeder, the conditions laid down in Articles 8 or 9 were not complied with at the time of the grant of the breeder's right, or

(iii) that the breeder's right has been granted to a person who is not entitled to it, unless it is transferred to the person who is so entitled.

(2) [*Exclusion of other reasons*] No breeder's right shall be declared null and void for reasons other than those referred to in paragraph (1).

Article 22 Cancellation of the Breeder's Right

(1) [*Reasons for cancellation*] *(a)* Each Contracting Party may cancel a breeder's right granted by it if it is established that the conditions laid down in Articles 8 or 9 are no longer fulfilled.

(b) Furthermore, each Contracting Party may cancel a breeder's right granted by it if, after being requested to do so and within a prescribed period,

(i) the breeder does not provide the authority with the information, documents or material deemed necessary for verifying the maintenance of the variety,
(ii) the breeder fails to pay such fees as may be payable to keep his right in force, or
(iii) the breeder does not propose, where the denomination of the variety is cancelled after the grant of the right, another suitable denomination.

(2) [*Exclusion of other reasons*] No breeder's right shall be cancelled for reasons other than those referred to in paragraph (1).

Chapter VIII The Union

Article 23 Members The Contracting Parties shall be members of the Union.

Article 24 Legal Status and Seat

(1) [*Legal personality*] The Union has legal personality.
(2) [*Legal capacity*] The Union enjoys on the territory of each Contracting Party, in conformity with the laws applicable in the said territory, such legal capacity as may be necessary for the fulfillment of the objectives of the Union and for the exercise of its functions.
(3) [*Seat*] The seat of the Union and its permanent organs are at Geneva.
(4) [*Headquarters agreement*] The Union has a headquarters agreement with the Swiss Confederation.

Article 25 Organs The permanent organs of the Union are the Council and the Office of the Union.

Article 26 The Council

(1) [*Composition*] The Council shall consist of the representatives of the members of the Union. Each member of the Union shall appoint one representative to the Council and one alternate. Representatives or alternates may be accompanied by assistants or advisers.
(2) [*Officers*] The Council shall elect a President and a first Vice-President from among its members. It may elect other Vice-Presidents. The first Vice-President shall take the place of the President if the latter is unable to officiate. The President shall hold office for three years.
(3) [*Sessions*] The Council shall meet upon convocation by its President. An ordinary session of the Council shall be held annually. In addition, the President may convene the Council at his discretion; he shall convene it, within a period of three months, if one-third of the members of the Union so request.

(4) [*Observers*] States not members of the Union may be invited as observers to meetings of the Council. Other observers, as well as experts, may also be invited to such meetings.

(5) [*Tasks*] The tasks of the Council shall be to:

(i) study appropriate measures to safeguard the interests and to encourage the development of the Union;

(ii) establish its rules of procedure;

(iii) appoint the Secretary-General and, if it finds it necessary, a Vice Secretary-General and determine the terms of appointment of each;

(iv) examine an annual report on the activities of the Union and lay down the program for its future work;

(v) give to the Secretary-General all necessary directions for the accomplishment of the tasks of the Union;

(vi) establish the administrative and financial regulations of the Union;

(vii) examine and approve the budget of the Union and fix the contribution of each member of the Union;

(viii) examine and approve the accounts presented by the Secretary-General;

(ix) fix the date and place of the conferences referred to in Article 38 and take the measures necessary for their preparation; and

(x) in general, take all necessary decisions to ensure the efficient functioning of the Union.

(6) [*Votes*] *(a)* Each member of the Union that is a State shall have one vote in the Council.

(b) Any Contracting Party that is an intergovernmental organization may, in matters within its competence, exercise the rights to vote of its member States that are members of the Union. Such an intergovernmental organization shall not exercise the rights to vote of its member States if its member States exercise their right to vote, and vice versa.

(7) [*Majorities*] Any decision of the Council shall require a simple majority of the votes cast, provided that any decision of the Council under paragraphs (5)(ii), (vi) and (vii), and under Articles 28(3), 29(5)*(b)* and 38(1) shall require three-fourths of the votes cast. Abstentions shall not be considered as votes.

Article 27 The Office of the Union

(1) [*Tasks and direction of the Office*] The Office of the Union shall carry out all the duties and tasks entrusted to it by the Council. It shall be under the direction of the Secretary-General.

(2) [*Duties of the Secretary-General*] The Secretary-General shall be responsible to the Council; he shall be responsible for carrying out the decisions of the Council. He shall submit the budget of the Union for the approval of the Council and shall be responsible for its implementation. He shall make reports to the Council on his administration and the activities and financial position of the Union.

(3) [*Staff*] Subject to the provisions of Article 26(5)(iii), the conditions of appointment and employment of the staff necessary for the efficient performance of the tasks of the Office of the Union shall be fixed in the administrative and financial regulations.

Article 28 Languages

(1) [*Languages of the Office*] The English, French, German and Spanish languages shall be used by the Office of the Union in carrying out its duties.

(2) [*Languages in certain meetings*] Meetings of the Council and of revision conferences shall be held in the four languages.

(3) [*Further languages*] The Council may decide that further languages shall be used.

Article 29 Finances

(1) [*Income*] The expenses of the Union shall be met from

(i) the annual contributions of the States members of the Union,

(ii) payments received for services rendered,

(iii) miscellaneous receipts.

(2) [*Contributions: units*] *(a)* The share of each State member of the Union in the total amount of the annual contributions shall be determined by reference to the total expenditure to be met from the contributions of the States members of the Union and to the number of contribution units applicable to it under paragraph (3). The said share shall be computed according to paragraph (4).

(b) The number of contribution units shall be expressed in whole numbers or fractions thereof, provided that no fraction shall be smaller than one-fifth.

(3) [*Contributions: share of each member*] *(a)* The number of contribution units applicable to any member of the Union which is party to the Act of 1961/1972 or the Act of 1978 on the date on which it becomes bound by this Convention shall be the same as the number applicable to it immediately before the said date.

(b) Any other State member of the Union shall, on joining the Union, indicate, in a declaration addressed to the Secretary-General, the number of contribution units applicable to it.

(*c*) Any State member of the Union may, at any time, indicate, in a declaration addressed to the Secretary-General, a number of contribution units different from the number applicable to it under subparagraph *(a)* or *(b)*. Such declaration, if made during the first six months of a calendar year, shall take effect from the beginning of the subsequent calendar year; otherwise, it shall take effect from the beginning of the second calendar year which follows the year in which the declaration was made.

(4) [*Contributions: computation of shares*] *(a)* For each budgetary period, the amount corresponding to one contribution unit shall be obtained by dividing the total amount of the expenditure to be met in that period from the contributions of the States members of the Union by the total number of units applicable to those States members of the Union.

(*b*) The amount of the contribution of each State member of the Union shall be obtained by multiplying the amount corresponding to one contribution unit by the number of contribution units applicable to that State member of the Union.

(5) [*Arrears in contributions*] *(a)* A State member of the Union which is in arrears in the payment of its contributions may not, subject to subparagraph *(b)*, exercise its right to vote in the Council if the amount of its arrears equals or exceeds the amount of the contribution due from it for the preceding full year. The suspension of the right to vote shall not relieve such State member of the Union of its obligations under this Convention and shall not deprive it of any other rights thereunder.

(*b*) The Council may allow the said State member of the Union to continue to exercise its right to vote if, and as long as, the Council is satisfied that the delay in payment is due to exceptional and unavoidable circumstances.

(6) [*Auditing of the accounts*] The auditing of the accounts of the Union shall be effected by a State member of the Union as provided in the administrative and financial regulations. Such State member of the Union shall be designated, with its agreement, by the Council.

(7) [*Contributions of intergovernmental organizations*] Any Contracting Party which is an intergovernmental organization shall not be obliged to pay contributions. If, nevertheless, it chooses to pay contributions, the provisions of paragraphs (1) to (4) shall be applied accordingly.

Chapter IX Implementation of the Convention; Other Agreements

Article 30 Implementation of the Convention

(1) [*Measures of implementation*] Each Contracting Party shall adopt all measures necessary for the implementation of this Convention; in particular, it shall:

(i) provide for appropriate legal remedies for the effective enforcement of breeders' rights;
(ii) maintain an authority entrusted with the task of granting breeders' rights or entrust the said task to an authority maintained by another Contracting Party;
(iii) ensure that the public is informed through the regular publication of information concerning
– applications for and grants of breeders' rights, and
– proposed and approved denominations.

(2) [*Conformity of laws*] It shall be understood that, on depositing its instrument of ratification, acceptance, approval or accession, as the case may be, each State or intergovernmental organization must be in a position, under its laws, to give effect to the provisions of this Convention.

Article 31 Relations Between Contracting Parties and States Bound by Earlier Acts

(1) [*Relations between States bound by this Convention*] Between States members of the Union which are bound both by this Convention and any earlier Act of the Convention, only this Convention shall apply.

(2) [*Possible relations with States not bound by this Convention*] Any State member of the Union not bound by this Convention may declare, in a notification addressed to the Secretary-General, that, in its relations with each member of the Union bound only by this Convention, it will apply the latest Act by which it is bound. As from the expiration of one month after the date of such notification and until the State member of the Union making the declaration becomes bound by this Convention, the said member of the Union shall apply the latest Act by which it is bound in its relations with each of the members of the Union bound only by this Convention, whereas the latter shall apply this Convention in respect of the former.

Article 32 Special Agreements Members of the Union reserve the right to conclude among themselves special agreements for the protection of varieties, insofar as such agreements do not contravene the provisions of this Convention.

Chapter X Final Provisions

Article 33 Signature This Convention shall be open for signature by any State which is a member of the Union at the date of its adoption. It shall remain open for signature until March 31, 1992.

Article 34 Ratification, Acceptance or Approval; Accession

(1) [*States and certain intergovernmental organizations*] *(a)* Any State may, as provided in this Article, become party to this Convention.

(b) Any intergovernmental organization may, as provided in this Article, become party to this Convention if it

(i) has competence in respect of matters governed by this Convention,

(ii) has its own legislation providing for the grant and protection of breeders' rights binding on all its member States and

(iii) has been duly authorized, in accordance with its internal procedures, to accede to this Convention.

(2) [*Instrument of adherence*] Any State which has signed this Convention shall become party to this Convention by depositing an instrument of ratification, acceptance or approval of this Convention. Any State which has not signed this Convention and any intergovernmental organization shall become party to this Convention by depositing an instrument of accession to this Convention. Instruments of ratification, acceptance, approval or accession shall be deposited with the Secretary-General.

(3) [*Advice of the Council*] Any State which is not a member of the Union and any intergovernmental organization shall, before depositing its instrument of accession, ask the Council to advise it in respect of the conformity of its laws with the provisions of this Convention. If the decision embodying the advice is positive, the instrument of accession may be deposited.

Article 35 Reservations

(1) [*Principle*] Subject to paragraph (2), no reservations to this Convention are permitted.

(2) [*Possible exception*] *(a)* Notwithstanding the provisions of Article 3 (1), any State which, at the time of becoming party to this Convention, is a party to the Act of 1978 and which, as far as varieties reproduced asexually are concerned, provides for protection by an industrial property title other than a breeder's right shall have the right to continue to do so without applying this Convention to those varieties.

(b) Any State making use of the said right shall, at the time of depositing its instrument of ratification, acceptance, approval or accession, as the case may be, notify the Secretary-General accordingly. The same State may, at any time, withdraw the said notification.

Article 36 Communications Concerning Legislation and the Genera and Species Protected; Information to be Published

(1) [*Initial notification*] When depositing its instrument of ratification, acceptance or approval of or accession to this Convention, as the case may be, any State or intergovernmental organization shall notify the Secretary-General of

(i) its legislation governing breeder's rights and

(ii) the list of plant genera and species to which, on the date on which it will become bound by this Convention, it will apply the provisions of this Convention.

(2) [*Notification of changes*] Each Contracting Party shall promptly notify the Secretary-General of

(i) any changes in its legislation governing breeders' rights and

(ii) any extension of the application of this Convention to additional plant genera and species.

(3) [*Publication of the information*] The Secretary-General shall, on the basis of communications received from each Contracting Party concerned, publish information on

(i) the legislation governing breeders' rights and any changes in that legislation, and

(ii) the list of plant genera and species referred to in paragraph (1)(ii) and any extension referred to in paragraph (2)(ii).

Article 37 Entry into Force; Closing of Earlier Acts

(1) [*Initial entry into force*] This Convention shall enter into force one month after five States have deposited their instruments of ratification, acceptance, approval or accession, as the case may be, provided that at least three of the said instruments have been deposited by States party to the Act of 1961/1972 or the Act of 1978.

(2) [*Subsequent entry into force*] Any State not covered by paragraph (1) or any intergovernmental organization shall become bound by this Convention one month after the date on which it has deposited its instrument of ratification, acceptance, approval or accession, as the case may be.

(3) [*Closing of the 1978 Act*] No instrument of accession to the Act of 1978 may be deposited after the entry into force of this Convention according to paragraph (1), except that any State that, in conformity with the established practice of the General Assembly of the United Nations, is regarded as a developing country may deposit such an instrument until December 31, 1995, and that any other State may deposit such an instrument until December 31, 1993, even if this Convention enters into force before that date.

Article 38 Revision of the Convention

(1) [*Conference*] This Convention may be revised by a conference of the members of the Union. The convocation of such conference shall be decided by the Council.

(2) [*Quorum and majority*] The proceedings of a conference shall be effective only if at least half of the States members of the Union are represented at it. A majority of three-quarters of the States members of the Union present and voting at the conference shall be required for the adoption of any revision.

Article 39 Denunciation

(1) [*Notifications*] Any Contracting Party may denounce this Convention by notification addressed to the Secretary-General. The Secretary-General shall promptly notify all members of the Union of the receipt of that notification.

(2) [*Earlier Acts*] Notification of the denunciation of this Convention shall be deemed also to constitute notification of the denunciation of any earlier Act by which the Contracting Party denouncing this Convention is bound.

(3) [*Effective date*] The denunciation shall take effect at the end of the calendar year following the year in which the notification was received by the Secretary-General.

(4) [*Acquired rights*] The denunciation shall not affect any rights acquired in a variety by reason of this Convention or any earlier Act prior to the date on which the denunciation becomes effective.

Article 40 Preservation of Existing Rights This Convention shall not limit existing breeders' rights under the laws of Contracting Parties or by reason of any earlier Act or any agreement other than this Convention concluded between members of the Union.

Article 41 Original and Official Texts of the Convention

(1) [*Original*] This Convention shall be signed in a single original in the English, French and German languages, the French text prevailing in case of any discrepancy among the various texts. The original shall be deposited with the Secretary-General.

(2) [*Official texts*] The Secretary-General shall, after consultation with the interested Governments, establish official texts of this Convention in the Arabic, Dutch, Italian, Japanese and Spanish languages and such other languages as the Council may designate.

Article 42 Depositary Functions

(1) [*Transmittal of copies*] The Secretary-General shall transmit certified copies of this Convention to all States and intergovernmental organizations which were represented in the Diplomatic Conference that adopted this Convention and, on request, to any other State or intergovernmental organization.

(2) [*Registration*] The Secretary-General shall register this Convention with the Secretariat of the United Nations.

Resolution on Article 14(5) The Diplomatic Conference for the Revision of the International Convention for the Protection of New Varieties of Plants held from March 4 to 19, 1991, requests the Secretary-General of UPOV to start work immediately after the Conference on the establishment of draft standard guidelines, for adoption by the Council of UPOV, on essentially derived varieties.

Recommendation Relating to Article 15(2) The Diplomatic Conference recommends that the provisions laid down in Article 15(2) of the International Convention for the Protection of New Varieties of Plants of December 2, 1961, as Revised at Geneva on November 10, 1972, on October 23, 1978, and on March 19, 1991, should not be read so as to be intended to open the possibility of extending the practice commonly called "farmer's privilege" to sectors of agricultural or horticultural production in which such a privilege is not a common practice on the territory of the Contracting Party concerned.

Common Statement Relating to Article 34 The Diplomatic Conference noted and accepted a declaration by the Delegation of Denmark and a declaration by the Delegation of the Netherlands according to which the Convention adopted by the Diplomatic Conference will not, upon its ratification, acceptance, approval or accession by Denmark or the Netherlands, be automatically applicable, in the case of Denmark, in Greenland and the Faroe Islands and, in the case of the Netherlands, in Aruba and the Netherlands Antilles. The said Convention will only apply in the said territories if and when Denmark or the Netherlands, as the case may be, expressly so notifies the Secretary-General.

Index

Cambridge Intellectual Property and Information Law

Titles in the Series (Formerly Known as Cambridge Studies in Intellectual Property Rights)

Brad Sherman and Lionel Bently *The Making of Modern Intellectual Property Law*
Irini A. Stamatoudi *Copyright and Multimedia Products: A Comparative Analysis*
Pascal Kamina *Film Copyright in the European Union*
Huw Beverly-Smith *The Commercial Appropriation of Personality*
Mark J. Davison *The Legal Protection of Databases*
Robert Burrell and Allison Coleman *Copyright Exceptions: The Digital Impact*
Huw Beverly-Smith, Ansgar Ohly and Agnès Lucas-Schloetter *Privacy, Property and Personality: Civil Law Perspectives on Commercial Appropriation*
Catherine Seville *The Internationalisation of Copyright Law: Books, Buccaneers and the Black Flag in the Nineteenth Century*
Philip Leith *Software and Patents in Europe*
Geertrui Van Overwalle *Gene Patents and Clearing Models*
Lionel Bently, Jennifer Davis and Jane C. Ginsburg *Trade Marks and Brands: An Interdisciplinary Critique*
Jonathan Curci *The Protection of Biodiversity and Traditional Knowledge in International Law of Intellectual Property*
Lionel Bently, Jennifer Davis and Jane C. Ginsburg *Copyright and Piracy: An Interdisciplinary Critique*
Megan Richardson and Julian Thomas *Framing Intellectual Property: Legal Constructions of Creativity and Appropriation 1840–1940*
Dev Gangjee *Relocating the Law of Geographical Indications*
Andrew Kenyon, Megan Richardson and Ng-Loy Wee-Loon *The Law of Reputation and Brands in the Asia Pacific Region*
Annabelle Lever *New Frontiers in the Philosophy of Intellectual Property*
Sigrid Sterckx and Julian Cockbain *Exclusions from Patentability: How the European Patent Office Is Eroding Boundaries*
Sebastian Haunss *Conflicts in the Knowledge Society: The Contentious Politics of Intellectual Property*
Helena R. Howe and Jonathan Griffiths *Concepts of Property in Intellectual Property Law*
Rochelle Cooper Dreyfuss and Jane C. Ginsburg *Intellectual Property at the Edge: The Contested Contours of IP*
Normann Witzleb, David Lindsay, Moira Paterson and Sharon Rodrick *Emerging Challenges in Privacy Law: Comparative Perspectives*
Paul Bernal *Internet Privacy Rights: Rights to Protect Autonomy*
Peter Drahos *Intellectual Property, Indigenous People and Their Knowledge*
Susy Frankel and Daniel Gervais *The Evolution and Equilibrium of Copyright in the Digital Age*
Edited by Kathy Bowrey and Michael Handler *Law and Creativity in the Age of the Entertainment Franchise*

Sean Bottomley *The British Patent System and the Industrial Revolution 1700–1852: From Privileges to Property*
Susy Frankel *Test Tubes for Global Intellectual Property Issues: Small Market Economies*
Jan Oster *Media Freedom as a Fundamental Right*
Sara Bannerman *International Copyright and Access to Knowledge*
Andrew T. Kenyon *Comparative Defamation and Privacy Law*
Pascal Kamina *Film Copyright in the European Union, Second Edition*
Tim W. Dornis *Trademark and Unfair Competition Conflicts*
Ge Chen *Copyright and International Negotiations: An Engine of Free Expression in China?*
David Tan *The Commercial Appropriation of Fame: A Cultural Critique of the Right of Publicity and Passing Off*
Jay Sanderson *Plants, People and Practices: The Nature and History of the UPOV Convention*

Lightning Source UK Ltd.
Milton Keynes UK
UKOW06n0630160717
305396UK00013B/203/P